Environmental Ecology

ENVIRONMENTAL ECOLOGY

By

Dr. P.R. Yadav

Lecturer
Department of Zoology
D.A.V. College
Muzaffarnagar (U.P.)

&

Dr. Shubhrata R. Mishra

Department of Botany
Vikram University
Ujjain (M.P)

D P H

DISCOVERY PUBLISHING HOUSE
NEW DELHI-110002

Published by:

Namit Wasan

DISCOVERY PUBLISHING HOUSE PVT. LTD.
4383/4B, Ansari Road, Darya Ganj
New Delhi-110 002 (India)
Phone : +91-11-23279245; 23253475; 43596065
E-mail : discoverybooksindia@gmail.com
discoverypublishinghouse@gmail.com
namitwasan9@gmail.com
web : www.discoverypublishinggroup.com

First Published: **2004**

Reprinted: **2021**

ISBN: 978-81-7141-753-7

Environmental Ecology

Printed at:
Infinity Imaging Systems
Delhi

Preface

The present title Environmental Ecology deals extensively with the elements, components, natural resources, physical, chemical, geological and biological aspects of environment. The last thirty years have witnessed a vast expansion of both interest and knowledge in environmental sciences. There has been a widenning public awareness of the importance of environmental interactions with particular focus on such issues as the impact of pesticides upon food chains and the loss of global biodiversity owing to habitat destruction.

The present title is intented to be accessible to any one with an interest in or responsibility for specific environmental concerns and hazards. Thus it is aimed at a broad range of readers. This book differs from the other books in its central concern with "statement about the future" as made by a wide range of scientists. The subject areas covered are extensive and specialists in every field will find materials that can be improved and extented.

The book is an effort to summarize the basic concept and principles of the subject, to present the elementary factual information with which a person to be competent in the field should be familiar, and to show how these principles and facts may be applied in a practical way to the interests and welfare of man.

The author has freely consulted various articles, discussion notes, reviews and extracts from the various scientific journals in the preparation of the present book, in able to make it comprehensive and upto date.

Though every care has been taken by the printer, publisher and me, it is quite likely that some errors might have found their way into the book but I hope these are of very insignificant nature. However, sugestions to improve book and pointing out of errors and mistakes, if any, will gratefully be appreciated.

The author expresses grateful to her friends and colleagues whose constant inspiration have initiated her in bringing out this book.

Special thanks are given to Mr. Wasan and staff of M/s Discovery publishing House for their whole hearted co-operation in the publication of this book.

Author

Contents

1

Symbiotic Relationships

One of the most striking features of marine organisms is the large number of close association that can be observed among many different species. These are a series of association that are not predator-prey or herbivore-plant relationships in which one member consumes the other. These close relationships between unlike species generally seem to be either unharmful to either member or, more likely, to be beneficial to one or both. *Symbiosis* is the name given to such associations and means an interrelationship between two different species Symbiotic relationships are found in the terrestrial environment as well as in the aquatic, but appear to be disproportionately common, extensive, and well developed in the marine environment, such that they merit separate coverage in a book such as this.

Definitions and Coverage

Symbiotic relationships cover a broad spectrum of associations from random, casual, or facultative associations through more and more obligatory groupings that benefit one or both members to finally those that are parasitic. Such differences in the degree of association have led to the subdivision of symbiosis into more narrowly defined groupings. The term *commensalism* is often used to refer to an association that is clearly to the advantage of one member while not harming the other member. *Inquilinism* is a special subdivision of commensalism, in which an animal lives in the home of another, or in its digestive tract, without being parasitic. *Mutualism* is that form of symbiosis in which two species associate together for their mutual benefit. In a mutualism relationship the partners are often called *symbionts.* In a commensal or inquiline relationship, however, the partner gaining advantage is called the *commensal* and other the *host.* Parasitism generally refers to an association in which one species lives in or upon another and draws nourishment from that species at the expense of, or to the detriment of, the other.

In other words, it is an association in which the advantage is solely to one member at the expense of the other Parasitism is an association that seems to be extremely common, in both marine and terrestrial environments. It will not be discussed in this chapter, since it contains few features unique to the marine environment. Basically, there are two broad groupings of nonparasitic symbiotic associations in the sea: those between algal cells and various invertebrate animals and those between various animals, both vertebrate and invertebrate. We shall consider these in that order.

Symbiosis of Algae and Animals

All symbiotic relationships in the sea between plants and animals are between unicellular algae or their parts and a wide variety of marine invertebrate animals. These symbiotic relationships have been shown to be most common in tropical waters, but they also are prevalent in temperate oceans. They do, however, appear to be absent, or virtually so, from polar waters. Furthermore, the associations are, for obvious reasons, restricted to very shallow subtidal or intertidal areas or to the uppermost layers of the pelagic realm where sufficient light is present.

Types and Composition of the Associations

There are basically two types of symbiotic association between algae and invertebrates. The more common is to have the entire functioning algal cell associated with the invertebrate animal. The second is to have only the functioning chloroplasts from the algal cell incorporated into the tissues of the invertebrate body. The algal cell symbionts have been typically classified into groups on the basis of their colour. *Zooxanthellae* is the name given to those cells appearing brown, golden or brownish yellow in colour, *Zoochlorellae* to those that appear green. A third, smaller group appears blue or bluish green and has been called *Cyanellae.* These colour groups, however, do not distinguish among the actual algal species involved. Zooxanthellae, the most common in the seas of the world, are primarily species of dinoflagellates but include a few diatoms and cryptomonads. As Schoenberg and Trench have

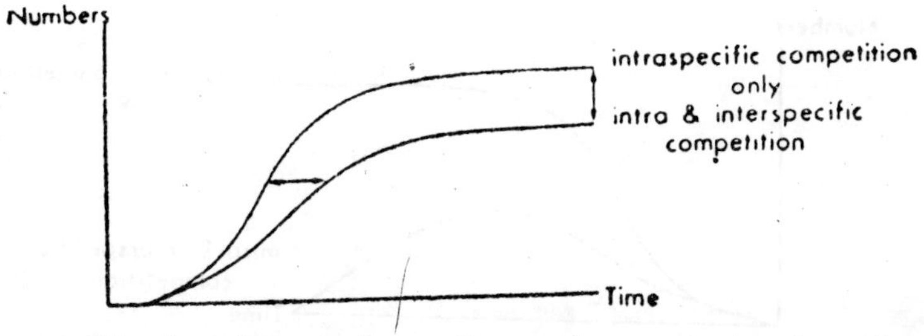

Fig. 1.1. Effect of introducing an element of interspecific competition into the population.

noted, the most common zooxanthellae species is the dinoflagellate *Symbiodinium microadriaticum.*

The zoochlorellae are much less common in the sea, but dominate symbiotic associations in fresh water. Among the marine associations, zoochlorellae include the alga *Platymonas convolutae* and unknown species of Chlorophyceae. Cyanellae are all blue-green algae, but the present, the identity of the species involved remain unknown. They are most common as symbionts in sponges and in planktonic diatoms. The algae all occur inside the bodies of the animals and most are found either within vacuoles inside individual tissue cells or else in various body spaces between or within tissue layers. Generally, the algal cells are restricted to certain tissues or areas of the host, and they grow and reproduce within the invertebrate without undergoing digestion by the host.

Table 1.1. Summary of the Types of Associations Between Algae and Marine Invertebrates

Algae Symbiont Group	*Algal Taxon*	*Invertebrate Animal Host Taxa*
Zooxanthellae	Dinoflagellata	Protozoa, Porifera Cnidaria, Platyhelminthes Mollusca
	Bacillariophyceae	Platyhelminthes *(Convoluta convoluta)*
	Cryptomonadida	Protozoa
Zoochlorellae	Pyramimonadales	Platyhelminthes (*Convoluta rescoffensis*)
	(?) Chlorophyceae	Cnidaria *(Anthopleura sp.)*
Cyanellae	Cyanophyceae	Porifera, Protozoa
Chloroplasts	Chlorophyceae	Mollusca (Sacoglossa)

Modified from D.C. Smith, Symbiosis of algae with invertebrates, Oxford Biology Reader 43.16 pp. Copyright Oxford University Press.

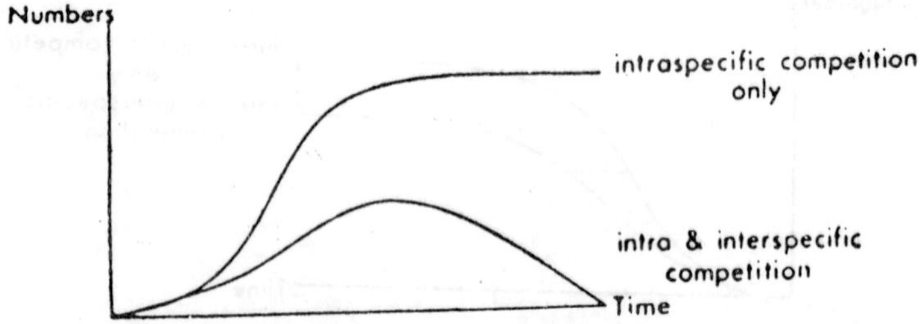

Fig. 1.2. As Fig. 1.1, but strength of competition has increased such as to cause extinction.

Among those associations that involve only the chloroplasts and the invertebrate body, the chloroplasts are usually derived from the cells of larger green algae, which are first ingested by the animal in feeding and subsequently are transferred from the digestive tract of the animal to other tissues. Algal symbiosis, either cellular or chloroplast, is widespread among various invertebrate groups. Symbiotic associations occur in Protozoa, Porifera, Cnidaria, Platythelminthes, Mollusca, and Echiura. Within these phyla, McLaughlin and Zahl report about 130 genera having algal symbionts. It is most common among Protozoa and Cnidaria.

Origin of the Association

Although we will probably never know the exact origin of algal-invertebrate symbiotic relationships, it is possible to outline some suggestions. In all the phyla mentioned above that contain symbiotic algae or chloroplasts, the final phase of digestion is intracellular, where individual cells take up particles from the stomach. It is thus possible to conceive of the origin of this association through the ingestion by cells of the digestive tract of either intact algal cells or chloroplasts, taken in by the animal during feeding. If, then, the algal cells were resistant to digestive action or the animal lacked enzymes to digest plant cellulose, the basis for a new association would be laid. This association could then evolve into a more obligatory relationship, provided that the new association conveyed an enhanced survival value to both symbionts. Since these associations, in fact, are common, they must have some selective value.

Distribution of Algae-Invertebrate Associations

Although there is not space or time to list and discuss all the species in which algal symbionts occur, it is of use to discuss briefly the various groups in which this association is prevalent. Beginning with Protozoa, symbiotic relationships are found in all of the epipelagic planktonic Radiolaria where the zooxanthellae occur in the outermost forthy layer. They also are found in a number of planktonic Foraminifera and even in marine ciliates. Symbiosis is not common among sponges (Porifera), but has been recorded in a few species, zooxanthellae in *Cliona* and cyanellae in *Demospongia.*

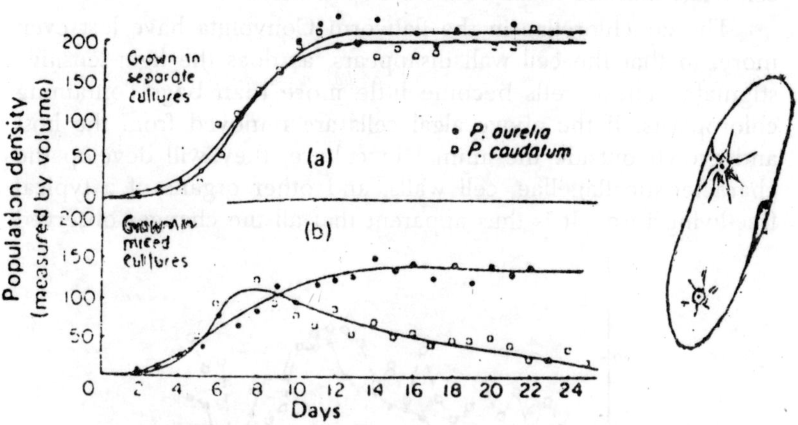

Fig. 1.3. Growth of Paramecium population (P. caudatum, P. aurelia) in separate culture (a) and mixed culture (b).

Among the Cnidaria, symbiosis becomes extremely common, especially among tropical cnidarians. Virtually all tropical, shallo water anemones, soft corals, sea fans, whips, and stony corals have symbiotic zooxanthellae in their tissues. Even certain of the tropical jellyfish contain zooxanthellae. In temperate seas, the incidence is not as great, though many anemones such as *Anthopleura* retain the symbionts. Whereas a few symbiotic relationships occur in Ctenophora, Annelida, Echinodermata, and Tunicata, the remaining major reservoirs of symbiotic relationships are found in the Platyhelminthes and Mollusca.

Among marine flatworms, the classic case is that of the genus *Convoluta,* which has been extensively investigated. Symbiotic relationships in the Mollusca are generally restricted to certain marine gastropods of the order Sacoglossa, where the animals retain only the chloroplasts, and to the bivalve family Tridacnidae.

Modifications Resulting from the Association

The symbiotic association between the algae and the invertebrates is generally a very close one which has resulted in rather significant anatomical and physiological changes in both the algal cells and the various invertebrate hosts. Perhaps the most profound changes that occur as a result of the association are found in the algal cells. Most marine symbiotic algae are dinoflagellates. The symbionts, however, have lost their locomotory flagellae and the characteristic grooves around the body. Furthermore, they have cell walls that are much reduced in thickness.

The zoochlorellae in the flatworm Convoluta have lost even more, in that the cell wall disappears, as does the light-sensitive stigmata. These cells become little more than bags containing chloroplasts. If the above algal cells are removed from the host and grown outside the animal in culture, they will develop the characteristic flagellae, cell walls, and other organs of a typical free-living form. It is thus apparent that all the changes observed

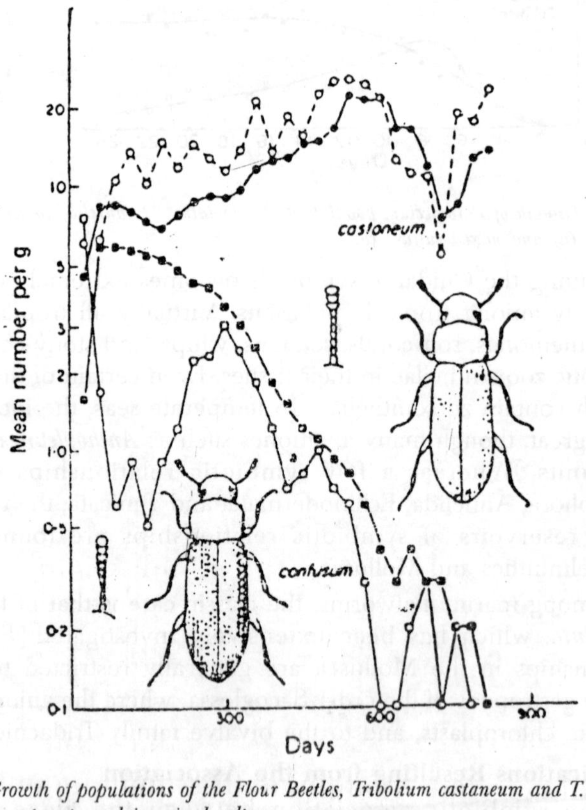

Fig. 1.4. Growth of populations of the Flour Beetles, Tribolium castaneum and T. confusum in separate and in mixed culture.

are a direct result of the symbiotic association. In the case of the animals, the changes vary with the type of organism and with the degree of interdependence established between the symbionts.

The single universal modification is that all of these invertebrates live in very shallow water, where they can obtain adequate light so that the algae can carry on photosynthesis. Perhaps the least amount of modification occurs in the lower invertebrates, such as Protozoa and Porifera, where the algae occur as simple inclusions in the cytoplasm of the animal. No special anatomical modifications are observed. Among the Cnidaria, definite modifications begin to become apparent. The algae occur in marine Cnidaria in the innermost of the two cell layers, the gastrodermis. The numbers of zooxanthellae in different corals vary, and those that have the most seem to have reduced tentacle size, indicating that they are less able to capture zooplankton food.

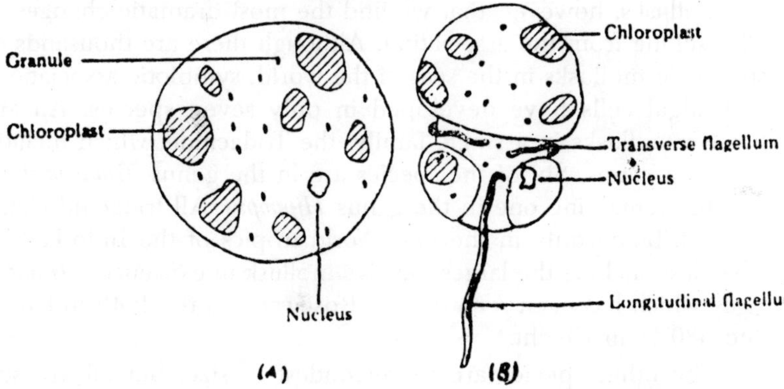

Fig. 1.5. Free-living (B) and symbiotic (A) zooxanthellae cells.

Among certain soft corals of the family Xeniidae the digestive regions of the animal are reduced and the animals are not responsive to animal food. In the jellyfish *Cassiopeia,* a striking modification is behavioural. These animals, rather than swimming in open waters as do most jellyfishes, lie upside down on the bottom in shallow tropical waters, exposing their oral arms to the light in order to illuminate their algae. The oral arms are also much enlarged and expanded to provide more area of habitation by the algal cells. Truly a strange way of life for a jellyfish! Among flatworms, the most studied case of symbiosis is that of *Convoluta roscoffensis,* a small animal inhabiting sand beaches on the Brittany coast of France.

The algae inhabiting *Convoluta is Platymonas convolulae,* which also occurs free-living in the same area. Convoluta exhibit perhaps the most profound behavioural and life history modifications, but little in the way of anatomical changes. *Convoluto roscoffensis* live burrowed into the sand of beaches near the upper reaches of the tidal zone. Whenever the tide is in, the animals are found buried. When the tide recedes, the animals move up onto the surface of the sand, where they spread out to expose their symbionts to the light. When the tide begins to return, the vibrations trigger a reburrowing response so that the animals are safely under the sand again before the rising tidal waters are able to sweep them out.

Even more significantly, as Smith notes, these animals apparently do not feed as adults. Furthermore, if the young worms do not ingest the algal symbionts upon hatching, they will not complete development and will die even if they feed. It is among the mollusks, however, that we find the most dramatic changes of all resulting from the association. Although there are thousands of species of mollusks in the seas of the world, symbiotic associations with algal cells have developed in only seven species. All are bivalve mollusks in a single family, the Tridacnide, which include the giant clams. Six of the species are in the genus *Tridacna* itself and the remaining one in the genus *Hippopus.* All tridacnid clams are distributed only in the Old World tropics of the Indo-Pacific. They also include the largest bivalve mollusk in existence : *Tridacna gigas,* which has been recorded by Rosewater to reach 4ft in length and 580 lb in weight.

The other species are more modest in size, but all are still large in comparison with most other bivalve species. These tridacnid clams are inhabitants of coral reef areas, where they are found abundantly in the shallow, sunlit water. They usually have a very uncharacteristic position for a bivalve mollusk, in that they lie either on the surface of the bottom or bored into coral or coral rock with the opening between the valves facing up toward the surface of the water. The valves usually gape widely, and within this opening can be seen an extensive, brightly coloured tissue layer. It is in this tissue layer that the symbiotic zooxanthellae are found. The brightly coloured tissue exposed within the gape is siphonal tissue; the colour results from the interaction of various pigments deposited there. The reason for the bright colors is that they act to protect the tissues of the clam from the damaging effects of sunlight, while passing enough light to allow the zooxathellae photosynthesize. What is truly remarkable about tridacnids, however, is the tremendous change that their bodies have undergone from that of a typical clam in order to accommodate this symbiotic association.

Bivalves generally rest with their foot either embedded in the substrate or held against it. In this position, a normal clam has its hinge uppermost. Obviously, this would not do for a tridacnid, since such a position would not permit the tissue and zooxanthellae to be illuminated. The result, as Yonge has described, is that the entire tridacnid has undergone a tremendous rotation with respect to the foot, so that the hinge comes to lie on the underside next to the foot, and the opening of the shell faces upward. At the same time, the siphons and siphonal tissue underwent an expansion and grew and extended themselves, covering the length of the upward-facing opening and providing the expanded area for occupation by the zooxanthellae. As a result of this rotation and expansion, one of the tridacnid's shell-closing muscles was lost. Because of these

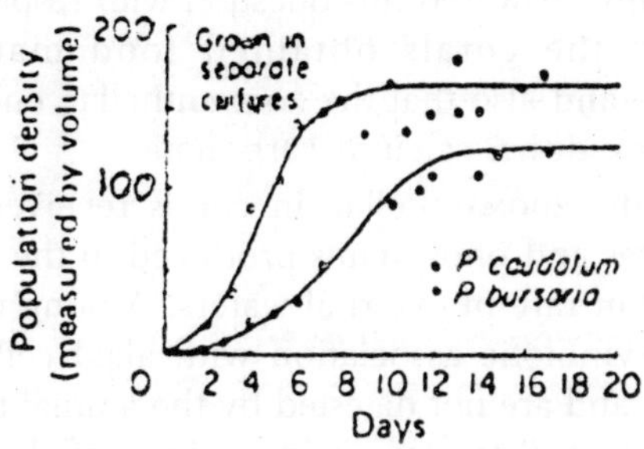

(b)

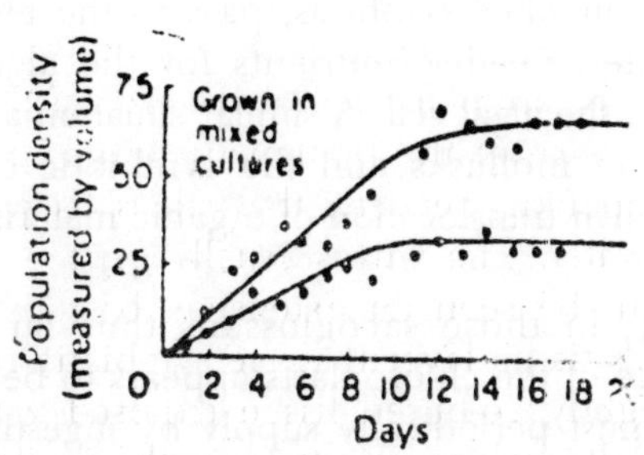

Fig. 1.6. Coexistence of Paramecium caudatum and P. bursaria in mixed culture (b) population curve for the two species in separate culture are shown in (a).

profound anatomical changes, the tridacnids are markedly different in body orientation from any other bivalves. These changes can only be attributed to the association with the symbiotic algae.

Symbiotic relationships with chloroplasts have been reported by Greene in the marine gastropods of the order Sacoglossa. Animals of this order regularly consume the contents of algal cells, and some species have evolved the ability to retain the chloroplasts in a functioning position on the dorsal surface. The only modifications so far observed are that they do not feed very often and the tropical forms provide a screen of lightabsorbing material above the chloroplasts to cut down the light intensity.

Value of the Association

Since symbiotic associations with algae are so common invertebrates, and since they often result in profound anatomicals or behavioural changes in the partners, it seems only natural to assume that they must have some positive value to each of the partners. What might be the value of such associations? We have already partially answered this question with respect to corals where we saw that the corals obtained food materials from the zooxanthellae and also that the zooxanthellae enhanced the ability of corals to lay down calcium carbonate.

In turn, the zooxanthellae in corals received nutrients in the form of nitrates and phosphates produced in the metabolic process of the coral, but rare in external waters. Among most invertebrates that have a symbiotic association with algal cells, the cells retain their integrity and are not digested by the animal to obtain nutrients. Thus, the nutrients that pass are in the form of chemical compounds. Energy containing molecules, such as glycerol, produced by the zooxanthellae in photosynthesis, pass to the animal, and nitrates and phosphates, needed nutrients for the algae, pass from the invertebrate to the algal cell. A similar situation also occurs between the sacoglossan mollusks and the symbiotic chloroplasts where Green has shown translocation of organic material from chloroplast to animal.

However, in those sacoglossans thus far investigated, the duration of life of the chloroplasts appears to be much shorter and the animals must periodically supply by ingesting the contents of algal cells. Among the giant clams, however, Yong has suggested that, in addition to reciprocal transference of nutrients, the animals may also actively digest the zooxanthellae cells. This would seem to be counter productive, especially considering the great anatomical changes that these animals have undergone in order to properly provide for their algal guests. Apparently, however, the clams can discern between healthy zooxanthellae cells and those that are degenerating or senile. As a result, the calm transports only the senile or degenerating cells from the outer mantle blood spaces, where they photosynthesize, into the deeper tissues, where they are consumed by the blood cells.

In this manner, then, the clams retain the symbiotic relationship, and at the same time, they cull the unfit algal cells to obtain additional nutrients otherwise lost when the cell dies. Other values may result from the association. As we have noted, the most of these associations are tropical; these waters are classically low in plant nutrients. It may well be that the algae benefit by the association through access to a larger and more reliable source of nutrients in the form of the metabolic products of the animal (NO_2, PG_4, CO_2) than they would obtain from the open water. Tropical waters also are lower in oxygen than temperate waters, because water at higher temperatures holds less oxygen. Since the photosynthetic process produces oxygen, it may be that the animals, especially in the crowded conditions of the shallow waters of the tropics, gain through additional amounts of oxygen produced by the symbiotic algae.

Establishment and Transmission of Zooxanthellae

Whereas each generation of sacoglossans must actually feed on algae to obtain their supply of symbiotic chloroplasts, perpetuation of the symbiotic association with zooxanthellae from generation to generation of hosts is accomplished differently among the different groups. Basically, two major routes are available. Either the zooxanthellae cells must be passed on directly from the parent to the eggs or larvae, or each new generation must reinfect itself a new from algal cells in the surrounding environment. Passage of algal symbionts directly to the next generation via the egg of larvae seems to be the method used by most Cnidaria.

In most corals, for example, the zooxanthellae enter the eggs at some time prior to their release from the parent. It is also assumed, though without much evidence, that the giant clams also transmit the zooxanthellae through the eggs. It has also been suggested the clams receive them from the surrounding corals. For other hosts, the perpetuation of a symbiotic relationship depends upon the reinfection of each generation. In such cases, the algal symbionts must be obtained a new from the environment. Among Protozoa and Porifera, this is the common means of transmission. Each new generation obtains the algal cells through ingestion of the free-living form of the symbiont. Surprisingly, this is also the case in *Convoluta roscoffensis,* where the association is highly dependent one. In this case, the algae-free larvae ingest the free-living *Platymonas convolutae.*

In order to ensure that the larvae will be infected, however, there has evolved a chemical that attracts the alga to the egg cases of Canidaria so that when the young hatch and begin to feed, the algae are present. Under such conditions, the new generation is virtually assured of reinfection. A final possibility for transmission is the ingestion by the potential host of food organisms themselves with symbiotic algae. Although this has not as yet been shown to be important, it remains an area requiring study.

Symbiosis Among Animals

We are concerned here with those special associations in which members of different species are regularly associated with each other in nonparasitic relationships. Such symbiotic associations are widespread in the sea, primarily in the crowded reaches of the epipelagic and shallow subtidal zones. As with algal symbiosis, such relationships appear to be somewhat more common or spectacular in the tropics, but are certainly also common in temperate seas.

Types of Associations

Symbiotic relationships among marine animals cover a broader spectrum that the strictly mutualistic associations we have seen among plants and chloroplasts and marine invertebrates. The simplest type of associations are commensal, whereon the "guest" lives on another organism "host," or in or upon some construction of that organism, such as a tube or burrow. Such associations are similar to epiphytic relationships seen among terrestial plants, wherein the epiphyte simply uses the other plant as a substrate without actually taking sustenance from it. In these cases, the commensal usually gains in some measurable way from the association and the host is not seriously inconvenienced. Marine commensals that live on or upon other invertebrates are called *epizoites.* Those that live inside other animals but are not parasites called *endozoites.*

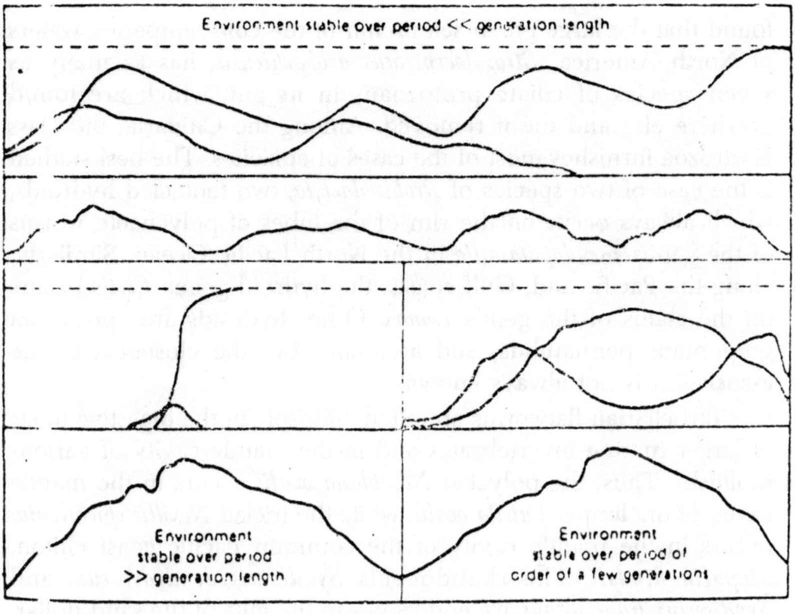

Fig. 1.7. Ideal course of competition between two species as regulated by the relation between generation length and the period over which the environment may be taken to be stable.

Epizoites are extremely abundant in marine waters and many are probably not true commensals; the relationship in the result of organisms that normally settle on the substrate settling at random on the outside of a slow-moving or sessile invertebrate. These will not be considered further here. Other epizoites are highly specific, and a symbiotic relationship is certainly the case, with the "guest" somehow seeking out the correct "host." These symbiotic epizoites are the most abundant group of commensals and are spread among the phyla Protozoa, Cnidaria, Entoprocta, Annelida, Arthropoda, and Mollusca. Many invertebrates harbour specialized ciliate protozoans, either on their external surfaces, internally in the digestive tract, or among the gills. For example, the vorticelled ciliate *Ellobiophyra donacis* is restricted to the gills of the clam *Donax vittatus*, while the collared ciliate *Lobochona prorates* is found on the telson of the isopod crustacean *Limnoria tripunctata*. Many other examples are also known.

Endozoic commensal ciliates are also common, especially in the digestive tracts of larger animals. Thus, for example, Beers found that the large green sea urchin of the cold temperate waters of North America, *Strogylocentrotus drobachiensis,* has as many as seven species of ciliate protozoans in its gut, which are found nowhere else and die if removed. Among the Cnidaria, the class Hydrozoa furnishes most of the cases of epizoites. The best studied is the case of two species of *Probisodactyla,* two tentacled hydroids, which always occur on the rim of the tubes of polychaete worms of the genus *pseudopotamilla* in the North Pacific Ocean. Similarly, along the Pacific and, Gulf coasts, the hydroid genus *Clytia* occurs on the clams of the genus *Donax.* Other hydroids are epizoic on gorgonians pennatulids, and ascidians, but the closeness of the association is not always known.

Turbellarian flatworms are often endozoic in the digestive tracts of larger marine invertebrates and in the mantle cavity of various mollusks. Thus, the polyclad *Notoplana ovalis* occurs in the mantle cavity of the limpet *Patella oculis,* while the triclad *Nexillis epichitonius* occurs in the mantle cavity of the common Pacific coast chiton, *Mopalia hindsii.* The rhabdocoels *Syndesmis dendrastorum* and *Syndisyrinx franciscanus* are endozoites in the guts of the sand dollar, *Dendraster excentricus* and the sea urchins *Strongylocentrotus franciscanus* and *S.purpuratus,* respectively. The phylum Entoproeta is a small, little-known group, most of which, as Nielsen has documented, are epizoic on other marine invertebrates, particularly polychaete worms, where they inhabit respiratory current areas.

Among the Annelida, a number of polychaete worms are epizoites with definite host specificity. The best studied is the hesinoid *Ophiodromus pugettensis,* which occurs on the ambulacral grooves of the bat star, *Patiria miniata* on the Pacific coast. However, the whole group of so-called "scale worms" are all very common epizoites on various other echinoderms and mollusks. Among the class Crustacea, the epizoic forms occur mainly within three groups: copepods, amphipods, and decapod crabs.

Copepods associated with other marine animals are usually highly evolved and modified parasites, but there are a few that are simply commensal, scampering over the outer surfaces of various invertebrates. Such an example are various species of the genus. *Hemicyclops,* which move on the surface of the other, larger crustaceans. Still others, like *Paranthessius,* are found in the mantle cavity of various bivalves. One of the most specialized epizoic associations is that among several species of barnacles and various large whales. Two families of barnacles are found only on whales and several are species-specific. Thus *Cryptolepas rachianecti* occurs only on the California gray whales. *Eschrichtius gibbosus,* and *Conchoderma auritum* attaches to *Coronula diadema,* which, in turn, is attached to the skin of whales.

A very curious form of epizoic relationship occurs between the amphipods of the suborder Hyperiida and the larger jellyfish. Hyperids are almost always epizoic or endozoic on jellyfish. Also epizoic or jellyfish are certain juveniles of various benthic crab species. Most mollusk epizoites are found among certain groups of clams. Generally, these are small clams that attach via a foot or byssus to the outside of various other invertebrates. Thus, *Montacuta ferruginosa* is found on the heart urchin, *Echinocardium cordatum; Orobitella rugifera,* attached via a byssus on the underside of the burrowing shrimp *Upogebia stellata*; and *Mysella pedroana* on the legs of the sand crab *Blepharipoda occidentalis.*

Among the gastropods found on other invertebrates the genus *Crepidula* is one of the most common. The above is certainly not an exhaustive list of the various epizoic and endozoic symbiotic associations, but gives an idea of the extent and diversity of such associations. Many, many more could be listed, but would not serve the purpose of this text. A second type of symbiotic association is that of organisms associated with animals forming a tube or burrow. In this case, the commensal or guest occupies the tube or burrow constructed by the host, but is not necessarily in close association with the body of the host. These commensals use the tube or burrow as a refuge from predation. They may or may not also tap the food source of the host. This type of symbiotic relationship ranges from one simply fortuitous, as in the case of an animal diving into any convenient burrow to avoid a predator, to examples of obligatory associations, where the commensals are not found other than in such tubes or burrows and are dependent upon them for continued existence.

In the former category are certain small gobiid fishes such as *Clevelandia ios,* which take refuge in any available tube or burrow but emerge to feed. Obligate tube dwellers include many small crabs of the family Pinnotheridae, known as "pea crabs." Many species of this family are known throughout the world, where they inhabit the tubes and burrows of polychaete annelids and other invertebrates, sometimes in species-specific associations. One of the best known examples of symbiosis with tube dwellers is that of the echiurid worm *Urechis caupo* on the Pacific coast of North America. This animal creates a permanent U-shaped burrow in which can be found four or more commensals. At the upper end is the fish *Clevelandia ios,* which merely uses the tube as a refuge. Further in the tube and closely associated with the worm are the scale worm *Hesperonoe adventar* and the pinnotherid crab *Scleroplax granulata. Hesperronoe adventor* seems restricted to *Urechis* burrows and lives against the worm, snatching food from it. Although the crabs are restricted to tubes, they may be found in the burrows of other animals as well.

Another commensal is the small clam *Cryptomya californica,* which has very short siphons and inserts them into the burrow of *Urechis,* thus enabling it to live lower in the substrate than the length of its siphons would normally permit. Occasionally the shrimp *Betaeus longidactylus* may occur in the burrow. A similar situation prevails with the tube-building polychaete *Chaetopterus.* This genus is virtually cosmopolitan, building tough parchment tubes. These tubes are also inhabited by various species of pinnotherid crabs and scale worms, the species differing depending on the geographical locality. Another type of association concerns those invertebrates that live in the mantle cavities of various mollusks. This appears to be a very common site for commensals to occupy. Thus, the nemertine *Malacobdella grossi* lives in the mantle cavity of various clams on the Pacific coast. Pinnotherid crabs occupy the mantle cavity of numerous bivalves, including oysters, where they are considered a pest.

Polychaete worms inhabit mantle cavities of chitons and limpets. In the tropics, shrimps of the family Palaemonidae are often found in mantle cavities of large bivalves such as *Pinna, Atrina,* and *Tridacna.* The remaining types of associations are those that appear to be mainly mutualistic rather than strictly commensal, as the aforementioned. The organisms display varying levels of behavioural and physiological modifications for the association. One of the most obvious but not well understood associations is that which occurs between various crabs and sea anemones.

Many hermit crabs all over the world have anemones attached to their shells, and in some genera, such as *Dardanus,* the presence of anemones seems to be universal. Still other crabs, such as *Hepatus, Munidopagurus macrocheles,* and *Stenocionops furcata,* are found with

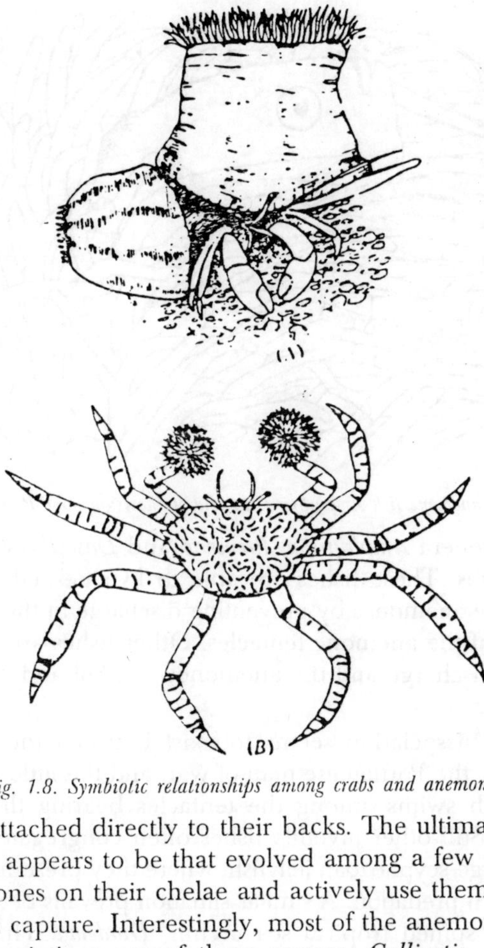

Fig. 1.8. Symbiotic relationships among crabs and anemones.

anemones attached directly to their backs. The ultimate symbiotic relationship appears to be that evolved among a few crabs, which carry anemones on their chelae and actively use them for defence and/or food capture. Interestingly, most of the anemones involved in these associations are of three genera, *Calliactis, Paracalliactis,* and *Adamsia,* species of which are rarely found elsewhere than with crabs.

These associations must be mutualistic, because whenever the crab changes shells or molts its carapace, the anemone is usually also transferred. This is effected by the manipulatory movements of the crab, but requires a degree of cooperation on the part of the anemone in loosening its grip on the shell or carapace and allowing itself to be physically moved. Another type of association is that between other Cnidaria and fishes. Throughout the Indo-West Pacific in the coral reef areas, there is a striking association of small

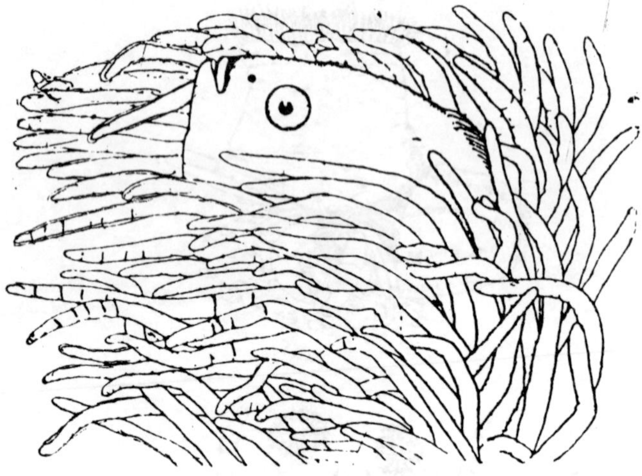

Fig. 1.9. The anemone fish Amphiprion sp. and its host anemone Radianthus sp.

fishes of the genera *Amphiprion, Premnas,* and *Dascyllus* with various large anemones. These fishes are able to live nestled among the tentacles of the anemones by preventing discharge of the formidable nematocysts of the anemone tentacles. Other fishes are not able to prevent this discharge and the anemones can kill and eat fishes of a similar size.

A similar association seems to exist between the dangerous siphoriophore, the Portuguese man-of war, and the little fish *Nomeus gronovii,* which swims among the tentacles bearing the powerful nematocysts. Still other juvenile fishes often congregate under the bells of the large scyphozoan jellyfish, where they presumably obtain protection from predation. A similar situation prevails between fishes and the long spined tropical sea urchin, *Diadema.* These urchins are circumtropical in shallow water and have enormously long, thin spines, which easily penetrate flesh. At least two tropical fishes, *Aeoliscus strigatus* and *Diademichthys deversor,* have adapted themselves to live among these spines, presumable for protection. They also look like sea urchin spines and so are camouflaged.

There is also the association of fish with other fishes, usually large, predaceous fishes. Pilot fishes and remoras are always found with other, larger fishes, other marine vertebrates such as turtles, or even any moving, inanimate objects. The final type of association is that known as "cleaning behaviour." As noted this is an association in which various species of fishes and shrimps actively attract large fishes to themselves for the purpose of cleaning them of various ectoparasites.

Origin of the Association

With such a diverse array of relationships as we have noted, it does not seem likely that there is any single cause for the origin of all the different types of associations. However, one factor common to all associations is that they have arisen in those areas of the ocean that are the most crowded with life. It would, therefore, appear reasonable that epizoites in particular could have become established due to the great competition for space among the settling invertebrate larvae. As a result, many settled upon other invertebrates and this then later evolved, in some cases, into more obligatory relationships in which those individuals that were commensal gained some advantage over the free-living forms. This is borne out by noting that many epizoites or burrow-inhabiting species are not strongly attracted to any given host species, while others are highly specific.

Similarly, many of the tube and burrow inhabitants, and those found in the mantle cavity of various mollusks, also probably had their origin in the search for space. After all, tubes, burrows, and mantle cavities really represent a certain amount of potentially occupiable, semienclosed space. The initial association of crabs and anemones may also well have begun when anemones began to take up residence on shells that were already occupied by hermit crabs.

Alternatively, the association could have arisen out of the habit of various crabs of picking up materials from the bottom and applying them to their carapaces to aid in camouflage. Many of the commensals that are epizoites or burrow-or tube-dwellers, especially the polychaeta worms, tend to be from groups that show strong *thigmotaxis;* that is, they respond positively to stimulation by contact with other, solid bodies. Such a behavioural response would naturally preadapt them to move into tubes or burrows or into various cavities in the bodies of larger animals. Once so established selection would favour the development of chemosensory clues enabling these animals to find such burrows, tubes, or cavities, and more specific associations would follow. Among the fishes and Cnidarians, it seems that the various types of symbiotic relationships have their origin in the protection from larger predators that such relationships confer.

It does not appear that this relationship stems from crowding, especially between epipelagic fishes and Cnidarians provided the symbionts could avoid having the nematocysts discharged, the tentacles of anemones and jellyfishes with their nematocysts represent a secure, formidable barrier to a larger predator. However, in the case of large sharks, pilot fishes, and remoras, the association probably originated in the search for food in the epipelagic. Pilot fishes and remoras probably were originally attracted by the amounts of food that were dropped in the feeding by the larger fishes, and they could have managed to establish themselves either by judiciously avoiding falling prey themselves to the sharks, or by being small enough to not be recognized as food by the larger fishes. Cleaning behaviour is a very complex relationship involving many fish species, and its origin and ecological functions have been confounded by various alternate hypotheses and factual inaccuracies as Gorlick et al. have discussed.

Modifications Due to the Association

As with the algal-animal symbiosis, the animal-animal relationships can result in anatomical, physiological, and behavioural modifications to one or both of the partners. The fewest modifications occur among the epizoites, especially those that are facultative epizoites and are found also in nonsymbiotic situations. Oftentimes, these animals have no anatomical modifications attributable to the symbiotic association, but those that are found with specific hosts usually have developed the means for *recognition* of the host from a distance. Such ability to recognize hosts is even more highly developed among obligate symbionts of various types. What is the basis for such recognition? Investigators such as Davenport have shown that it is more likely a chemical unique to the host. What is released into the water.

The commensal has, in turn, developed chemical sensory receptors, which detect the chemical and enable it to "home" in on its host. If such symbionts are tested in the laboratory in an apparatus designed to give them a "choice," they show a remarkable ability to choose the "host" on which they are found from among other closely related forms. Such attraction by chemical means has been demonstrated for several scale worms commensal with various starfish species; for the peculiar fishes, *Carapus,* which live in sea cucumbers; and for the polynoid *Hesperone adventor* and its host, *Urechis caupo.* Yet another modification of commensals and one that also aids in host recognition, is the ability of the larvae or young to select the appropriate substrate.

In many epizoite species that settle onto their hosts from the plankton, metamorphosis and subsequent development do not occur unless the larva is in contact with the appropriate host. The substrate discrimination abilities of some of these symbionts are truly remarkable. Some can apparently recognize substrate surface texture at the molecular level! It is not surprising, then that whale barnacles can settle not only on whale skin, but only on certain species of whales. However, in this case, it is likely that a chemical unique to the host is also present. Similarly, the clam *Modiolaria* is able to recognize the structure of tunicin, the major chemical component of the external layer of sea squirts with which it lives. A basic modifications of tube dwelling symbionts is that they are often smaller or thinner and flatter than their free-living relatives. Thus, the pinnotherid, or "pea" crabs, are among the smallest of the various marine crabs.

The shrimps that inhabit the mantle cavities of bivalve mollusks are also among the smallest of the shrimps. Similarly, the scale worms are often very flattened in comparison with other polychaetes. The fishes, such as *Clevelandia ios,* that are commensal in tubes or in the internal spaces of clams and sea cucumbers, *Apogonichthys* and *Carapus,* respectively, are very thin in order that they may fit into these restricted areas. The pearl fishes, *Carapus,* show perhaps the greatest modifications in that they have lost their scales and pelvic fins and have shifted the anal opening far forward under the head. This latter change apparently is to ensure that defecation will occur outside the body of the sea cucumber. Furthermore, as Arnold has shown, these fishes show behavioural modification in that they enter into the cucumber tail first. Such extensive modifications suggest a long history of adaptation.

Among those anemones found associated with crabs, a number of behavioural modifications occur. In the first place, the anemones must respond positively to manipulations by the crab when it rushes to transfer the anemone to another shell or from its shed carapace to a new one. This it does by releasing its grip on the old shell so that the crab can remove it with its claws. This requires a higher degree of nervous integration than we have come to believe could be accomplished by the extremely primitive, brainless nervous system of the anemone! We are thus currently ignorant of an understanding of this process among those crabs that carry anemones in their chelae.

The anemones again show a behavioural adaptation in that they allow themselves to be carried in the chelae and do not close up. In these cases, however, the crab also shows changes in that the chelae are modified to carry the anemones rather than gather food. Behaviourally, the crab extends the anemones when threatened or when gathering food; it uses its second pair of legs to transport food to the mouth, rather than the chelae, since the latter carry the anemones. Perhaps the most drastic changes undergone by the anemone member of these crab-anemone systems is that exhibited by *Adamsia palliata*, which lives with the hermit crab, *Eupagurus prideauxi*. Here, the anemone steadily expands its basal disc to completely encircle the vulnerable abdomen of the hermit crab, thus eliminating the need for the crab to change its shell as it grows. The anemone then expands to allow for the growth of the crab, meanwhile orienting itself so that the oral disc and mouth are ventral to the crab, thus enabling it to obtain food from the crab. The greatest morphological changes among fishes associated with other fishes occur in the remoras, where the animals have the dorsal fin modified to form a large sucker, which enables them to remain attached to their hosts.

Pilot fishes and the anemone fishes have no special morphological modifications. Among anemone fishes, however, there are definite behavioural and biochemical modifications. Studies have shown that anemone fishes of the genus *Amphiprion* must undergo a period of "acclimatization" to an anemone before they can, with impunity, dive into its tentacles. During this period, the fish is not capable of preventing nematocyst discharge by the anemone. During the acclimatization, Schlichter has demonstrated that the fish goes through a series of first brief and then longer encounters with the anemone, during which time, the fish coats itself with anemone mucus, thus tricking the anemone to discern the fish as itself. Hence, there is no nematocyst discharge from the anemone. When this coating is complete, the acclimatization period is over and the fish is able to dive in among the tentacles without causing nematocyst discharge. In addition to this fascinating coating process, anemone fishes have additional behavioural modifications.

In contrast to most fishes, which swim away when approached by humans or other large potential predators, anemone fishes swim out of their anemones toward the potential predator! This serves to attract the predator to them, at which time, they do an abrupt about-face and dive into the anemone. Presumably, the behaviour would tend to make an unwary potential predator chase them and subsequently be caught and consumed by the anemone. In order to further attract attention to themselves, anemone fish also have vivid color patterns, contrasting markedly with their background. In contrast to the anemone fishes, which develop immunity to the nematocyst of their host, the fishes associated with the Portuguese man-of-war do not acclimate, nor do they ever obtain complete immunity from the nematocyst.

They instead appear to spend their liver playing a perpetual game of Russian roulette with the tentacles of the siphonophore, constantly moving so as to avoid contact with the lethal nematocyst batteries! The same is true for the juveniles of various fishes that shelter below the umbrellas of the large jellyfish. Among cleaner fishes, certain morphological modifications can be noted, such as pointed, narrow snouts and forcepslike teeth, adaptations for picking up the small ectoparasites. Both cleaning shrimps and fishes also

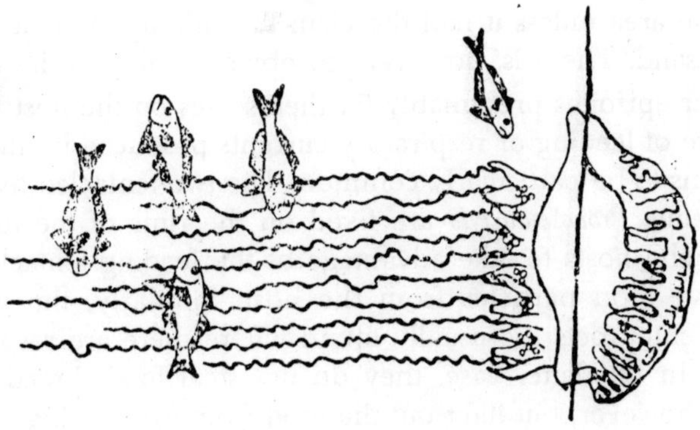

Fig. 1.10. Portuguese man of war, Physalia, with the fish Nemeus gronovii sheltering among the tentacles.

tend to have very bright and contrasting colors, which make them stand out against the background and advertise their presence.

Certain cleaning shrimps, notably the Pederson shrimp, *Periclimenes pedersoni,* have extraordinarily long antennae, which they wave about to further advertise their presence to the fishes. Cleaning behaviour relationships also impose behavioural modifications on the fishes cleaned. These animals must seek out and present themselves at the "cleaning station" and then remain motionless while the cleaner moves over their bodies. They must also open their mouths and spread their gills and opercula to allow the cleaner to enter the mouth and gill area. Of course, at the same time, the larger fish being cleaned must refrain from sudden inhalations, which might accidentally ingest the smaller cleaner!

Value of the Association

These symbiotic associations have obvious value to one or both of the members, and the relative value of the relationship to each varies depending upon the association. The major value of the symbiotic relationship to the epizoites would seem to be with the epizoite rather than the host organisms. For some epizoites, the value of the association lies in the fact that the host represents the only available suitable substrate in an otherwise unsuitable area. For example, the hydroid *Clytia bakeri,* which is epizoic on the clam *Donax gouldi* in sand areas in southern California, could not live in the area unless it had the clam to settle upon, as it cannot exist on sand. There is, however, no obvious value to the clam.

Other epizoites presumably fix themselves on the host to take advantage of feeding or respiratory currents produced by the host. This seems to be particularly common. The two-tentacled hydroids of the genus *Proboscidactyla* are fixed on the rims of the tubes of their annelid hosts to take advantage of the feeding tentacles and to remove food particles from the host. Similarly, the various epizoitic entoprocts are usually found where there are respiratory currents. In this latter case, they do not steal food directly from the host, however, but filter out the food from the moving current. Perhaps most epizoites benefit by the protection afforded them by the larger host. Thus, the various flatworms occurring in the guts and mantle cavities of their larger hosts are protected from predation and, in the case of intertidal forms, from possible desiccation. Surely, the hyperiid amphipods found on the large scyphozoan jellyfish would be much more subject to predation if free living in the plankton.

Finally, the epizoite may also gain the advantage of movement. This advantage may be of particular significance if the epizoite is itself sessile and the host freely moving. Commensals that inhabit tubes gain primarily protection from predation. This, for example, is the main value to the goby *Clevelandia ios* in the tubes of *Urechis caupo*. Other commensals, however, such as the scale worms gain not only protection but also food as they snatch morsels from their tube-dwelling hosts. Commensals in the mantle cavities of bivalve mollusks benefit from the protection of the bivalve shells, but also are able to employ the feeding and respiratory currents of the clam for their own needs.

In a few cases, these symbionts may tend toward parasitism in that they may elect to consume pieces of the gills or mantle of their bivalve host. Among the fishes associated with large jellyfishes, the Portuguese man-of-war, and the large sharks, it seems apparent that the whole value of the association is to the fishes in the form

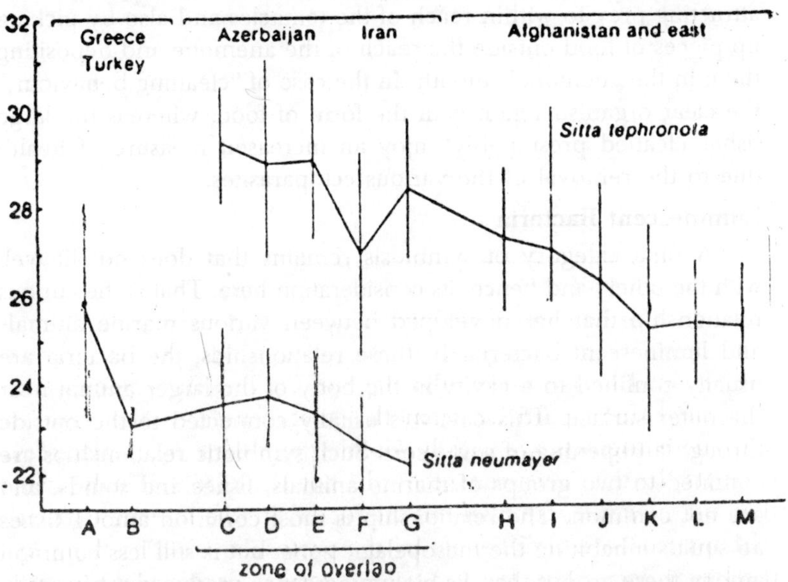

Fig. 1.11. Geographical ranges and character displacement in the Bill-length of the Western rock Nuthatch (Sitta neumayer) and Eastern rock Nuthatch (S. tephronota). Vertical lines represent the range of beak lengths in each local population.

of protection from predation in an area notorious for great predation pressure. The pilot fish and remoras are also protected from predation by their association with the large predaceous sharks, but they may also obtain food when the shark is feeding. The fishes associated with sea-urchins and living in the respiratory trees of sea cucumbers are protected by their association, but their presence presumably has no correlative advantage to their host. Among the mutualistic associations, the values of the association are more pronounced and, as the term applies, extend to the host as well.

In the various crab and anemone associations, the crabs, as hosts, gain the advantage of the nematocyst batteries of the anemones, either as protection against predators or else as an offensive weapon to capture food. The anemones, on the other hand, while still sessile, gain the advantage of movement; some, such as those on hermit crabs, may also be able to obtain food fragments lost during the feeding process of the crab. A similar situation prevails between anemone fish and their anemones. The fishes gain mainly protection, but they benefit their partners by attracting prey to within reach of the tentacles and also by picking up pieces of food outside the reach of the anemone and depositing them in the anemone's mouth. In the case of "cleaning behaviour," the clear organisms gain is in the form of food, whereas the large fishes cleaned presumably enjoy an increased measure of health due to the removal of the various ectoparasites.

Luminescent Bacteria

A final category of symbiosis remains that does not fit well with the others and hence, its consideration here. That is the curious relationship that has developed between various marine animals and luminescent bacteria. In these relationships, the bacteria are usually confined to a cavity in the body of the larger animal near the outer surface. This cavity is usually connected to the outside through an opening of some sort. Such symbiotic relationships are confined to two groups of marine animals, fishes and squids, and are not common. The relationship is most common among fishes ad squids inhabiting the mesopelagic zone, but is still less common among these groups than its bioluminescence produced intrinsically without bacteria. The various suggested values for the production of light by mesopelagic animals have been discussed in a Chapter.

The relationship of bacteria and the fishes or squids is a mutualistic one in which the bacteria obtain food from the larger animal.

In turn, the fishes or squid use the light produced by the bacteria for various defensive and/or offensive purposes, as discussed in a Chapter. The light produced by bacteria is usually continuous, and as a result, the fishes and squids often develop rather elaborate modifications to control the light. These anatomical developments can include the production of a reflecting layer behind the cavity to reflect the light outward, often lenses to focus and concentrate the light, and most often, a screen or shade on the outside which can be raised or lowered to either "turn on" or "turn off" the light. The bacteria show no corresponding changes in behaviour or morphology as a result of the association. The origin of these elaborate associations is not well understood, and the bacteria apparently are not passed from one generation of fishes or squid through the eggs. It would appear that each generation somehow must reinfect its light organs from the external environment.

2

ECOLOGICAL FACTORS

The environmental factors may be defined as the whole complex of climatic, edaphic, and biotic factors that act upon an organism or an ecological community and ultimately determine its form and survival. On the other hand organisms react to differences or changes in their environment in several characteristic ways, either by trying to avoid harmful situations or by being able to adjust physiologically, within their genetic limits, to adverse factors. In ecological studies the total environment is said to be made of some nonliving components as well as some biotic or living components, thus making the nonliving environment and living environment respectively. The nonliving environment includes the nonliving factors both physical and chemical, which influence the life of organisms. They are temperature, light anergy, water, atmosphere, including atmospheric gases and air current etc., fire, gravity, topography, (the configuration of the earth's surface), soil, pH, and nutrients, etc. The living environment includes the living or biotic factors, i.e., animals and plants. How organisms are influenced by their environment is determined largely by the following principles:

Shelford's Law of Tolerance

According to the law or tolerance, organisms have an ecological minimum and maximum, with a range in between which represents the limits of tolerance. Some subsidiary principles to the law of tolerance may be stated as follows:

(i) The limits of tolerance for reproductive individual seeds, eggs, embryos, and larvae are usually narrower than nonreproductive adult plants or animals. For example, adult blue crabs and many other marine animals can tolerate brackish water or fresh water with a high chloride content, but their larvae cannot live in such waters and hence they cannot reproduce in rivers and are not established there permanently.

(ii) When conditions are not optimum fur a species with respect to a particular environmental factor, the limit of tolerance may be reduced with respect to other factors. For example, according to Penman (1950), when soil nitrogen is limiting, the resistance of grass to drought is reduced.

(iii) Organisms with wide ranges of tolerance for all factors are likely to be most widely distributed.

(iv) Organisms may have a wide range of tolerance for one factor and a narrow range for another.

To describe a species with a narrow range of tolerance for a particular factor, we use the prefix steno–, for those with a wide range we use the prefix eury–. The terms stenothermal and eurythermal refer to temperature tolerance. Similarly, stenohaline and euryhaline, stenophagic and euryphagic stenoecious and euryecious are a few other pairs of words used in respect to salinity, food, and habitat respectively.

Liebig's Law of the Minimum

This law stales that the growth of a plant is dependent on the amount of food stuff which is presented to it in minimum quantity. This law has been extended by Taylor (1934), According to him, the functioning of an organism is controlled or limited by that essential environmental factor of combination of factors present in the least favourable amount But according to Odum, it seems best to restrict the concept of the minimum to chemical materials (oxygen, phosphorus, etc.,) necessary for physiological growth and reproduction, and to include other factors and the limiting effect of the maximum in the law of tolerance. However, two subsidiary principles must be added.

The first is a constraint that Liebig's law is strictly applicable only under steady-state conditions, i.e., when inflow of energy and materials balances the outflow. The second important consideration is factor interaction. Thus, high concentration of so me substance, or the action of some factor other than the minimum one, may modify the rate of utilization of the latter. Sometimes organisms are able to substitute partly a chemically closely related substance for one that is deficient in the environment. Thus, where strontium is abundant, molluscs are able to substitute it for calcium to a partial extent in their shells.

Combined Concept of Limiting Factors

According to Odum, the presence and success of an organism or a group of organisms depends upon a complex of conditions. Any condition that exceeds the limits of tolerance is said to be a limiting condition or a limiting factor. Among others, temperature is very often a limiting factor. Thus, organisms are controlled in nature by (i) the quantity and variability of materials for which there is a minimum requirement, and physical factors which are critical, and (ii) the limits of tolerance of organisms themselves to these and other components of the environment.

Salinity and Osmotic Pressure

The most deterring ecological barrier we know is the salt content of sea water. Only very few plant and animal species can thrive in a range of salinity extending from sea water to fresh water. Moreover, within this continuum of salinity, a particular degree of change is not necessarily accompanied by a corresponding change in the organisms. In a certain range a relatively pronounced difference in salinity (for example, 25% vs 15%) produces only a slight difference in number of species, whereas in the vicinity of 8 percent and at the boundary between fresh and brackish water there is a sudden dramatic effect. These effects of water salinity and the associated osmotic pressure deserve closer attention. We shall take animals as an example. All the animals and plants alive today originated in the ocean.

In order for them to invade and conquer terrestrial and freshwater habitats, they had to undergo adaptive modifications if their original states–at a price, of course. We must keep this fact in mind whenever we are discussing plants and animals of fresh water and land. To begin, let us consider the primary marine animals–species which throughout their phylogenesis have occupied no environment other than the ocean. The osmotic pressure of their intercellular fluid is very slightly higher than that of the open sea; it corresponds to 35 percent total salinity. The fluid in the cells has the same osmotic potential, and differs from that outside the cells only in ionic composition. Whereas intercellular fluid–in the water-vascular system of echinoderms, for instance, or in the blood vessels of annelids–closely resembles sea water, the cells contain considerably higher amounts of potassium ions.

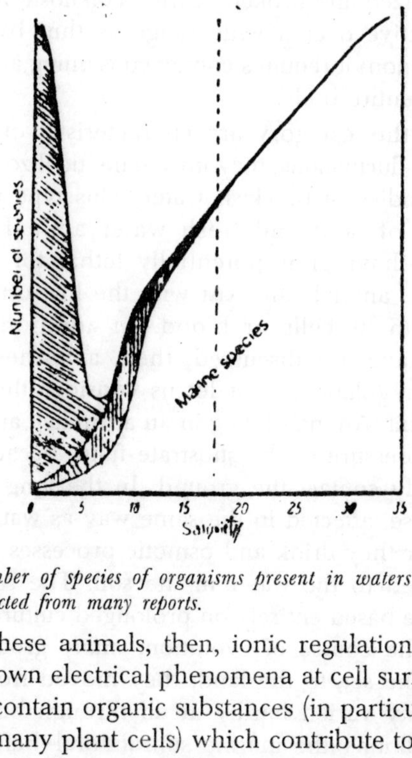

Fig. 2.1. The number of species of organisms present in waters of different salinity, reconstructed from many reports.

Even in these animals, then, ionic regulation occurs; it leads to the well-known electrical phenomena at cell surfaces. Moreover, the cells also contain organic substances (in particular amino acids, and sugar in many plant cells) which contribute to maintenance of osmotic pressure. Animals of this type can endure only minimal fluctuations in the osmotic pressure (that is, the salinity) of their environment. Their distribution is thus restricted to the oceanic plankton or to deep regions of the sea. The majority of benthonic animals–echinoderms, tunicates, crustaceans, annelids, molluscs, and cnidarians–are in this category. Animals living in shallow water, tidal zones and estuaries must be able to tolerate fluctuations in salinity.

Changes in the osmotic pressure or the milieu, which have an immediate effect on the intercellular fluid and thence on the cells, must be compensated. Without such compensation, reduction of environmental salinity would cause water to flow into the animal and its cells, which would swell arid burst. Conversely, increase in salinity results in shrinking of the cells. These processes can be counteracted only by active alteration of the osmotic pressure in the cells and in the intercellular fluid. Water is expelled and ions are actively taken up. Within the cells osmotically effective amino acids are synthesized and broken down. "Poikilosmotic" behaviour of this sort effective over a wide range, is thus by no means a passive phenomenon; it requires energy-consuming activity of single cells and of the entire body.

Animals in this category are characteristic of regions with moderate salinity fluctuations; they invade the tidal zone, the mouths of rivers, and bodies of brackish water. This type represents the point of origin of land and fresh water animals. But purely poikilosmotic behaviour is potentially lethal for both lines of development. No animal can exist with the low osmotic pressure of fresh water in its cells or blood. In addition to the ionic regulation we have just discussed, then, a further mechanisms necessary–osmoregulation. First let us consider the evolution of the terrestrial habit. Animals living in an air milieu are less exposed to the osmotic pressure of the substrate fluid, because only small areas of the body contact the ground. In the long term, though, they are of course, affected in the same way as water animals, by way of the water they drink and osmotic processes in the parts of the body exposed to the water in the soil. The results we shall now consider are based entirely on prolonged culture experiments. When, during the course of evolution, many groups of animals migrated from the sea to the land, they moved from a milieu in which the salinity varied hardly at all through one with highly unconstant conditions–the marine supralittoral–before they again reached a stable, though quite different, situation inland.

Most crustaceans initially protect themselves from the wide fluctuations in the coastal region by retaining an osmotic pressure higher than that of the external medium (hypertonicity regulation). Then the beach forms acquire an additional capacity for hypotonicity regulation; they are able to keep the internal milieu at an osmotic pressure lower than that of the environment. Such constancy of the internal milieu is typical of most of the coastal crustaceans living above the water line (Uca, Ocypode, Carcinus, Talitrus, Orchestia, Ligia). The terrestrial hermit crabs (Coenobita) and some brachyurans also appear to keep their internal milieu constant. In this case, however another method is used. Evidently these animals require both fresh and sea water; if only one kind is available, the internal osmotic pressure rises or falls and the animals eventually die. The time of death depends on many other factors (molting is particularly hazardous). When land hermit crabs drink, then, they sometimes choose fresh water and sometimes sea water. Their internal pressure can vary over a wide range, but can be continually readjusted if water of the appropriate salinity is consumed in time. An entirely different approach enabled gastropods to conquer the supralittoral zone.

On flat soft beaches they too develop hypertonicity regulation; its acquisition can be followed within genera. For example, Limapontia capitata exhibits no osmoregulation at all, whereas a species living higher up the beach, Limapontia depressa, can regulate its osmotic pressure. Gastropods have responded to high salinity by developing a poikilosmotic mechanism effective over a wide range; in this case there is no hypotonicity regulation. By these means, gastropods of the ocean beaches can tolerate extremely wide fluctuations of the pressure in the internal medium and in the cell fluid.

In Ovatella and Assiminea, pressures corresponding to a salinity between 6 percent and 90 percent (sometimes up to 100 percent) have been measured. This situation requires highly specific adaptation of the enzymes, which must remain functional despite great differences in their intracellular milieu. By contrast, gastropods on rocky coasts apparently never perform osmoregulation. When conditions are unfavourable they shut themselves off from the outside world by closing the operculum. Of course, adaptation to changing salinity conditions cannot be obtained free of cost. In the case of homoiosmotic animals such as Ligia or Orchestia platensis the price is obvious. In both low- and high-salinity water they must perform hard osmotic work, so that it comes as no surprise that with respect to growth and rate of reproduction they exhibit a distinct optimum in the approximately isotonic range. The same is true of poikilosmotic species. Synthesis and breakdown of the osmotically effective substances costs energy.

Even the presence of enzymes that operate with a wide osmotic range (Sarkissian, 1974) evidently represents a drain on energy. Poikilosmotic species, then, also pay for their ability to tolerate a wide range of salinity with very different rates of development and reproduction. It seems clear that when organisms are under such stress the strategy adopted is not to increase energy consumption (which would mean eating greater amounts of food, and this; is normally not available) but to cover the cost by slower growth, a lower rate of reproduction, and less resistance to generally harmful environmental factors. The poikilosmotic mussels in the inner Baltic Sea, for example, are considerably less able to tolerate cold than those in the North Sea–although in the Baltic selection for high cold tolerance would be expected. Desert plants and lichens are particularly sensitive to chemicals in their environment. Only two of the four types were able to evolve further, into true land animals.

Terrestrial hermit crabs and snails on rocky coasts need at least occasional access to water of marine salinity. Their evolution therefore ends in the supralittoral. Once the other terrestrial crustaceans and gastropods had evolved into genuine land animals, the salinity at which body fluids could be maintained by maximum regulation fell significantly. For terrestrial arthropods it corresponds to about 10 percent, and for soft-skinned land animals to about 6 percent. At the same time, the ability of arthropods to tolerate high salinity was greatly restricted by loss of hypotonicity regulation. Phylogenetically young land animals (isopods) can still exhibit such regulation, and can thus tolerate high environmental osmotic pressure. This physiological feature is useless in the present habitat of these animals, and can only be understood in the historical context. In the evolutionary tree of soft-bodied land animals, the poikilosmotic branch is reduced: salinity tolerance is strictly limited. The pattern that thus developed–in arthropods and land vertebrates (which originated in fresh water), a fixed limit for regulation corresponding to 10 percent with no hypotonicity regulation, and in soft-bodied land animals (gastropods and annelids) a 6 percent limit with poikilosmotic behaviour limited to a very small branch–gives the starting point for the recolonization of the ocean by land animals. Certain animals were predisposed for this return to the sea–desert animals that drink from salt lakes and must excrete salt, and animals from environments with aberrant chemistry (such as liquid-manure pits).

Another predisposing property, evidently, is alteration in the set level of tonicity during the course of the year, as a protection against cold. The comparatively massive advance of terrestrial animals (insects, mites) into arctic oceans argues for this interpretation. In the, transition from land to sea arthropods redevelop hypotonicity regulation. Only the Collembola appear to extend the poikilosmotic branch of evolution which otherwise includes soft-skinned forms–amphibians gastropods (Succinea) and enchytraeids. The increase or osmotic pressure in (he blood or the only truly marine frog (Rana cancrivora, eastern Asia) is brought about not by salts but by dissolved urea. By such modifications secondary marine animals are initially capable of living in either sea or fresh water. They need no salt: they can either tolerate or excrete it. A real dependence on marine salinity arises by loss of hypertonicity regulation, as has happened in the beach fly Coelopa. Once this has occurred, the animal is cut off from both land and fresh water; examples in this category include marine nematocerans (most species of Clunio) and mites (Halacaridae). These arthropods, which returned to the sea very long ago, exhibit hardly any trace of their old homoiosmotic characteristics; they have secondarily become almost poikilosmotic.

Moreover, they have secondarily raised the regulated internal tonicity to such a level that they are iso-osmotic in the water of the open ocean, and thus are exposed to only slight physiological stress. Another route away from the ocean, simpler to understand and not associated with such a diversity of types, is that leading to fresh water by way of brackish water. Salinity does fluctuate in brackish-water regions, but these fluctuations are much smaller than those in the supralittoral. No species travelling this route has developed hypotonicity regulation. Again, a typical feature is lowering of the internal tonicity; in fact, this is much more pronounced than in terrestrial animals. This progression can be readily followed in species of Gammarus from the ocean, brackish water and fresh water.

Soft-skinned freshwater animals (clams and snails) have the lowest internal .osmotic pressures of all animals. Because of this low set tonicity level remigration into the ocean presents difficulties. Bony fish (Teleostei) in the ocean must continually excrete salt: they have developed hypotonicity regulation as an adaptation to marine salinity. They have also raised their regulated tonicity as compared with bony fish in fresh water, but not enough to make salt excretion unnecessary. And this regulation uses up energy, with severe consequences. Fish in the arctic freeze at higher temperatures than does the medium surrounding them. Selachians (sharks and rays)–like Rana cancrivora and Latimeria– have followed the poikilosmotic route. They raise their internal pressure to levels equivalent to those of the environment by retaining urea. This feature is an extremely strong indication of the freshwater origin of selachians–if, indeed, the high urea production does not actually imply a semiterrestrial ancestry.

Freshwater arthropod adapted to marine life by hypotonicity regulation, acquired in a stepwise manner. A good example of a typical brackish-water arthropod on its way back to the sea is the caddis fly Limnophilus affinis. But only relatively few species returned from fresh water to the open ocean; examples include the bug Halobates and the nematoceran Cricotopus. No arthropod that originated in fresh water has been found to have secondarily raised its regulated tonicity level. The internal osmotic pressure of Clunio, high in comparison to that of other chironomids such as Chironomus and Cricotopus, may well be associated with the terrestrial origin of this genus. Soft-skinned freshwater animals have achieved this return journey only to a slight extent the chief examples to mention are the marine rotifers. Nothing is known about how they maintain water balance.

The river gastropod Theodoxus has advanced into brackish water up to fairly high salinities; it exhibits a slight increase in poikilosmotic capacity as compared with its limnetic relatives. The routes of migration, then, can lead from sea to land, land to sea, sea to fresh water and the reverse–and there are many others. We shall mention only one, which involves salt lakes and deserts.

Salt lakes and highly saline lagoons in arid maritime regions are colonized from the land from the Ocean and from fresh water. It is not difficult to distinguish animals of marine and limnetic origin on the basis of differences in internal osmotic pressure. Insects that came from land or fresh water develop a marked hypotonicity regulation, though the maintained tonicity is significantly increased. This process can readily be followed within the genus Chironomus, which has progressively invaded small bodies of water of increasing salinity. There are many ways by which animals can migrate from these saline environments to brackish-water regions, where the inventory of species is very similar. We have seen that a given environmental factor can elicit quite different physiological responses among animals coexisting in a particular habitat. These responses can be understood only in relation to the ecological history of the animals. The same considerations hold true for ecological factors other than salinity.

There is evidently little connection between the development of physiological functions and the phylogenesis of the animals in the Sense of evolution of higher forms of life. Rather, physiological functions appear to arise chiefly in relation to the changes of habitat that Occur during phylogenesis. This interpretation is generally accepted in the case of excretion of nitrogenous metabolic waste. The complicated processes by which ammonia is converted to larger molecules in land animals are determined by the animal ecological history. And the same is true of sensory functions.

Dolphins evolved the capacity to orient by ultrasound, and it may be that this new adaptation, "economical" in terms of the price paid, made possible their great success as inhabitants of the sea-a habitat where other animals in a more favourable osmotic situation and or with gill respiration are much better off energetically. One could say that the dolphins can "afford" higher energy consumption because of a "newly invented technological trick." In this discussion we have emphasized the costs incurred by animals that avoid intraspecific competition by moving to different habitats.

The uptake of dissolved organic matter, which plays a decisive role in the nutrition of a great Dumber of marine invertebrates (Schlichter, 1975), is much more difficult under the osmotic conditions in fresh water indeed, it is impossible (Siebers and Bulnheim, 1976). In migrating to fresh water, marine animals gave up an extremely abundant source of energy-rich and qualitatively valuable food, leaving it to the microorganisms. Furthermore, hypertonicity regulation requires that the water drawn in by osmosis be continually, pumped out, and that there be continual active uptake of salts; against a concentration gradient.

Special organs for salt uptake and water elimination are necessary for existence on land and in fresh water. Energy is expended for the construction and operation of these organs. And there arc still other problems; The organs for salt uptake are evidently not so specific that they can recognize precisely the correct ions in the water. They "confuse" heavy metals with sodium and potassium ions, and if uptake of the latter is blocked by heavy metals the animal dies. The same problem arises when animals return to the ocean. Fish and birds and mammals in the ocean must continually excrete salt; they must develop specific organs for the purpose, as well as others to take in water to replace that lost by osmosis. From the viewpoint of energy consumption, fish and birds are inferior to the primary marine animals. As Dollo's Law of the irreversibility of evolution states, animals returning to this intrinsically "economically favourable" environment do not re-evolve the original physiological constitution.

It is worth making a cost-benefit analysis to learn how warm blooded animals have managed, despite these difficulties to he so successful in the ocean: but no such analysis has as yet been carefully done. The best solution to the problem of minimizing costs on the land has been developed by the vascular plants. Their roots extend into the moist and mineral-rich soil, their leaves into the dry air. Water evaporates continually from the leaves, generating a flow of sap through the plant, initially always from the roots to the leaves. Sap flow in this direction makes it possible to maintain, without cost, an osmotic pressure within the plant that is relatively high in comparison with the soil as well as a cost-free influx of water against the prevailing osmotic gradient. Both the movement of materials from the roots to the leaves and osmoregulation, then, are initially brought about in a purely physical way, by means of the differences in the vapour pressures of soil and air and in the structures of the plant below and above ground–all at no cost to the plant. Of course, water consumption is very high. Unlike animals, plants have no internal circulation of water.

In all our discussions of the effects of seawater salinity and osmotic pressure we have been thinking in terms of the normal composition of sea water. The proportions of the different salts are in fact uniform throughout the oceans and in all waters directly connected with the ocean. Saline habitats inland, by contrast, have an entirely different ionic composition. Thus they can usually not be colonized by primary marine forms, which are really dependent on salt and have correspondingly high concentrations of salt in their cells and body fluids. Only secondary marine organisms, not actually dependent on salt but simply capable of excreting it efficiently, can colonize such sites. This fact explains the great differences between the fauna and flora of inland saline habitats and those of marine habitats. A particularly striking characteristic is the absence of most of the true marine algae (only very pollution-tolerant forms can survive in some inland saline sites) and sensitive ocean animals such as all of the marine Cnidaria, echinoderms and molluscs.

The same considerations apply to plants. The primary marine species have tissue osmotic pressures identical to that of the surrounding medium. Plants in fresh water and on land must actively take in salts and protect themselves from the osmotic influx of water. Land plants have exploited this situation as described above, the water simply evaporates, and in the process sets up a current that simultaneously carries nutrients through the plant. If vascular plants return to saline habitats they are in the same predicament as animals, and must eliminate the excess salt. Salt glands develop in such plants, just as in animals. Land and freshwater plants are thus from the outset in a less favourable energetic position than marine plants. Adaptation to the new habitat exacts a price. Should these terrestrial plants return to the marine environment, they cannot simply realize the old advantages of life in the sea–they must now bear additional costs. If they are to maintain themselves in competition with marine plants they must become or remain competitive in a different manner.

This sort of optimization in a new habitat can occur in surprising ways. A secondary marine animal which relies on the continual excretion of salt can reduce the amount that must be excreted by selective feeding; when marine fish or birds are taken as food (as they are by many seals and dolphins) relatively little salt is consumed. An animal with such a diet need not secrete salt as vigorously as one that eats primary marine animals. A desert animal that selects, from the very salty plants available to it, the parts containing little salt, is in a favourable energetic position. For example, the kangaroo rat Dipodomys microps of the North American deserts scrapes off the less salty tissues of plants with specially shaped teeth. It consumes only as much salt as a normal herbivorous land animal (Kenagy, 1973). Related rodents make use of the water produced in their bodies when food is metabolized–so efficiently that they never need to drink, despite the dryness of their food.

Land-dwelling plants and animals face special problems with regard to water balance. Because of the evaporation of water into the air, their milieu in some respects corresponds to; a highly saline environment in which water is lost due to osmotic effects. We shall now discuss the methods by which water is conserved under these conditions, and consider the question whether it is possible to make use of the moisture in the air by taking it in

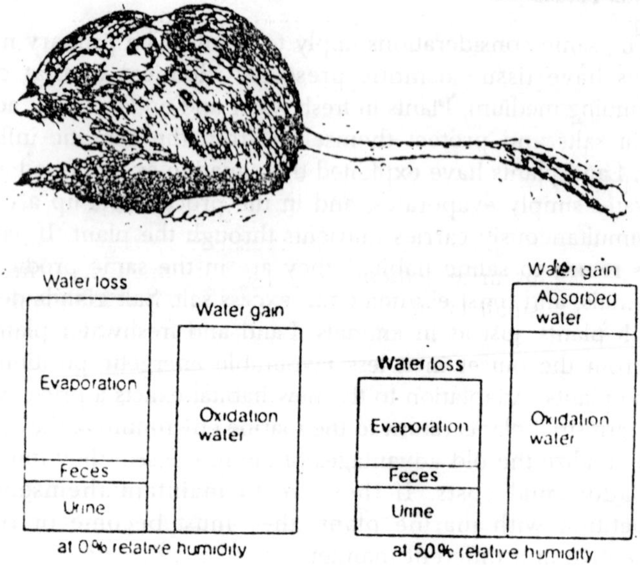

Fig. 2.2. Water balance in the kangaroo rat Dipodomys spectabilis at different relative humidities.

directly. It will become apparent that in the last analysis water balance is merely a special case of salt balance and osmoregulation. The problem of moisture, which occupies a central place in textbooks on terrestrial ecology, must be given less emphasis in a general presentation that includes marine ecology. It is therefore treated as a subtopic in this chapter.

First let us consider the strategies by which animals adapt to differences in relative humidity. Simple model experiments and theoretical considerations show that a small sphere of gelatine loses more moisture per unit time than a larger sphere under the same conditions. The small sphere soon dries up; the larger one retains moisture for a longer time. As a strategy of adaptation .to low-humidity regions, then, we would expect to find increase in the size of animals. But this increase brings a disadvantage; in that larger animals take longer to reach sexual maturity. Their reproductive rate is fundamentally lower than that of a small animal. Thus in regions where there is little or no danger of desiccation, smaller animals have a selective advantage over larger ones. From the cold deserts of the arctic to the arid deserts of the tropics, there is a distinct increase in mean body size of the insects sampled (small insects, if present, are active only at night, when humidity is high). On the other band, the number of individuals per unit area is greater in cold deserts than in hot deserts. What has been said of deserts applies similarly to other habitats.

Humid-cool regions are occupied, on the average, by smaller insects than dry slopes; maritime regions, such as the North Sea coast and the British Isles, on the average have smaller insects than continental regions like eastern Prusia or Hungary. But there is a catch in this arrangement–larger insects must pass through juvenile stages small enough to be in danger of desiccation. The danger is particularly great in arid regions. Many of the animals living here provide protection for their young or by actual brood-care behaviour, shift the emergence of the youngest stages to a time when the humidity is relatively high. They actively select particularly favourable (i.e., moist) elements in the habitat. The youngest aid thus smallest stages of crickets stay in the more many place near the bases of the plants; mature crickets prefer the open areas between the plants. A similar shift of locale occurs during the bug life cycle; from their original post at the bases of plants, the young stages migrate to the exposed tips. Another way to escape desiccation is to adopt the nocturnal habit.

Mammals lose more water by evaporation when the air temperature is higher than that of their bodies. Under such conditions, they must allow water to evaporate in order to keep the body cool. Some desert mammals save water by doing without such cooling; they are able to tolerate an increase in body temperature when the .air temperature is very high. The same species preserve the water ordinarily lost in respiration, by retaining it in the nose with a specific cooling mechanism On the other hand, remarkably few organisms have developed the ability to withdraw water vapour from the air. Such behaviour is know only among a very small number of animals (book lice) and lower plants. No vascular plants can do it, but many of them can utilize water suspended in the air as droplets–fog: for example. The occurs in the high deserts of the Andes.

Very high humidity is unfavourable to terrestrial organisms, in all temperature zones. It is obvious that plants should suffer, for they depend on transpiration. But we still do not know why grasshoppers, crickets, butterflies, beetles, and their larvae show practically no growth or development at relative humidities, above 80 percent.

TEMPERATURE

No environmental factor seems so easily measurable, and there is no factor of which we are so readily aware, as temperature. Because of its accessibility, a chaotic mass of temperature data has accumulated, most of which is unrelated to the ecological context within which it was published. In fact, it is difficult to measure temperature at the site of importance to the organism–in the organism itself; the temperature here (in the case of land organisms) bears little relationship to the meteorological temperature. The latter can serve only as a rough estimate.

Accordingly, it is considerably harder to establish a relationship between the distribution of an animal or plant and the temperature than, for example, to learn how distribution depend on the salinity of the available water. As an illustration of the difficulty all such judgements, consider the following example. In Bavaria, the part of West Germany with the most continental climate, the nightingale is, found in Only a very few Place where the climate is especially mild. It is of common occurrence in many other parts of Germany. About April 20 nightingale return from its wintering grounds; it cannot tolerate nighttime frost, and hard frosts are generally to be expected in Bavaria until the beginning of May. By contrast; the hoopoe is a regular inhabitant of many locations in Bavaria.

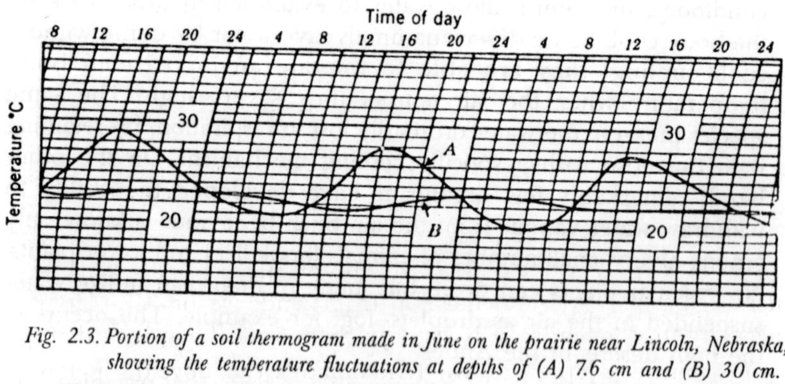

Fig. 2.3. Portion of a soil thermogram made in June on the prairie near Lincoln, Nebraska, showing the temperature fluctuations at depths of (A) 7.6 cm and (B) 30 cm.

It requires considerably higher temperatures than does the nightingale. It arrives later (not before the beginning of May, when the night frosts in general have stopped and the approaching continental summer brings very warm days and very small amounts of rain. In most other parts of Germany (apart from Baden-Wurttemberg) the hoopoe is considerably less common, because the summer temperatures are low and the precipitation greater. It is not clear whether the influence of temperature here is direct or is exerted by way of another factor such as food supply. Temperature affects all chemical processes as formulated in Vant'Hofts Law a 10°C increase in temperature accelerates a chemical reaction by a factor of 2-4. We say that this chemical reaction has a Q_{10} of 2-4.

The biochemical reactions of organisms are naturally subject to this law. Given that the temperature can fluctuate over a wide range during the course of a day, it is understandable that during evolution all organisms have developed mechanisms that liberate them, to a greater or lesser extent, from this temperature dependence. The warm-blooded animals have gone furthest in this respect, and will be discussed separately. The ectothermic organisms, with body temperatures that can change over a wider range (microorganisms, plants, and poikilothermic animals), seem at first glance to be obliged to follow even the most erratic changes in environmental temperature. But in fact they too have developed a lavish array of regulatory mechanisms, and in many respects even these organisms are evidently unaffected by temperature. The most illustrative example is the physiological clock of plants and animals, which runs with period of about 24 have regardless of the prevailing temperature.

It is self-evident that this should be so; a physiological clock that ran faster at high than at low temperatures would offer no advantage in the struggle for existence and would therefore never have become established in the solution. But this example demonstrates that temperature compensation is possible even in ectothermic organisms. If we want to know more about the general aspects of such compensation, we must ask how widespread these phenomena are an explore the underlying physiological mechanisms. It is astonishing the true compensation seems never to have become a property of the processes of development. Allectother microorganisms appear to develop in dependence on Vant'Hoffs Law. And this dependence is even more extensive than is evident.

Organisms adapted to low temperatures can develop at low temperatures, but their development occurs no more rapidly than would that of tropical organism if it were able to develop at that temperature. That is, adaptation to permanently low temperatures is brought at the cost of extending the period of development. The stoneflies and mayflies that live in mountain brooks need a while

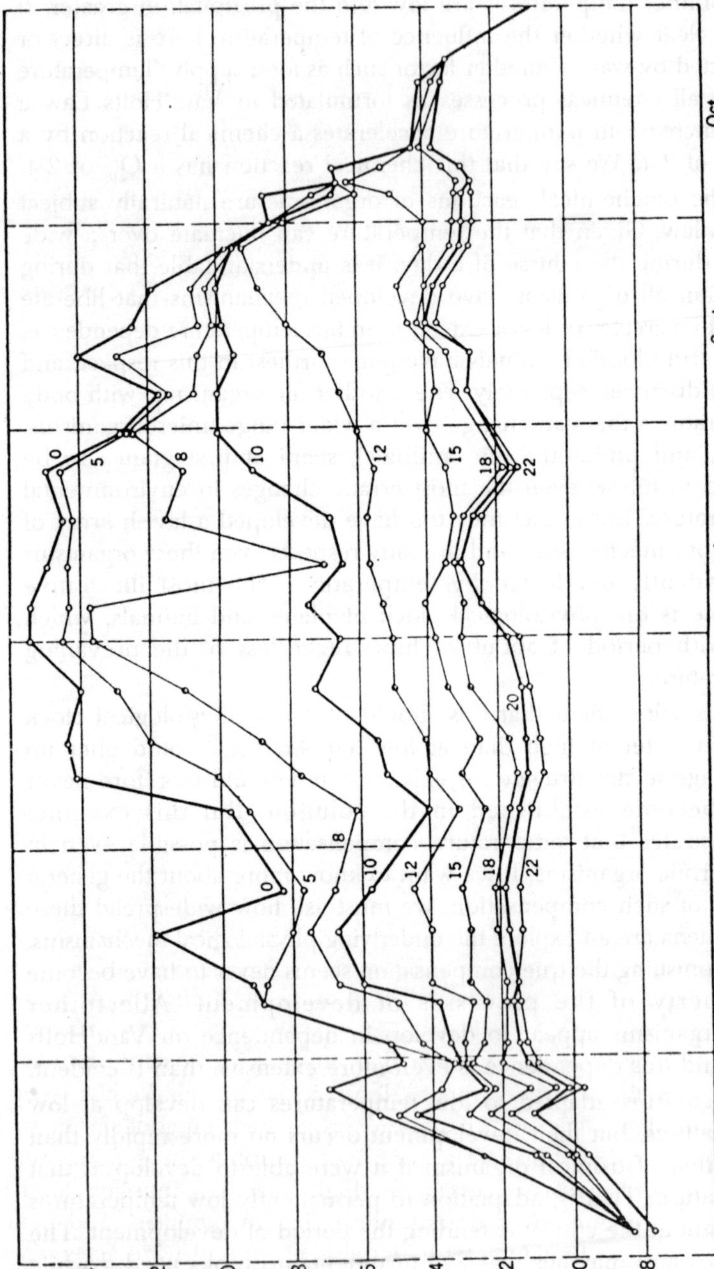

Fig. 2.4. *Seasonal change in water temperature at the indicated depths (in meters) in Lake Mendota, Wisconsin.*

year to grow to the size of a housefly but a housefly could not-grow at all at such temperatures. The antarctic fish Trematomus lives in an environment with a constant temperature of–1.6°C and requires about 10 years to reach the size of a small trout. And there are still other costs involved. An animal that has adapted to low temperature and is paying for this adaptation with a prolonged period of development at the same time loses the ability to exist at higher temperatures. All the mayfly larvae, stoneflies, and amphipods indigenous to mountain brooks die if exposed to the, temperature that any pond will reach during the summer, about 25°C.

Typical winter animals such as the snow scorpionfly Boreus cannot tolerate temperatures around 20°C for very long. The price for this adaptation, then, is high. The best proof is given by the fact that we find refrigerators useful. If during their millions of years of evolution microorganisms had developed vital functions that could proceed as rapidly at low as at high temperatures, we would not be able to keep food in simple refrigerators. From the outset, then, we would hardly expect to find truly temperature-compensated processes of development in ectothermic organisms. It is still not entirely clear why organisms were able to achieve independence of temperature in many functions, but never in those of growth and development. Comparative biochemical analyses have led to a current hypothesis that can be expressed in simplified form as follows:

The normal enzymes of an organism, are adjusted to operate best at temperatures in the range from 28°C to just over 30°C; here they are most effective and last for a long time. There are enzymes that can operate just as effectively at higher or lower temperatures as the "normal" enzymes do at 28°-30°C. However, these enzymes survive only briefly they soon break down and must continually be resynthesized. The possession of such enzymes costs energy. Functions involving these special enzymes can proceed at temperatures higher or lower than normal at the same rate as in the normal temperature range. However, because of the high energy requirement, it is not possible to provide all functions with such enzymes; there must be a system of priorities. In general, sense organs and organs necessary for flight reaction (muscles) fend to take precedence over organs subserving metabolism (and hence growth). This fundamental hypothesis has found wide acceptance, but cannot as yet be considered sufficiently well documented to be given the status of a theory.

Finally, by Vant'Hoff's Law, the curves we find for temperature dependence are exponential. That is, very slight temperature differences can have very large or small effects, we can imagine a species of which most individuals cannot reach maturity during a normal year their number decreases from year to year, until eventually one of the rare very warm summers makes it possible for the individuals to complete development. Now the size of the population leaps suddenly from very small to very large so that the species can persist during the following normal years.

Of course, one must bear in mind that it is not the temperature of the air but that of the organism itself that determines the rate of development. Body temperature can depart considerably from air temperature. In flight, the temperature of all the larger insects rises to more than 35°C because of the activity of the flight musculature. Bees can warm up their hive in this way; bumblebees, by beating their wings, keep their brood warm so effectively that the species can invade even arctic regions. Furthermore, a great many animals can heat themselves in the sun. Grasshoppers, in the cool of the morning, position themselves broadside to the sun's rays and as a result quickly become very warm. During the midday that they turn their heads to the sun so as to present the smallest possible surface, and absorb relatively little heat: This sunbathing behaviour may well be an absolute necessity for the survival of many insects in temperate and cool regions. It is characteristic of red ants when they first leave the nest in spring, of many lepidopterans in arctic and subarctic regions, and probably of a number of insect larvae.

Redbugs (Pyrrhocoridae) in spring, and later their larvae, use the sun's radiation to regulate their body temperature. Usually they take up a position at the foot of a tree to sun themselves. But if forced to stay there permanently, so that their body temperature is always relatively high, they develop too rapidly; the synchronization between stage of development and time of year is lost, and the animals die. Under natural conditions the bugs move back and forth between sunny and shady places, adjusting their movements to the changing conditions through the year. On the forest floor in Denmark, the development of the large brown weevil Hylobius abieties generally takes three years. But in clearing, where the ground is heated by the sunshine, development is complete in two years.

As a consequence, the damage this beetle does to coniferous forests is considerably greater where clear-cutting is practiced. The difference in duration of development corresponds to a southward shift of about 1,500 km (Bejer-Petersen, 1975). It appears that all ectothermic land animals begin to lose water when their body temperatures reach 34-35°C, and water loss increases sharply as the temperature continues to rise. This is probably a cooling mechanism (Jakovlev and Kruger, 1954; Jakovlev, 1956). In experiments on Mediterranean crickets the body temperature could not be raised above this level even by irradiation, whereas similarly irradiated dead animals reached a temperature of about 45°C within two minutes. Many of the physiological parameters that have been measured (transpiration, respiration the enzyme activity of various organs) indicate that in terms of the biochemistry of their enzyme complement all the eurythermic animals are adapted to an optimal temperature of around 27°C.

In many cases reptiles, too, can invade cold regions only because they use the sun's radiation to accelerate development. The adder and the lizard Lacerta vivipara have advanced further north than any other reptiles. Neither lays eggs; they bear living young. During the day they let their bodies get very hot in the sunshine, and the unborn young are heated as well. Plants can be injured if they are overheated by the sun. Under intense irradiation the dark bark of a tree can become very warm–a danger both in winter, when a sharp temperature gradient is produced between the lighted and shaded sides of the tree, and in summer, when the heating can be sufficient to interrupt sap flow. Injury so caused is evident at the edges of clearings and alone roads, especially in beeches. Green leaves are not heated very much by radiation; furthermore, their temperature can be brought appreciably below that of their surroundings by transpiration.

The same is, of course, true of animals; by high transpiration rates and specific behavioural mechanisms ectothermic animals can also become cooler than their surroundings. When desert birds bury their eggs in sand, in many cases it done to protect them from heat. For the same reason insects, spiders, and lizards seek out shady places. It is evident, then, that the temperature of an organism can fluctuate even more widely than the meteorological

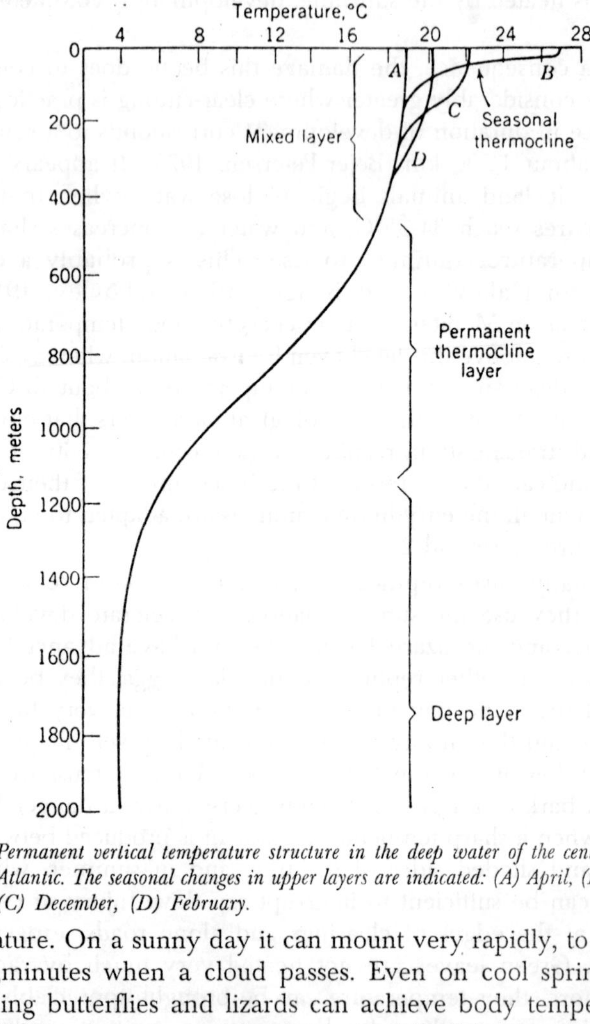

Fig. 2.5. Permanent vertical temperature structure in the deep water of the central North Atlantic. The seasonal changes in upper layers are indicated: (A) April, (B) August, (C) December, (D) February.

temperature. On a sunny day it can mount very rapidly, to plunge within ,minutes when a cloud passes. Even on cool spring days sunbathing butterflies and lizards can achieve body temperatures above 35°C, whereas at night they are chilled by frost. Can we make direct comparisons between such changing temperatures and what we know about the effects of constant temperatures? If we know the time (in hours or minutes) during which an organism is exposed to a particular temperature; and its rate of development under constant temperatures, we should be able to calculate how long it would need for development under a particular changing-temperature regime. As early as 1928 Kaufmann made such

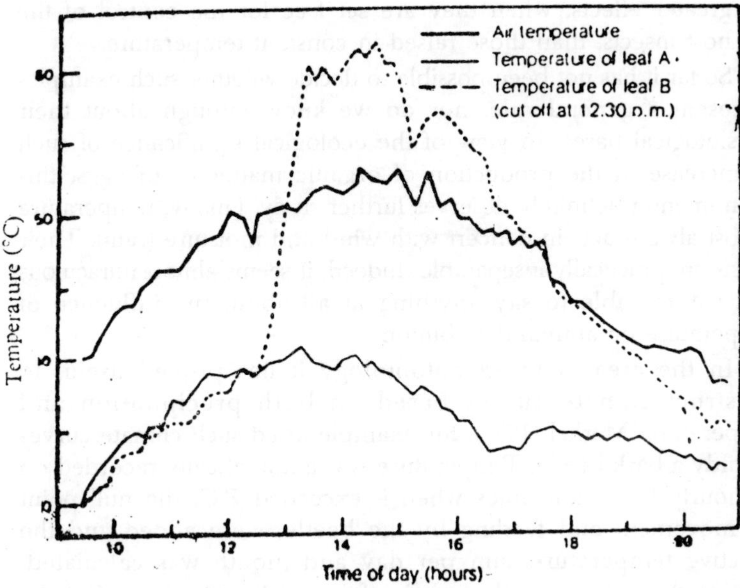

Fig. 2.6. The temperature of a leaf can be well below that of the air, as a result of transpiration and the associated cooling. A cut-off leaf is rapidly heated by sunlight, because of lack of water.

calculations. His conclusions have been confirmed recently by a large number of experiments. The calculations are complicated by the exponential relationship between growth had temperature, and by the existence of a threshold a null point, for development. Because of these features, growth and development in general proceed more rapidly under fluctuating temperatures than in a constant temperature equal to the mean, and the difference is most pronounced at low temperatures. However, there are a few notable phenomena that have not yet been satisfactorily explained.

1. The productivity of many plants is increased when the temperature fluctuates, Perhaps low nighttime temperatures act to prevent high respiratory losses. Such an explanation could also apply to the diurnal vertical migrations of planktonic animals, which when they are not feeding sink to the colder levels in the water.
2. In many of the ectothermic animal that have been studied the- rate of reproduction is greatly increased in fluctuating temperatures. Parasitic wasps of the genus Trichogramma raised under fluctuating-temperature conditions have appreciably

greater effects, when they are set free for the control of the host insects, than those raised in constant temperature.

So far it has not been possible to decide whether such examples represent a general rule, nor do we know enough about their physiological bases. In view of the ecological significance of such an increase in the production of organic matter or of eggs, this phenomenon definitely deserves further study. Finally, temperature almost always acts in concert with wind and moisture (rain). Their effects arc practically inseparable. Indeed, it seems almost miraculous that we are able to say anything at all about the influence of temperature on animal distribution.

In the area of forest entomology it has proved useful to construct climate curves based on both precipitation and temperature. Merkel (1977), for example, used such climate curves to study a bark beetle. Temperature was automatically recorded on an hourly basis; the times when it exceeded 7°C, the null point for movement and feeding by the beetle, were added and the effective temperature sum per day and month was calculated. Plotting these sums on the x axis of a coordinate system and the sums for precipitation on the y axis gives the climate curve. When there is heavy rainfall the curve rises steeply, and when the weather is dry it is nearly parallel to the x axis. By entering observations of stages in the life of the beetle in this graph, one obtains data specific to the measurement site which can be compared with results similarly obtained at other locations or in other years. Using this method, researchers have developed considerable insight into the dependence of insects on the climate in their habitats.

But even this procedure is not entirely satisfactory, because depending on the general state of the weather the beetles can require different integrated temperatures. For example, the large bark beetle Ips typographus does not appear until the temperature at the site where it has spent the winter rises above 7°C. By this time the surface of the soil has reached 10-20°C. The beetles then begin feeding under the bark; to do so, they require temperatures between 12 and 19°C. The temperature sums necessary for maturation of the gonads vary depending on the weather. The breeding flight of the mature beetle can occur only if the air temperature is at least 20°C. The body of the animal must have reached at least 23°C, which can happen at an air temperature of 20°C if the weather is sunny but requires 23°C air temperature under overcast skies.

Another approach to the difficult problem of temperature in terrestrial habitats is to determine the effective mean temperature (eT) by the method of Pallmann et al. (1940): This procedure is based on the fact that sucrose in, aqueous solution is broken down by hydrogen ions to form glucose and fructose.

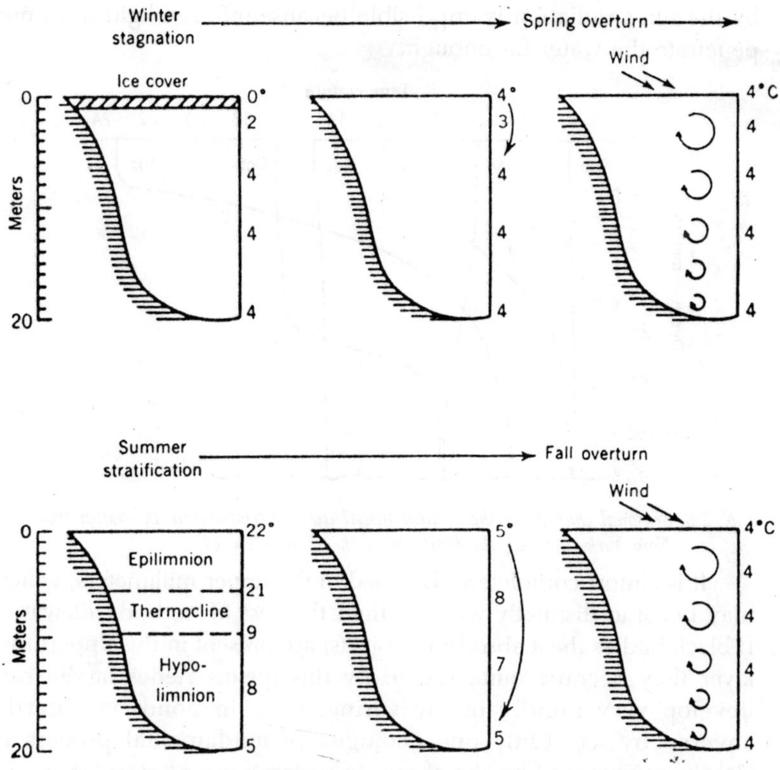

Fig. 2.7. Diagrammatic vertical section of a deep lake of the temperate region, showing the seasonal cycle of temperature.

In a buffered solution, with constant hydrogen-ion concentration, the reaction is temperature-dependent. The degree of sucrose inversion can readily be monitored polarimetrically, providing a convenient measure of the temperature situation in a particular period of time. The advantage of the method lies in the fact that the reaction rate, like the growth processes of ectotherms, rises exponentially with temperature. Sucrose inversion ought therefore to give a very precise, measure of temperature-dependent biological processes. But in spite of this obvious advantage the method has not been widely adopted, no doubt for fear of overgeneralization. That is, there are no really well documented relationships between organic life and the temperatures recorded in this way. Exemplary studies in which this method is compared with, others would be most informative. In water things are simpler. A general heating by the sun's radiation is impossible because infrared light does not penetrate the water far enough.

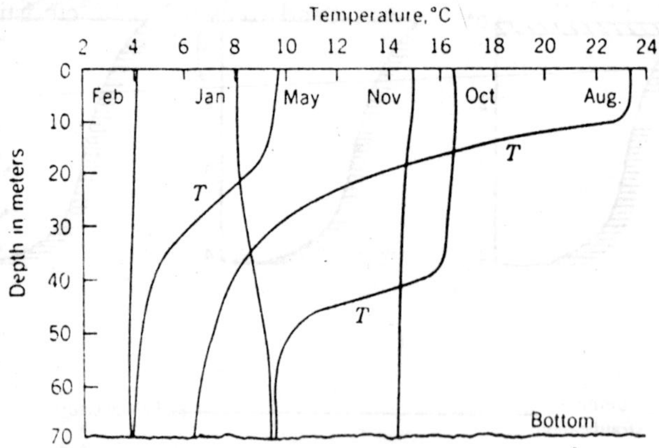

Fig. 2.8. Seasonal changes in the vertical distribution of temperature in coastal water off Now York, showing the positions of the thermocline (T).

It is almost completely absorbed in the upper millimeters, which may become distinctly warmer than the rest of he body of water. If black bodies (heat-absorbing objects) arc present in this uppermost layer they become quite warm. By this means Aedes larvae can develop very rapidly in the spring, even in ponds still partly covered by ice. Only one ecologically fundamental process is clearly not covered by the above considerations–photosynthesis in green plants.

Photosynthesis is not a simple biochemical process; to a considerable extent it involves photochemical events and is thus not so dependent on temperature. The Q_{10} of photosynthesis is less than 2, distinctly lower than that of biochemical reactions in general (with a Q_{10} of 2-4). Hence photosynthesis can operate at low temperatures, and at higher temperatures does not increase to the same degree as respiration or any of the functions of animals and microorganisms. Fig. illustrates a basic ecological phenomenon: in the moist, warm tropics dead organic matter is rapidly decomposed, whereas in cool regions decomposition occurs slowly. However, the production of organic matter by the green plants is not very different in the two locations. This fact was remarked upon by Darwin in his book about the voyage of the Beagle. In comparing the jungles of the Amazon and of Tierra del Fuego, he writes that the tropical rain forests appeared to him as a symbol of life, full of power and growth; the forests at the southern tip of the continent seemed like a symbol of death, full of dead tree trunks, branches and twigs. Forests with dead trees are characteristics of temperature and cool zones, because at low temperatures the dead wood remains intact for a long time.

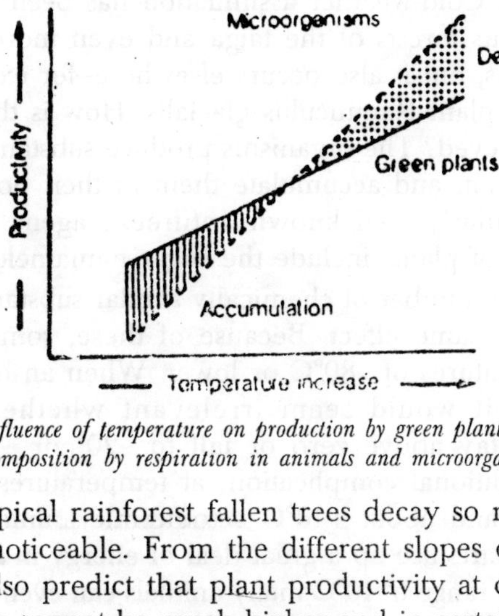

Fig. 2.9. The influence of temperature on production by green plants and on the processes of decomposition by respiration in animals and microorganisms.

In a tropical rainforest fallen trees decay so rapidly that they are hardly noticeable. From the different slopes of the two lines, we would also predict that plant productivity at consistently high temperatures cannot be much higher and in certain circumstances may be even less-than at temperatures in the intermediate range. At high temperatures the steady losses due to respiration are so large that the balance between photosynthesis and respiration may show a deficit. Again, of course, the temperatures of interest are not those in the meteorological reports. If the plants have access to enough water, massive transpiration can hold the leaf temperature well below that of the surrounding air. Warming by heat radiation, such as occurs in animals, plays no role in the, leaves of plants; essentially all the infrared radiation passes through the leaf oris reflected by the chlorophyll. So far we have been considering only the period in which plants and animals are active.

In climates with a cold winter the organisms face the problem of surviving freezing temperatures and sometimes temperatures well below the freezing point. Many animals can avoid exposure by moving to parts of the water or soil that do not freeze, but many others cannot do this; in the permafrost regions of the arctic such a strategy is entirely impossible. Very few organisms can tolerate freezing of their body fluid or of their cells. At the lowest temperatures which they can survive, nearly all organisms manage to keep their internal milieu in the liquid state. This fact, together with the photo chemical nature of photosynthesis, explains the ability of green plants to assimilate carbon dioxide with a positive balance even in winter. Cold-weather assimilation has been demonstrated in the coniferous forests of the taiga and even more strikingly in antarctic lichens, but it also occurs elsewhere–for example, in the high-mountain plant Ranunculus glacialis. How is this prevention of freezing achieved? The organisms produce substances that lower the freezing point, and accumulate them in their body fluids and cells. A particularly well known antifreeze agent in animals is glycerol, those of plants include the sugar hamamelose.

There are a number of chemically similar substances that tend to produce the same effect. Because of these, some beetles can tolerate temperatures of –80°C or lower. When an organism is so well adapted it would seem irrelevant whether the winter temperatures stay above zero or fall to –20° or –30°C. In fact there is an additional complication; at temperatures between the freezing point and about +10°C ectothermic animals and plants lacking green parts use up a great deal of energy in respiration; in the temperature range 6°-10°C many animals can even more actively about. But only predatory organisms can, under certain conditions, cover these losses, since the nutrients in animal tissue are readily utilizable. Herbivores and detritus feeders cannot eat enough to replace the lost energy at those temperatures.

It is considerably better for them if the winter temperatures remain well below the freezing point, so that none of their store all energy is used up. The paucity of species in maritime regions (for instance, Schleswig-Holstein or the British Isles) is in part explainable on this basis; the winter is not warm enough for feeding, but not so cold that animals can retain all of their stored energy. In this regard continental climates are more favourable. The great variety of insects in the northern United States and central Russia is partially due to this effect. On the other hand, the very warm summers of continental regions are naturally more favourable to ectothermic animals than the rather cool summers of maritime regions with the same average annual temperature. Bony fishes in the ocean face a special problem in surviving the winter. As mentioned previously, their osmotic pressure is lower than that of their surroundings. The water in the open ocean freezes at a temperature of −17°C, whereas a fish would freeze at only −1°C. Here we have yet another price paid for adaptation to a new habitat.

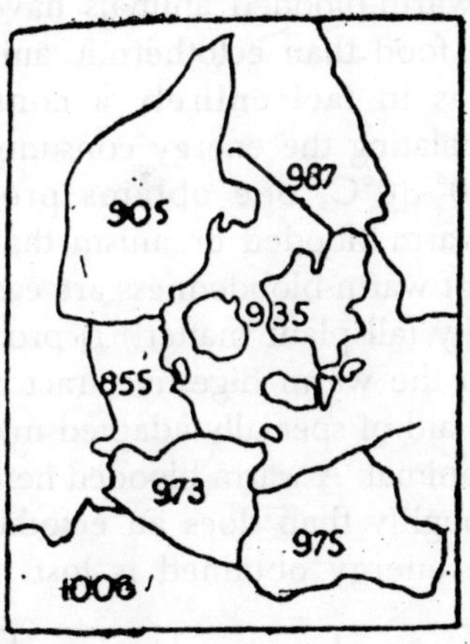

Fig. 2.10. The number of species of wild flowering plants in northern Germany and Denmark; it is lowest in the region with the most pronounced maritime climate. The regions with many species have areas of continental climate, so that they also include continental species.

Arctic and antarctic fishes solve the problem by moving to deeper levels during the winter or by raising the osmotic pressure of the blood to approximately that of the medium. In so doing, however, the evidently enter a sort of dormant state in which they ta e hardly any food. Species less well adapted in some situations–for example, when ice crystals are stirred into deep water by violent storms- die in massive numbers. Warm-blooded animals are in a quite different situation. These are organisms with a greatly elevated metabolic rate and concomitant high energy consumption and heat production. The heat is a side product of metabolism which may be given off from the body or, with suitable insulating mechanisms, can be retained. By changing the insulation properties of the body surface as environmental conditions change, such an animal can achieve a constant body temperature between 35° and 42°C depending on the species. When the external temperature is very high cooling mechanisms involving the evaporation of water come into play. As a result the animals are largely independent of the surrounding temperature; they can remain fully active even at low temperatures and colonize regions from which most ectothermic animals are excluded. Such regions include the antarctic continent and, to a considerable extent, the tundra of the far north (around latitude 80°N), where the temperature is so low and solar radiation so slight that ectothermic herbivores are practically negligible.

The price the warm-blooded animals have paid is that they require much more food than ectothermic animals of equal size. This requirement is in fact entirely a consequence of their temperature. Extrapolating the energy consumption per unit time of a crayfish at 20°-30°C, one obtains precisely the energy consumption of a warm-blooded organism that weighs the same. But the advantages of warm-bloodedness are easy to see. Food that is hard to utilize fully (all plant matter!) is processed a good deal more successfully in the warm digestive tract of such animals (in some cases with the aid of specially adapted microorganisms) than by any ectothermic animal. A warm blooded herbivore thus exploits its food more thoroughly than does an ectotherm. On the other hand, much of the energy obtained is lost in maintaining the

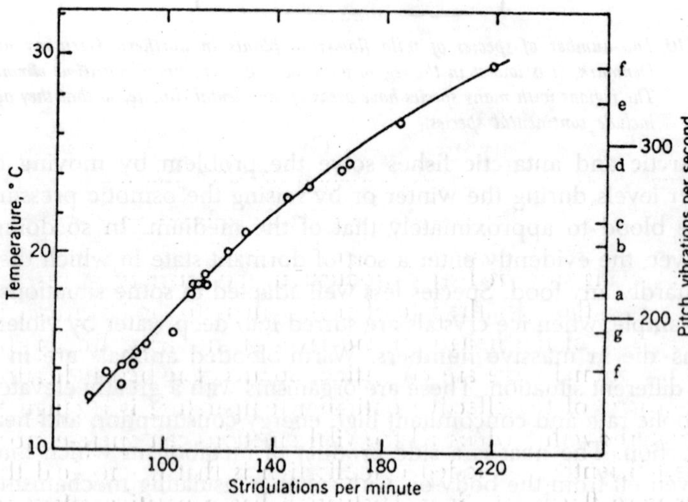

Fig. 2.11. Relation of stridulation rate of tree cricket Oecanthus to temperature and also to pitch. The note C is "middle C."

necessary temperature. In effect, then, an ectothermic organism produces mere animal matter than a warm-blooded animal does, for a given amount of plant matter consumed. It is not surprising that attempts have been made to reduce this high energy requirement.

The technique of insulation has been brought to perfection by some organisms-arctic animals consume hardly any more energy when the environmental temperature is low than when it is high. But warm blooded animals, like ectotherms, still face the problem of minimizing the loss of energy during the least favourable seasons. A warm-blooded animal that simply took shelter when the weather was bad would lose so much energy because of its ongoing high metabolism that it could survive only briefly. The conspicuous phenomenon of bird migration, and the comparable migrations of bats, show one way out of this dilemma. Another solution is periodically to reduce body temperature, as hibernators do. In the dormant state these animals can last out the unfavourable season without much loss of energy–but they lose the advantage of warm-bloodedness. Such experiments playa central role in ecology, however, so that they deserve a brief discussion here, in the context of temperature preferences. The first question to consider is whether the preferred temperature corresponds to the optimum. This is by no means necessarily the case. Behavioural research has shown that brooding geese presented with eggs of different sizes will choose one much larger than their own. Here the preferendum is far from the optimum. And we know that similar discrepancies can occur in the realm of ecology.

The Mediterranean cricket Gryllus bimaculatus has a temperature preferendum in the region of 34°C. But the optimum for this species is much lower, between 25° and 31°C. This range is so wide because the measured optima differ, depending on the function being analyzed–egg production, the quantity of reserve substances in the eggs, the rate of development to the imago stage, the final size of the imago, or the level of mortality. But in the life of the animal there are no distinct optima. The best indication one can give of the effective optimum temperature is a range of temperatures within which all the vital functions can operate more or less efficiently. An added complication is that in the field the temperature fluctuates. Such fluctuation has a positive effect on egg production. However, the crickets does not (as tiger beetles do) seek out locations at different temperatures at different times of day. Finally, the food supply can modify the effect of temperature. We may well ask whether animals live within their optimum range at all. To what extent may other factors encountered under field conditions force them into a range of temperatures we would not consider optimal? We cannot answer this question here, but will return to it in the discussion of competition.

Range of Temperature Tolerance

The different species of plants and animals vary greatly in the limits of temperature that they can tolerate. Plants and animals, which can tolerate very large fluctuations in temperature, are called as eurythermal, (e.g., cyclops, wall lizard, grass snake, toad, man, etc.) whereas others which can tolerate only a small variation in temperature are termed stenothermal (e.g., fishes, snails, coral reefs, etc.). In organisms all metabolic processes necessary for life begin at a certain minimum temperature and increase with rise in temperature until they reach the maximum level at a temperature called optimum temperature. Further rise in temperature beyond optimum brings about decrease in metabolic rate, until it ceases at a temperature called maximum temperature. This is of great ecological significance and limits the distribution of plants and animals.

Minimum temperature

The lowest temperature at which an organism can live indefinitely in an active state is termed minimum effective temperature. If an animal or plant is subjected to a temperature below the minimum effective limit it enters into a condition of inactiveness called chill coma. The activities of all animals and plants in chill coma can be restored by warming them to minimum effective temperature. The lowest temperature, at which the survival is possible, is called as minimum survival temperature. The minimum survival temperature depends very much upon time. For example, in an experiment the eggs and larvae of fruit fly Ceratitis capitata were killed after 7 weeks when kept at 7°C, after 3 weeks at 4°C, and after 2 weeks at 1°C. The minimum survival temperature is generally slightly above 0°C, because at 0°C the water freezes and this produces mechanical harm in the form of rapture of cell walls and in the stoppage of circulation. But certain animals and plants have certain adaptations, where by they can tolerate temperatures even below 0°C. For example, a green alga

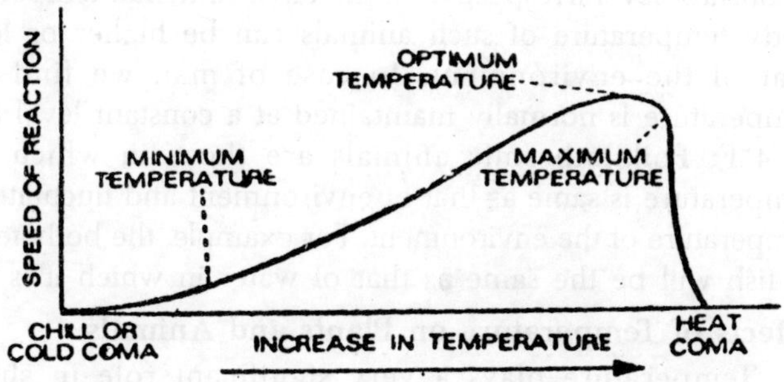

Fig.. 2.12. Graph showing effect of temperature on the physiological activity of animals.

Chlorella can withstand a temperature of–182°C for an hour without any harm. Similarly, Alaskan black fish can recover to normal activity after an exposure to–20°C for 40 minutes.

Optimum temperature

When the temperature is increased above the level of minimum temperature, there is also an increase in the physiological activities of animals and plants. The temperature at which the physiological activities of animals and plants are at the maximum is called as optimum temperature. This optimum temperature also varies in different species at different times.

Maximum temperature

The maximum temperature varies very much in different plants and animals. Generally the maximum temperature, which the animals can tolerate, is 48°C, but certain plants found in desert regions can tolerate a temperature of 70°C. The maximum temperature at which a species can live indefinitely in an active state is called as maximum effective temperature. If the temperature is raised above maximum effective temperature the animals or the plants enter into the heat coma, a state of inactivity, but will recover if soon cooled.

Homoiothermic and Poikilothermic Organisms

All organisms (animals and plants) can broadly be classified into two groups:

(a) warm-blooded or homoiothermic, and

(b) cold-blooded or poikilothermic.

The former category includes mammals and birds whereas the later includes reptiles, amphibians, fishes, and plants. Homoiothermic animals are those animals which maintain the body temperature at a constant level irrespective of the environmental temperature. The body temperature of such animals can be higher or lower than that of the environment. In case of man we find the body temperature is normally maintained at a constant level at 37°C or 98.4°F; Poikilothermic animals are those in which the body temperature is same as that of environment and fluctuates with the temperature of the environment. For example, the body temperature of fish will be the same as that of water in which it is found.

Effects of Temperature on Plants and Animals

Temperature plays a very significant role in shaping the structure, guiding physiological processes, behaviouristic patterns, and distribution of most of the species of plants and animals. The important effects of temperature on the plants and animals are as follows:

Effect on metabolism

All chemical reactions going on inside the body of plants and animals are collectively called as metabolism. Most of these reactions are under the control of -enzymes, which in turn are influenced by temperature. Increase in temperature, upto a certain limit, brines about increased enzymatic activity, resulting in an increased rate of metabolism. For example, the activity of enzyme liver arginase which acts upon agrigine (aminoacid), goes on increasing when the temperature is increased from 17°C to 48°C. But increase in temperature beyond this limit (48°C) brings about

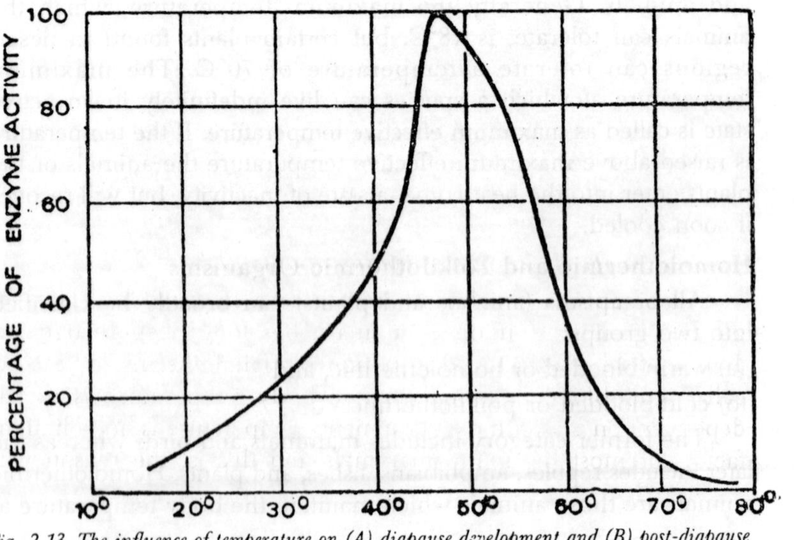

Fig. 2.13. The influence of temperature on (A) diapause development and (B) post-diapause morphogenesis in the embryo of the grasshopper, Austroicetes cruciata.

retardation in the activity of the enzyme. Accordingly there is reduction in the rate of metabolism dependent upon this enzyme.

Effect on sex-ratio

In certain animals, such as rotifers and daphnids, the sex-ratio is also affected by temperature. Under normal temperature daphnids give parthogenetic eggs which develop into females. In increased temperature daphnids give sexual eggs which after fertilization develop either into males into females.

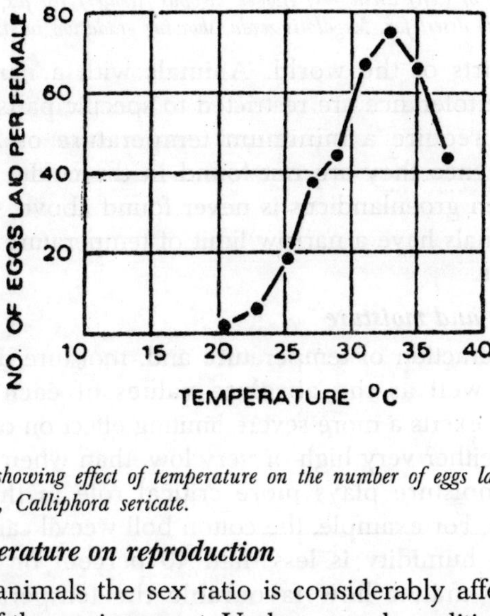

Fig. 2.14. Graph showing effect of temperature on the number of eggs laid per female in blow-fly, Calliphora sericate.

Effect of temperature on reproduction

In some animals the sex ratio is considerably affected by the temperature of the environment. Under normal conditions, daphnids produce parthenogenetic eggs which develop into females. But when the temperature of the environment is raised, they give rise to sexual eggs, which, after fertilization, may develop either into males or females.

Effect of temperature on distribution of animals

The importance of temperature as a limiting factor in the geographical distribution of animals, has long been recognized. Temperature affects the distribution of poikilotherms, or the so-called cold-blooded animals whose body temperature is mostly dependent on that of the environment. Temperature is not a limiting factor for most birds and mammals and this is, the reason why only birds and mammals occur in the coldest as well as in the

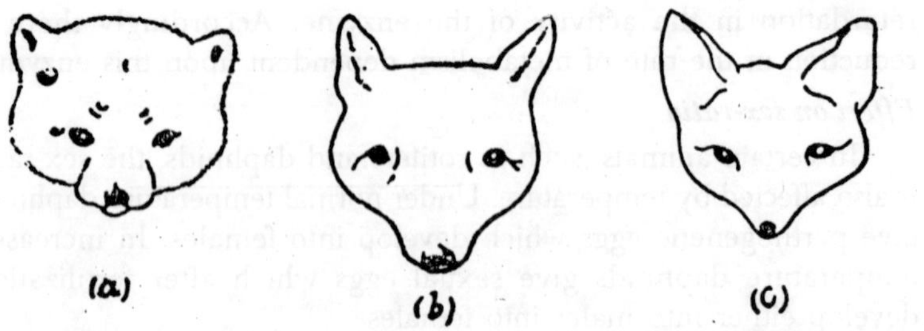

Fig. 2.15. Head of (left) arctic fox, Alopex lagopus, (center) red fox, Vulpes vulpes, and (right) desert fox, Megalotis zerda, showing gradation in size of ears.

warmest parts of the world. Animals with a narrow image of temperature tolerance are restricted to specific parts. For example, coral reefs require a minimum temperature of 21°C for their existence, hence they are not found in the colder regions of the globe. Pecten groenlandicus is never found above. 0°C, Generally aquatic animals have a narrow limit of temperature tolerance than land animals.

Temperature and moisture

The interaction of temperature and, moisture depends on the relative as well as the absolute values of each factor. Thus, temperature exerts a more severe limiting effect on organisms when moisture is either very high or very low, than when it is moderate. Similarly, moisture plays more critical role in the extremes of temperature. For example, the cotton boll weevil cannot develop if the relative humidity is less than 40 percent or more than 88 percent, no matter how favourable the temperature may be. Similarly, the animal remains dormant, regardless of humidity, if the temperature is lower than 10°C or higher than 39° humidity of 40 percent. But it develops in only 11 days if the humidity is between 60 to 65 percent. The two factors acting together produce the limitation, and may be referred to as a "limiting combination." According to Shelford 1951), limiting combinations probably operate under natural conditions more frequently than we realize.

Effect of temperature on animal behaviour and structure

The directive effect of temperature influences the animal behaviour to a considerable extent. Certain animals may be positive to one stimulus at one temperature and negative to the same stimulus at a different temperature. Temperature changes are also known to affect structural changes. Experiments show that

Drosophila may undergo structural modifications at high temperatures. The vestigial wings of one of the mutants will develop into normal wings at high temperatures. Evolutionary changes in the same fly may be induced by temperature. Some of the changes induced by radiation in Drosophila can, to some extent be, duplicated by temperature effects. Similarly, colour patterns in many insects can be induced or changed by regulating the temperature under which they develop. Many of the differences in the colour pattern animals of the same species living in different environments may be thus explained.

Other effects of temperature

If extreme, the effect of temperature may be lethal. There may be many more results of the effect of temperature. It has been found in Drosophila that temperature may affect the mechanism of heredity by affecting the position of genes and the behaviour of chromosomes. Mutations also occur under certain temperature. According to Jordan's rule fishes living in cold waters are said to have more vertebrae than those which occur in warm waters. Temperature also induces several other structural modifications in animals.

Effects of cold, heat and lethal temperatures

The thermal, tolerance is not the same in all animals. They can tolerate extreme temperatures only to a limited extent. Even with a very slow increase in temperature, we reach an absolute upper lethal limit for the particular animal beyond which it cannot adjust to a further increase. Similarly, the absolute lower lethal limit is that point beyond which no further decrease in temperature is tolerated.

Morphological and Physiological Adaptations to Meet Temperature Extremes

Animals and plants during the course of evolution have developed several ingenious methods to overcome the harmful effects of extremes of temperature. Some of the methods include:

Formation of Spores, Cysts, Seeds etc.

Some of the animals and plants produce cysts, eggs, pupae, spores, and seeds that can tolerate temperature extremes. Seeds of rye are active even at 0°C and can germinate at that temperature. Similarly, Amoeba, in encysted conditions, can tolerate even temperatures below 0°C. Encystment is very common in freshwater communities.

Removal of Water from Tissue

This is another method which enables the animals and plants to tolerate the extremes of temperature. Dried seeds, spores and cysts avoid freezing because there remains no liquid in them that can freeze. It has been observed that dry seeds can germinate even after exposure for 3 weeks to –190°C. Some of the bacterial cysts have high thermal resistance.

Dormancy

It is another special adaptation. The term dormancy includes two phenomena called hibernation and aestivation.

Hibernation

Over-wintering in a dormant state is known as hibernation. Animals which pass winter in a lethargic state are said to be hibernating. True hibernation consists of winter hypothermia accompanied by sleep. Hibernation occurs in poikilothermic animals as well as homoiothermic animals particularly micropterous bats (e.g., Myotis, Rhinolophus, Plectus, Macrotus, etc.), hedgehogs, ground-squirrels, and juming mice. This is characterised by reduced metabolic rate, low body temperature and reduced heart beat rate. In pioikilothermal animals the body temperature is equal to that of environmental temperature, whereas in mammals it is 1°C above the environmental temperature. On the approach of favourable conditions the animals come out of sleep and their body temperature rises.

Aestivation

Dormancy in summer when high temperature excessive dryness and/or shortage of food may occur is called as aestivation. It is very common among insects, some invertebrates, plants, lungfishes, amphibians and in certain mammals. Lungfish Protopterus, during summer, burrows into the mud, secretes a cocoon of slime around itself and lies dormant in this condition. In insects aestivation takes place in the form of diapause, during which growth and development are suspended, or greatly retarded. As we see dormancy is related with severe environmental conditions hence even insects belonging to same species undergo hibernation and aestivation during different periods of year in different parts of the world. This is beautifully illustrated by Coccinella septempunctata.

In London, it hibernates from January to May and from October to November, whereas, between the period ranging from June to September the normal development takes place. There is no aestivation period. In Moscow, it hibernates between January to April; normal development takes place in May, it aestivates in June, in July and August there is normal development, and from September to December it again hibernates. In Formosa, the development takes place from January to May and from October to December whereas aestivation period range, from June to September. There is no hibernation.

Homoiothermy

It is a special device for dealing with temperature extremes while in active conditions. Birds and mammals are able to maintain a constant body temperature (within certain limits) despite variations in temperature of the environment. In summer evaporation of water from their bodies takes place, this brings about cooling effect, and hence brings the body temperature to normal values. In cold weather the relative high temperature of their bodies is maintained due to insulating action of fur (mammal); feathers (birds) and fats (birds and mammals) and also with the help of certain physiological adjustments.

Thermal Migrations

This takes place in animals only. The journeys taken by animals that enable them to escape from extremely hot or cold situations are referred to as thermal migrations. The distance travelled by the animals during these journeys may vary from few centimeters to several hundred miles. Thermal migration in fishes have been studied by tagging methods. Desert animals move to shaded places to avoid the scorching heat, whereas, some of desert animals become nocturnal and thus avoid the heat of the day (*e.g.* desert reptiles and snakes). The frogs, turtles and other amphibious forms make short trips into or out of water and this provides cooling or warming as needed by the animal concerned. Burrowing animals, such as rodents, escape from high temperature by burrowing deep into the soil.

Light

Light is the factor that permits life to exist on earth at all. But it is difficult to distinguish among various habitats on this basis, for there is enough light to support photosynthesis over the earth's entire surface. Very few habitats–the ocean depths, the caves–do not receive enough light for plants to be productive. Everywhere else in the biosphere there is light in adequate amounts and with the correct spectral composition. It is only natural, then, that the quantitatively significant production of organic matter everywhere is in the final analysis based on the same mechanism–photosynthesis by means of chlorophyll. The different kinds of chlorophyll, as far as we know, hardly differ at all in productivity. We can consider them as a single entity. The chief natural sources of light are the sun, the moon, the stares and the bioluminiscent organisms.

However, all the energy of importance for organisms, under natural conditions, is derived directly or indirectly from the sun. Light is an environmental factor of great ecological and physiological significance. On light depends the synthesis of food by green plants, which in turn support all animal life on the earth. Light is a highly directional factor, its intensity, quality and duration all have influence on plants and animals. It is also responsible for directing and controlling various structural peculiarities and behaviour patterns in organisms and acts as a limiting factor in many activities of organisms.

Composition of Radiant Energy and Visible Spectrum

Solar energy consists of electromagnetic waves that extend from very short or high frequency wavelengths, to very long low frequency wavelengths. In solar energy high frequency short wave radiations, from 390 millimicrons (nm) downwards are ultravoilet rays, X-rays, gamma rays and cosmic rays. Low frequency long wave radiations from 700 millimicrons upwards are infrared, radar micro-waves and radio-waves (with frequency more than 1000 microns). A segment or part of solar energy, which can be perceived by the human eye, is called as visible light. This visible light is made up of a series of colours ranging from violet through blue, green, yellow, orange and red, constituting the visible spectrum. The frequency of wavelengths in visible light ranges from 390 to 700 millimicrons.

Light in Relation to Terrestrial Environment

The intensity of light reaching the earth's surface varies greatly. The variations in the intensity, quantity and quality arc caused by the angle of incidence, degree of latitude, amount absorbed and dispersed by atmosphere and a number of other climatic and topographical factors such as fog, clouds, suspended water drops and dust particles, etc. When the angle of incidence is smaller light rays have to travel by a longer distance through the atmosphere which causes reduction in intensity. Light intensity also becomes smaller and smaller with the increase in degree of latitude.

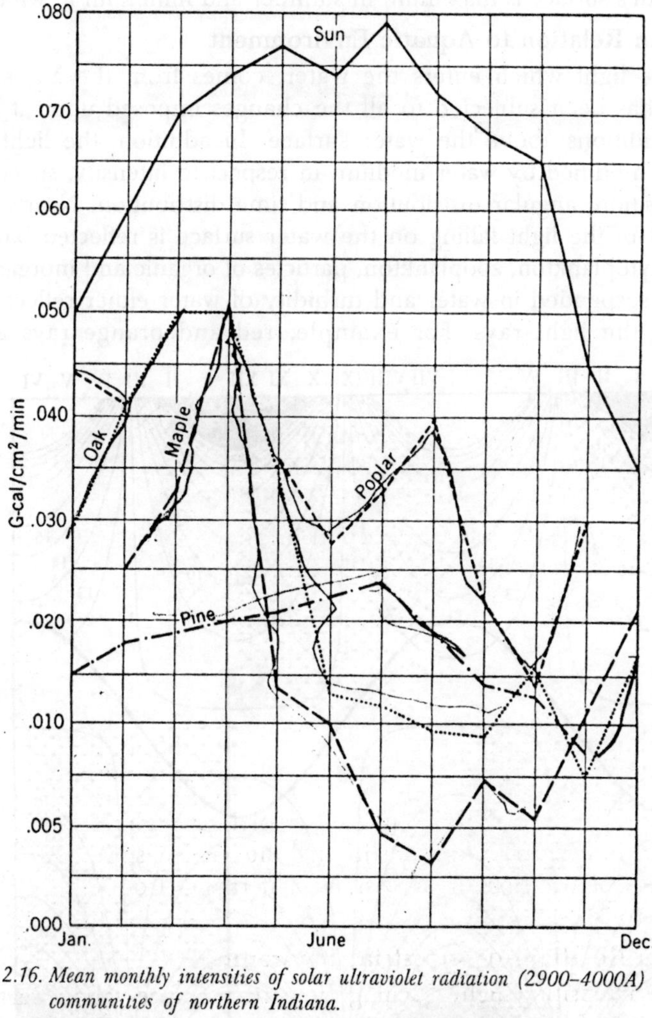

Fig. 2.16. Mean monthly intensities of solar ultraviolet radiation (2900–4000A) in forest communities of northern Indiana.

Atmospheric gases, such as nitrogen and oxygen, also absorb and disperse it small portion of light (mainly shorter wavelengths as ultravoilet) as it passes through the gaseous layer surrounding the earth. Other factors such as atmospheric moisture, clouds and suspended dust particles in the atmosphere exert great influence on the intensity of light. Topography and layers of vegetations such as forest canopy, also bring about some modifications in light intensity. For example, in a pine forest the amount of light reaching the earth's surface is maximum in summer and minimum in winter.

Light in Relation to Aquatic Environment

The light which enters the water, comes from the air, and hence has been subjected to all the changes imposed upon it by the conditions above the water surface. In addition, the light is further modified by water medium in respect to intensity, spectral composition, angular distribution and time distribution. Nearly to percent of the light falling on the water surface is reflected back. The phytoplankton, zooplankton, particles of organic and inorganic matter suspended in water and turbidity of water either reflect or absorb the light rays. For example, red and orange rays are

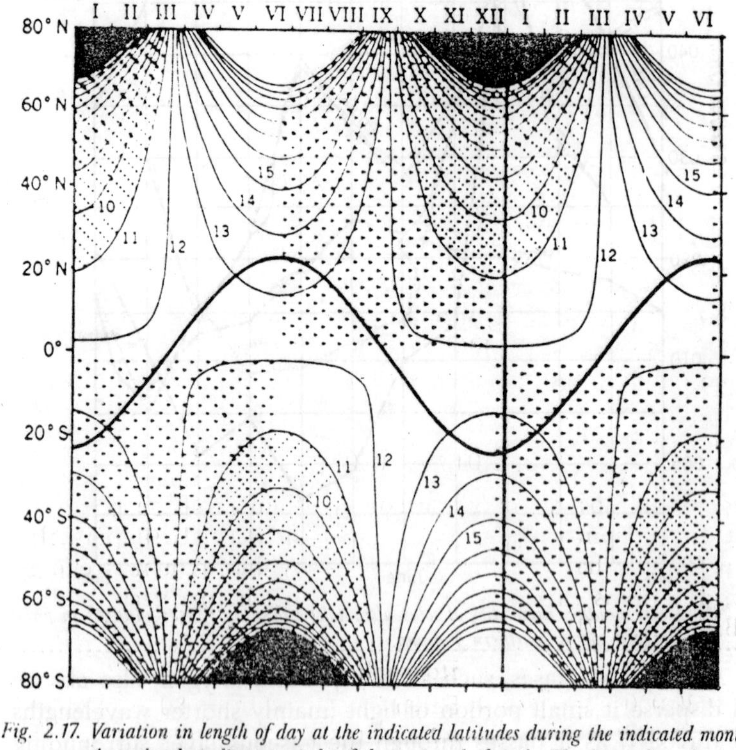

Fig. 2.17. Variation in length of day at the indicated latitudes during the indicated months. Arrowheads indicate periods of decreasing daylength. The heavy line shows the course of zenith sun.

completely absorbed upto the depth of 20 meters. Yellow rays may penetrate up to 50 meters where as green and blue rays penetrate 50 to 100 meters deep. As an initial consideration, we can take it that light energy is a factor in excess. Plants make use of but a small fraction of the incident radiation; the efficiency of light utilization is always less than 5 percent and usually 1 percent of the available energy (Chart 1); It follows that productivity within a habitat is usually not dependent on the amount of light–there is always enough.

Chart–1. Production by a Field of Maize, as an Example of the Calculation of Primary Production

Total dry weight of the maize from 0.4 ha (=1 acre) (10,000 maize plants)		6,000 kg
Ash (inorganic components) subtracted		– 300 kg
Total weight of organic components		5,700 kg
Their equivalent in glucose		6,700 kg
Plus organic matter lost by transpiration appropriate to the season (expressed as the glucose equivalent)		2,000 kg
Total weight of the glucose formed by 0.4 ha maize		8,700 kg
Energy required for synthesis of 1 kg glucose		3,800 kcal
Energy required for synthesis of 8,700 kg glucose	ca.	33×10^5 kcal
Total solar energy available to 0.4 ha % utilization of the available energy		$2{,}040 \times 10^6$ kcal

$$= \frac{33{,}000{,}000 \times 100}{2{,}040{,}000{,}000} = 1.6\%$$

Theoretically possible annual maximum yield (kg dry matter/ha, from de Witt as cited by Baeumer, 1971)	Location	Number of sheep theoretically supportable by this amount of energy (per ha per year)
25,000	Stockholm	68
30,000	Berlin	80
51,000	Purerto Rico	140
57,000	Tropical Australia	156

The level of production is dictated by temperature and water supply, as well as by the presence of minerals. Therefore it must be possible to calculate the level of production possible. Under the same conditions of soil, nutrient availability, and climate all plant communities must in theory exhibit roughly the same production, and observations have shown that in fact they do. Studies carried out in the Solling Project (supported by the German Research Foundation as part of the International Biological Programme) showed that the production of organic matter in a meadow, an area of coniferous forest, and an area of beech forest was quantitatively comparable.

This fact is hard to accept. We know, after all, that there are shade plants and sun plants. We know of plants that carry on photosynthesis during only part of the year–spring flowers, for example, which later die back into the ground. How can we reconcile such things with the above claims regarding uniform productivity and a single basic biochemical process? The only

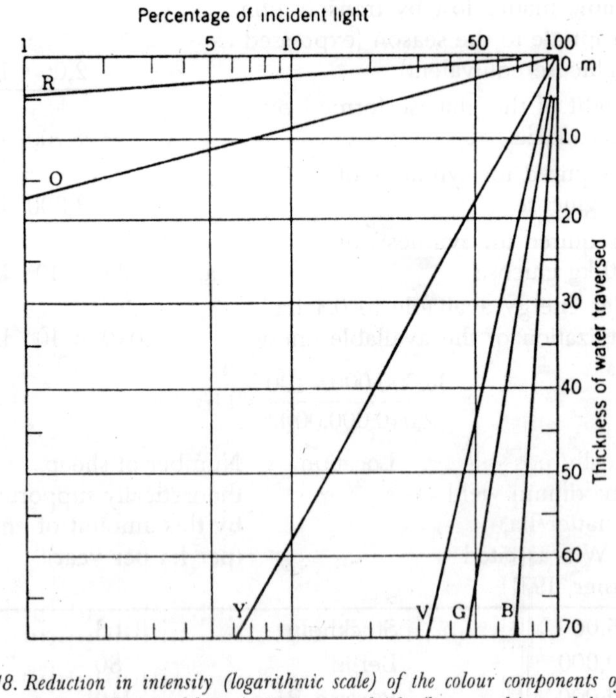

Fig. 2.18. Reduction in intensity (logarithmic scale) of the colour components of sunlight (indicated by initial letters) at increasing depths (linear scale) in a lake of optically pure water.

difference between shade leaves and sun leaves is that the former contain more chlorophyll; the same is true of Shade plants as compared with sun plants. The greater amount of chlorophyll compensates for the lower relative light intensity. Greater amounts of chlorophyll are obtained at a price; there is room for them in the leaf only if the thick epidermis and cuticle are eliminated. Therefore shade plants are very vulnerable to drought. In their normal habitat this is irrelevant, because shade is practically always associated with ample moisture.

The essential point here is that the mechanism of photosynthesis is identical to that of the sun plants. The assertion that productivity of different plant communities is similar under similar conditions does not refer to single species within the communities; no one could maintain that productivity is the same among different species. Differences at this level are inevitable in view of the variations in length of growing season, which is controlled by other factors. The comparison here is between one plant community composed of many species and another of equally complex composition. The overall production in such a community results from the activity of different plants occurring at different times; but because the underlying biochemical system is the same in all cases, the total production of the community per unit area and per unit time is the same. To recapitulate: the uniformity in the productivity of different plant communities under the same conditions is ultimately based on the fact that the amount of chlorophyll exposed to the light is independent of the species composition of the community.

The modern high-yield varieties developed for agricultural purposes produce no more organic matter than the wild varieties, but what is produced is differently distributed. Instead of an extensive root system and strong stems more is produced of the part usable by humans–the kernels of grain. There is one exception in principle to the uniform-productivity rule–one which leads to a slightly but distinctly increased productivity. Whereas the first product of normal photosynthesis is a triose, a sugar with three carbon atoms (hence the term "C_3 plants" for those with this form), in a number of plants the CO_2 taken up is first coupled to phosphoenolpyruvate–a C_3 molecule–to form oxalacetate, a C_4 molecule. Since the first product of photosynthesis here is a C, molecule, these are called C, plants. Later the oxalacetate is split up again; the carbon dioxide becomes available to the normal photosynthetic apparatus and is further processed in the normal pathway of photosynthesis, described

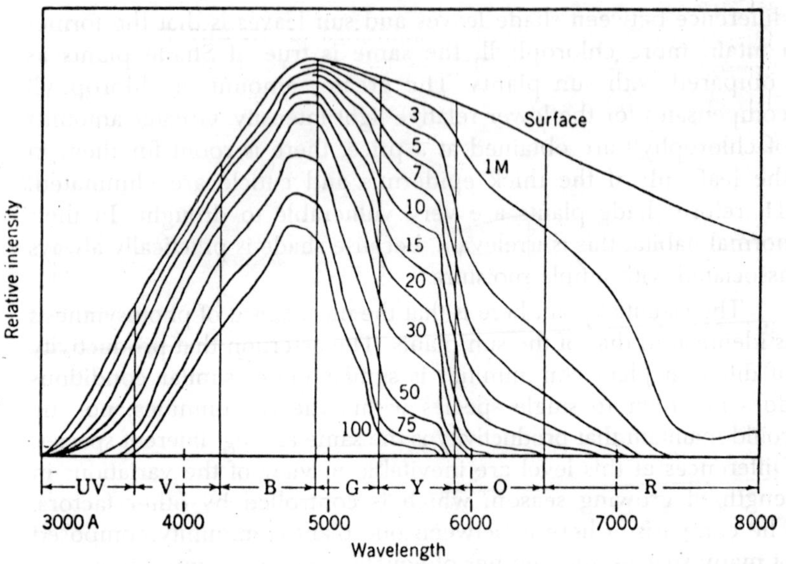

Fig. 2.19. The spectral distribution of solar energy at the earth's surface and after modification by passage through the indicated meters of pure water. Similar light conditions would be found in the clearest ocean and lake waters.

previously. This procedure uses up more energy than normal photosynthesis, and it is probably for this reason that C_4 plants are restricted to regions where solar radiation is very intense. But it conveys advantages.

The limiting factor for the plants is carbon dioxide. Phosphoenolpyruvate (PEP) carboxylase has a higher affinity for carbon dioxide than the ribulose-diphosphate (RuDP) carboxylase that acts as the acceptor for the CO_2 molecule in normal photosynthesis. As a result, more carbon dioxide can be fixed per unit time and productivity can be greater than that possible with normal photosynthesis. Examples of plants with this kind of photosynthesis can be found among the grasses and dicotyledons (e.g., Atriplex) in very sunny regions which (usually) have an irregular water supply. The familiar crops maize and sugar cane are in this group, as well as aggressive, wide-spreading tropical weeds because the difference between the two photosynthesis types is not clear-cut but gradual (all plants can bind slight amounts or carbon dioxide to PEP), a number of attempts are currently being made to find and cultivate varieties of other crop plants with C_4 photosynthesis. The superiority of C_4 plants over C_3 plants is based on yet another principle: C_3 plants have so-called photorespiration. That is, they lose considerable quantities of organic matter by respiration even during the day-in fact, the daytime respiration level is about 5 times as high as that at night.

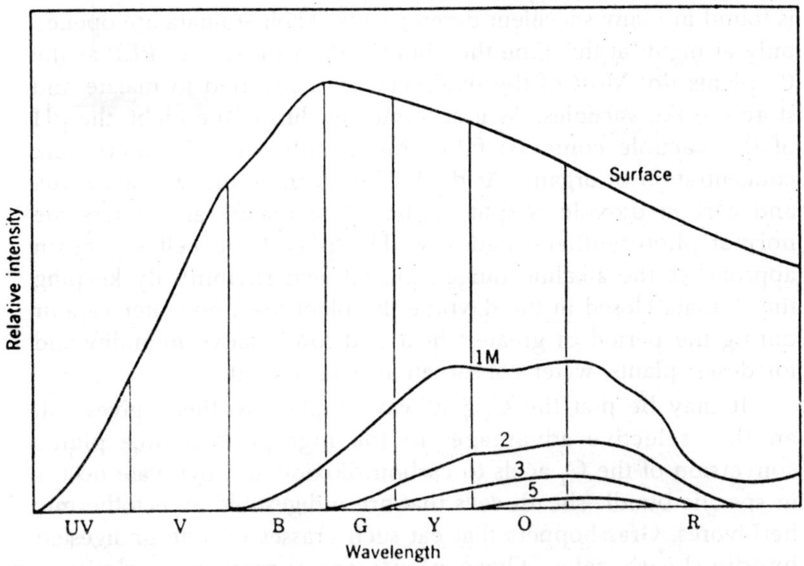

Fig. 2.20. Reduction in intensity and shift in spectral composition of light in heavily stained Rudolph Lake, Wisconsin.

The enzymes involved are not located in the mitochondria, but in the very small peroxysomes. For a long time there has that RuDP carboxylase, which couples carbon dioxide to the RuDP molecule, often "confuses" oxygen with carbon dioxide, especially at high oxygen concentrations. When this happens the ribulose diphosphate is oxidized to form phosphoglycerate and phosphoglycolate. The latter is oxidized to glyoxylic acid; the hydrogen peroxide produced in this reaction is converted to water by peroxidase. During photorespiration–the oxidation of glycolate–the plant obtains no energy; the process represents a pure waste. The entire pathway is probably explicable only as a "historical relict." It originated at a time when the oxygen concentration was still significantly lower and that of carbon dioxide higher than today. The C_4 plants, with their higher specific affinity for carbon dioxide, have solved the problem of CO_2 deficiency, but the C_3 plants have not.

On the other hand, the C_4 plants require greater amounts of energy for their actual photosynthesis–a higher level of solar radiation per unit time. For this reason they cannot spread into regions too far from the equator (Ehleringer, 1978). A very similar modification, but complicated by introduction of a timing factor, is found in many succulent desert plants. Their stomata are opened only at night; at this time they bind carbon dioxide to PEP as the C_4 plants do. Most of the oxalacetate is converted to malate and stored in the vacuoles. As it accumulates during the night, the pH of the vacuole contents: falls dramatically with the increasing concentration of organic Acids. In the morning the stomata close, and carbon dioxide is split off from the malate and enters the normal photosynthetic pathway, The pH of the cell sap again approaches the alkaline range (diurnal acid rhythm). By keeping the stomata closed in the daytime the plant loses no water vapour during the period of greatest heat and low relative humidity-and for desert plants, water conservation is important.

It may be that the C_4 pathway of photosynthesis gives still another selective advantage. In the high-performance plants conversion of the C_4 acids to carbon dioxide and pyruvate occurs in specific bundle-sheath cells that are indigestible by ectothermic herbivores. Grasshoppers that eat such grasses excrete undigested bundle-sheath cells. These plants are therefore a relatively unprofitable source of food for such animals (Caswell et al., 1973, 1977; warmblooded herbivores have no-difficulty in digesting the bundle-sheath cells). The pronounced daily rhythm of acidity in the succulents may also provide protection from plant-eating animals, but this possibility has not yet been documented.

As far as ecology is concerned, a crucial aspect of productivity is the site of incorporation of the organic matter produced by a vascular plant-does the new substance take the form of nectar, seeds, leaves, roots or wood? Evidently the different plants vary in this regard; annuals differ from perennials. In addition, there are endogenous control mechanisms, most conspicuous and self-explanatory in biennial plants. And within a given species "resource allocation" depends very much on external factors.

A well-fertilized and well-watered plant develops only a small root systems, whereas under water stress or conditions of mineral scarcity a larger, and usually deeper, root system is elaborated. Flowers and seeds account for a smaller proportion of the total growth of a plant when water is abundant than when it is in short-supply. This may be an absolute rule for some species; these form flowers and seeds only under conditions of water stress. Finally, the growth of a plant is critically affected by the amount of light received (photo morphogenesis, the regulatory hormone phytochrome).

These considerations lead us to the effects of light on animals. For animals, light is almost never a source of energy; it is most significant with respect to orientation of an animal in space (via the eyes) and in time (the alternation between light and darkness acts as a timing signal for the daily and annual rhythms of animals). Light thus appears not to be a factor necessary for animal metabolism, so that one would expect there to be no problem in raising animals in permanent darkness if other conditions were favourable. Such experiments have been done with only a few species, and most of them have failed. Drosophila, terrestrial chironomids and cicadas, for example, have very high mortality rates in maintained darkness. No one knows why. As a model of the ecological effect of light in the context of the physiology of metabolism, consider Vitamin D. This is taken up as 7-D-hydrocholesterol; in the skin, by the action of short-wavelength sunlight, it is converted into cholecalciferol. This conversion does not occur if insufficient light reaches the skin. When the supply of Vitamin D is inadequate the skeleton is incomsymptoms of rickets in children. Rickets is found predominantly at northern latitudes, where the winters are long and dark and UV radiation is weak even in summer.

In tropical regions with intense sunlight the illness is essentially unknown. Dark-skinned humans living in the north are particularly subject to rickets, because most of the limited radiation available is absorbed by their skin pigment and cannot be used for conversion of the precursors into the vitamin. Conversely, people with light skins exposed to the strong tropical sunlight are in danger of excessive ossification. The problem of insufficient irradiation can be simply solved by taking vitamin pills, but so far no means have been found by which overproduction of Vitamin D can easily be prevented. This example suggests that the distribution of animals may be closely related to light, but there is as yet practically no confirming evidence. It also appears plausible that animals of valleys and the high mountains might differ in their radiation requirements

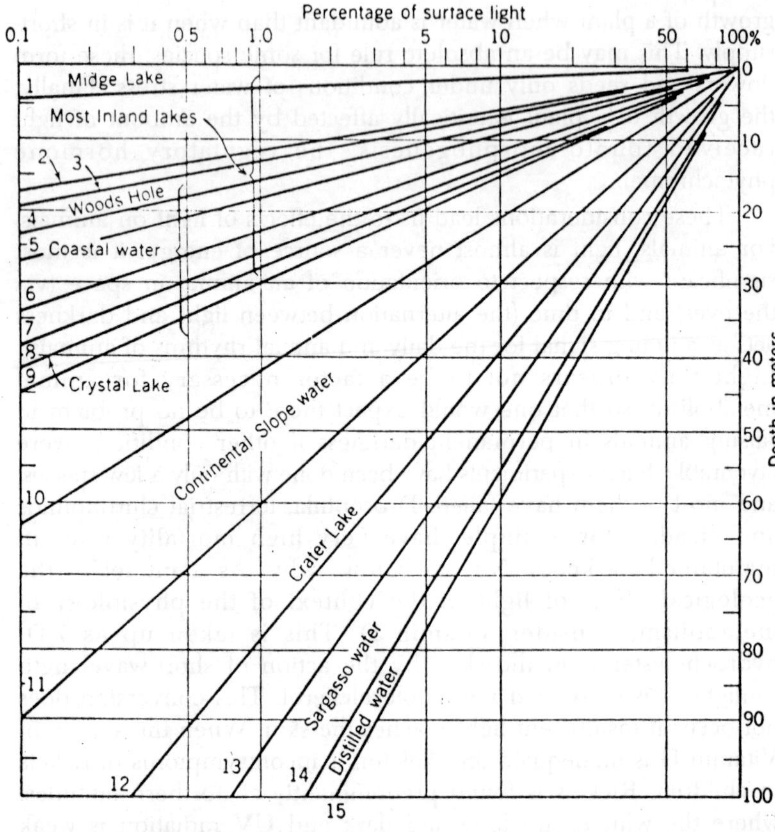

Fig. 2.21. Comparison of the transparencies of representative natural waters. Curves based on the average extinction rate for the yellow-green (5000–6000A) component of daylight.

and tolerance. Such a possibility has often been proposed, but again evidence is almost entirely lacking. Gluck (1979), however, has shown that there is a clear correlation between the locations where birds breed successfully and the amount of light falling on their nests. He found that the nests of chaffinches, goldfinches, and hawfinches were located where the total irradiation was relatively high, whereas greenfinches, serins, and linnets choose nest sites with a distinctly smaller amount of light. Some animals–unlike plants–are active at night. The cause of noctural activity is frequently the higher relative humidity that prevails at night. In this case, then, light acts solely as a timing signal.

Effects of Light on the Plants

Light plays a very important role in plant life. Nearly all the aspects of plant life, like structure, form, shape, physiology, growth, reproductions, development, local distribution, etc., are controlled by it. Based on light factor alone, plants can be divided into two groups: shade tolerant species called sciophytes and shade intolerant species i.e., heliophytes. Sciophytes .have lower growth rate and lower respiration rate. The role of light on plant life may be summarized as follows:

Effect on photosynthesis

The most important role of light is in photosynthesis. Photo synthesis is the only efficient means by which energy from abiotic world is trapped for utilization by the biotic world. The trapping of solar energy is mainly accomplished by means of chlorophyll. Radiant energy, particularly the blue, violet (wave length 440 nm) and short red with wavelength 655 nm, is absorbed by chlorophyll a and b. Chlorophyll a is most directly concerned in photosynthesis, for only this form is capable of transforming energy to power carbohydrates building reactions.

$$6CO_2 + 12H_2O \xrightarrow{\text{Light energy}} \underset{\text{Glucose}}{C_6H_{12}O_6} + 6H_2O$$

Light intensity effects the primary production in autotrophic layers. Thomas (1955) derived the relationship of light intensity to photosynthesis in terrestrial and aquatic ecosystems and showed that there is linear increase upto a certain optimum level, called light saturation, followed by decrease at very high light intensities.

Effect on vegetative shape of the plant

The intensity and direction of light have a great effect on the shape of the plant. Light inhibits the production of auxins or growth hormones as a result of which the shape and size of the plant are influenced. Plants grown in insufficient light or in the total absence of light produce maximum amount of growth hormones, as a result of which they are elongated with weak pale yellow stem, and very few branches. The tissues are poorly developed and there are no supporting structures like xylem and phloem. The leaves ale represented by scales. Roots are also poorly developed.

Effect on differentiation of plant tissue

Light affects the formation of certain specific chemical substances that effect the differentiation of specialized cells and organs. This can be illustrated by the development of palisade tissue in leaves and growth of storage organs and roots. It has been seen that leaves of herbs and shrubs that grow regularly in deep shade consist entirely of spongy tissue. Cells that normally form palisade tissue in the sun develop into spongy parenchymatus cells in shade. Conversely, spongy cells under strong illumination develop into palisade. The heliophytes grown under strong illumination have thick palisade tissue (length palisade tissue) or double layer or both. Intercellular spaces in heliophytes are smaller while in sciophytes they are larger. Leaves of the same plant also differ in the amount of palisade depending upon the amount of light they receive.

Effect on the production of chlorophyll

Light is necessary for the formation of plant pigments. A primary response of the plant to light is the production of chlorophyll. Plants with plastids produce chlorophyll only in light and the chlorophyll practically disappears in continued darkness. However chlorophyll does not develop, in response to light, in bacteria, fungi. Thalesia, Cuscuta and some other parasites and saprophytes. In these the power to make chlorophyll has never been attained or has been lost in consequence of parasitic or saprophytic habits. Seedlings of most conifers, young fern fronds, some mosses and one-called algae are other exceptions as these become green (develop chlorophyll) in the absence of light.

Effect on number and position of chloroplasts

Light has a very strong influence on the number and position of chloroplasts. The upper part of the leaf which receives full sunshine has larger number of chloroplasts which are arranged in line with the direction of light. These chloroplasts arrange themselves in a number of layers and screen each other from the full effect of radiant energy. In leaves of plants which grow under shade, chloroplasts are very few in number and are arranged at right angle to the light rays, thus increasing the surface of absorption.

Effect on leaf structure

The leaf structure undergoes greatest modifications in response to light than any other organ of the plant. Differences in the thickness of leaves grown in the sun and the shade are usually very clearly marked. Thickness of the leaf increases with the increasing light intensity. The form of the leaf is largely determined

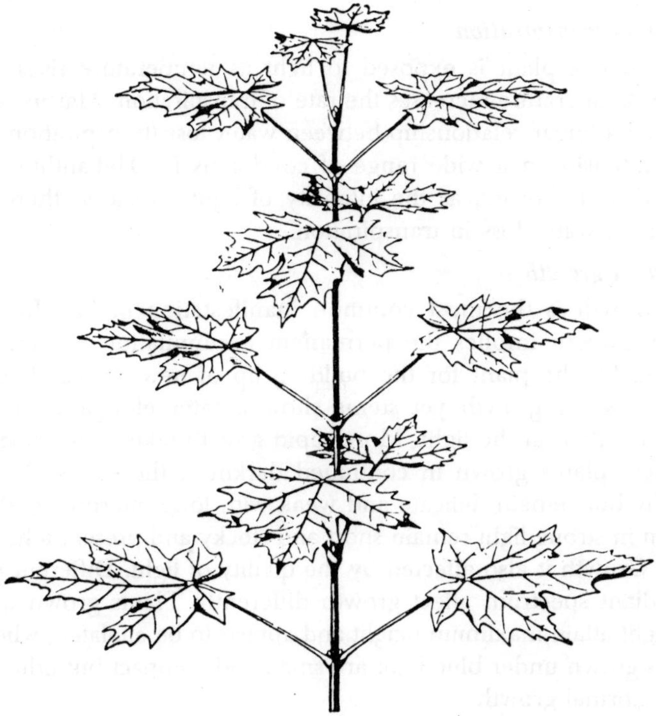

Fig. 2.22. Alternating position and horizontal orientation of leaves on shoot of Norway maple, resulting in maximum exposure to sunlight.

by the action of the light upon the chloroplasts and consequent changes in the form of cells that contain them. Light also effects the differentiation of palisade tissue and spongy parenchymatus tissue in the leaves, (as discussed earlier in details). Spongy cells tend to produce an extension of the leaf at right angle to the incident light, while palisade cells extend the leaf in the direction of the incident light. Hence leaves that contain excess of spongy tissue are relatively broader while those in which palisade is more are relatively thicker.

Effect on stomatal movement

Stomatal movement is greatly modified by light. In nearly all the plants stomata open in the presence of light, when other conditions are favourable. However, in unfavourable conditions, the influence of light is very different. For example, when water supply becomes low, stomata may close even in the presence of light.

Effect on transpiration

When a plant is exposed to light its temperature rises. This rise in temperature increases the rate of transpiration. Martin (1935) derived a linear relationship between water loss (transpiration) and light intensity in a wide range of conditions for Helianthus (sunflower) and showed, as the intensity of light increased there was increased water loss in transpiration.

Effect on growth

Growth is the most common manifestation of life. In strict sense, growth implies the permanent accumulation of materials utilized by the plant for the building up of new tissues. Light is necessary for growth yet stems show greater elongation in the darkness than in the light. More rapid growth takes place at night. In dicot plants grown in continued darkness, the stems elongate rapidly but remain delicate and weak with long internodes. Plants grown in strong light remain short and stocky and present a healthy look. Growth it also affected by the quality of light. Different parts of radiant spectrum affect growth differently. Plants grown under red light attain maximum height and appear to be etiolated, whereas plants grown under blue light are small and compact but otherwise show normal growth.

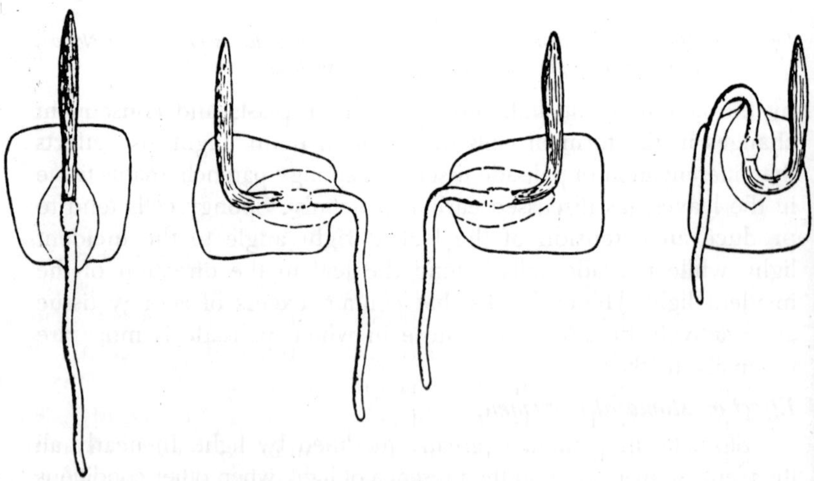

Fig. 2.23. Negative geotropism of shoots and positive geotropism of roots of four kernels of maize that have germinated in different positions.

Effect on development of flowers, fruits and seeds

Intensity of light has very pronounced effect on the development of flowers, fruits and seeds. Diffused light or reduced light promotes the development of vegetative structures and causes delicacy. Vegetative crops like potato, carrots, turnips and beets give highest yield in regions with high percentage of cloudy days. Tobacco plants are grown under artificial shading to get larger leaves needed for making cigars. Intense light favours the development of flowers, fruits and seeds. The great profusion of flowers in grasslands throughout the growing seasons in largely due to strong light. Similarly larger number of bright coloured red flowers is found in alpine meadows where the light is very intense. The countries which produce highest yield of grains and fruits, in the world, are only with nigh percentage of sunny days.

Effect on the distribution of plants

According to Krebs (1972), light does not acts as a limiting factor in geographical distribution of plants. Light does, however, play an important role in determining local distribution of plants. Plant species become adapted to live in a certain kind of habitat and in this process evolve a series of characteristics *(e.g.,* shade tolerant and shade intolerant spp.) which prevent them from occupying other habitats. Grime (1966) "Suggested that light may be one of the major components directing these adaptations. However, the local distribution of plants is affected by the duration and intensity of light. In sea, for example, the vegetation differs at different depths as light 'intensity goes on decreasing. Similarly, on land, the vegetation in polar regions is different from other regions.

Effects of Light on Animals

The visible light represents a small fraction of the whole radiant Spectrum. The shares many of the properties of other wavelengths specially of those just longer or just shorter than it. Light is one of the complex environmental factors that has far reaching effects on the animals. The visible light acts as an important limiting factor for many multisided activities of the animals. It effects a number of biological processes like pigmentation, reproduction, development, growth, locomotion, migration, etc and some other structural and behavioural characteristics of animals. The important effects of light on the animals may be discussed as follows:

Effects of light on metabolism

Visible light is indispensable for photosynthesis. The chlorophyll of Euglena like organism furnishes the mechanism by which CO_2 is combined with H_2O to form carbohydrates and to store the energy obtained from the sunlight

$$6CO_2 + 6H_2O + \text{sunlight} \rightarrow C_6H_{12}O_6 + 6O_2$$

However, the above formula genera used to explain the formation of a carbohydrate does not give an idea of the intermediate steps and is only a sum of many reactions, but it is certain that light plays an essential role in the process even though some steps may be completed in the dark. The rate of photosynthesis varies somewhat with different wave lengths of light. In terrestrial ecosystems the quality of light does not vary enough to have an important differential effect on the rate of photosynthesis.. But as light penetrates water, the reds and blues are filtered out and the resultant greenish light is poorly absorbed by chlorophyll. However, phycoerythrings (supplementary pigments) of the red algáe enable them to utilize this energy and to live at greater depths. The relationship of light intensity to photosynthesis in both aquatic and terrestrial plants and Euglena like organism follows the same general pattern of a linear increase upto an optimum or light saturation level, followed in many instances by a decrease at very high intensities.

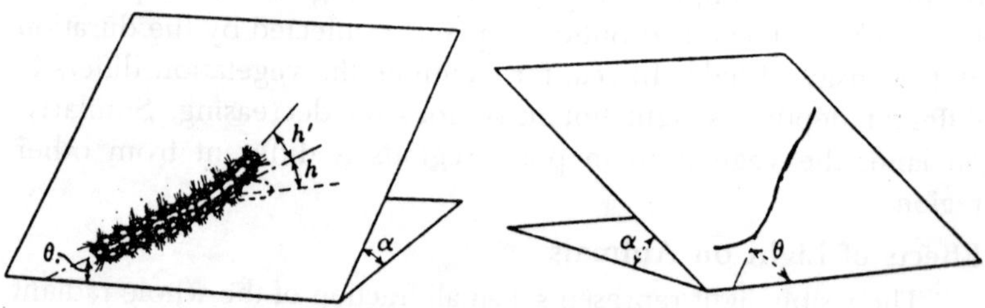

Fig. 2.24. Arrangement of experiment to measure the geotactic orientation at angle θ of caterpillar Malacosoma placed on plane inclined at angle α. Lateral movement of head indicated by h and h′.

Effects of light on colour and structure

Light induces chemical changes resulting in the formation of pigments. The absence of light results in the loss of colour among the cave animals. The role of pigmentation and protective colouration are well known in terrestrial animals. Many inhabitants of caves, where light of course has no ecological significance, generally have vestigial eyes, or they may be absent. Numerous animals of the deep sea are blind or have reduced eyes and some have telescopic eyes. Thus it becomes clear with these examples, that light affects certain structures, especially the eyes. Among green plants light is required for the production of chlorophyll in the chloroplasts. Plants grown under insufficient light do not develop their normal green colour. On the other hand, excessive light causes destruction of chlorophyll.

Effects of light on development

Salmon larvae undergo normal development only when they get sufficient light. In the absence of light their development is not normal and mortality is the result.

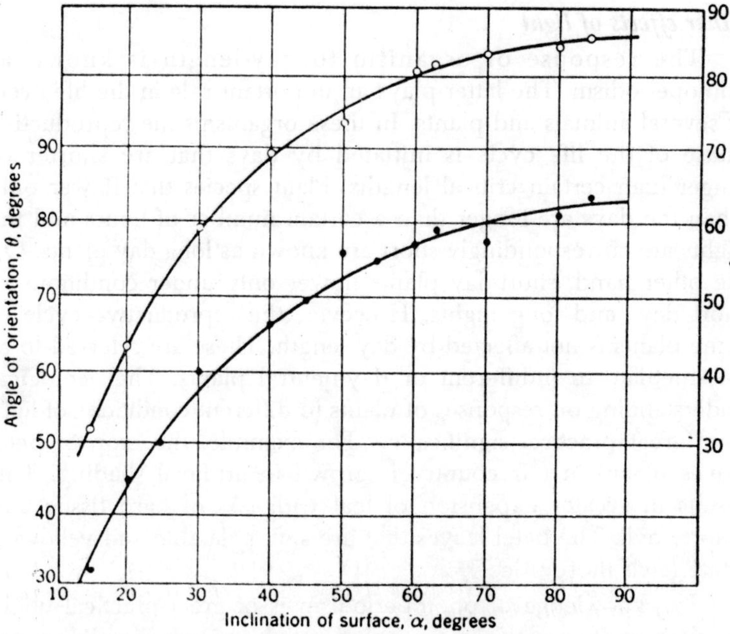

Fig. 2.25. Relation between the angle of inclination of plane (α) and the mean angle of orientation (θ) of caterpillar creeping upon it. Open circles are means for one individual; black circles are means for all individual tested.

Effects of light on locomotion and orientation

In certain lower animals the speed of locomotion is considerably influenced by light. This phenomenon is known as photokinesis.

For example, according to Welsh (1932), the larvae of pinnotheres move fast in the increased light intensity. When the intensity of light is decreased, they move slowly. Locusts immediately stop their flight if the sun is hidden by the clouds. Light also plays a significant role in orienting the growth or locomotion of animals and plants.

Orientation is brought about either by the differential growth or movement of parts of the organism or by a change in the direction of locomotion of the whole organism. If the orientation of plants is toward gravity, it is called geotropism. On the other hand, the orientation of locomotion of motile organisms is explained by the terms geotaxis, phototaxis etc. In the aquatic environment the buoyant action of water often reduces the effect of gravity so that primary orientation is toward light or current.

Other effects of light

The response of organism to day-length is known as photoperiodism. The latter plays an important role in the life cycle of several animals and plants. In these organisms the reproductive phase of the life cycle is initiated by days that are shorter or longer than certain critical lengths. Plant species that flower only when the days are longer than a certain number of hours and the nights are correspondingly short are known as long-day plants. On the other hand, short-day plants flower only under conditions of short days and long nights. However, the reproductive cycle of some plants is not affected by day length. These are referred to as intermediate to indifferent or day-neutral plants. The ecological understanding on responses of plants to different conditions of light is of great practical significance. For example, the crop of betel leaves (paan) in our country is grown in artificial shading. This results in greater expansion of leaf with less of hard tissue and chlorophyll. The betel leaves that are soft, palatable and yellowish white fetch more price.

The knowledge of photoperiodism is of great practical utility in selection of species and season of its cultivation depending upon the plant parts which are economically important. Photoperiodism similarly plays an important role in the life cycles of many animals. For example, wing production in aphids, metamorphosis in mosquitoes, etc are influenced by day-length. Trout that ordinarily spawn in December can be induced to lay their eggs in August by artificially changing the day length. According to Farner (1950),

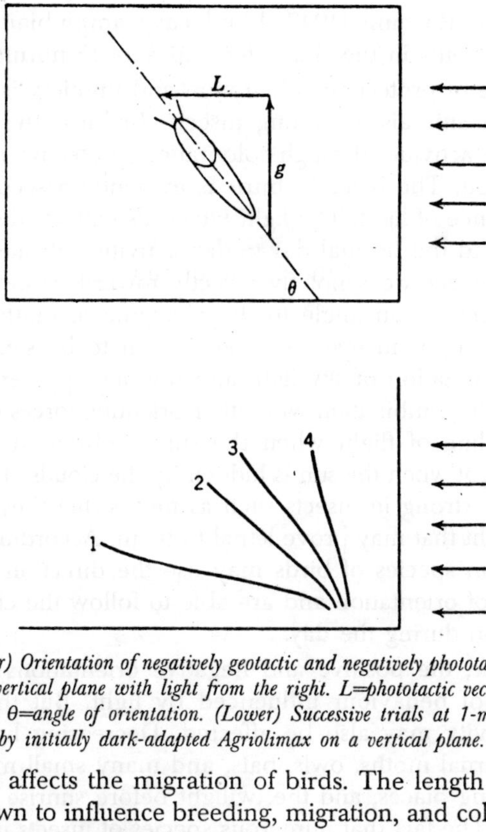

Fig. 2.26. (Upper) Orientation of negatively geotactic and negatively phototactic Agriolimax on a vertical plane with light from the right. L=phototactic vector, g=geotactic vector, θ=angle of orientation. (Lower) Successive trials at 1-minute intervals made by initially dark-adapted Agriolimax on a vertical plane.

the light also affects the migration of birds. The length of the day has been shown to influence breeding, migration, and colour change in many birds and mammals.

Effects of light on terrestrial organisms

Light affects the distribution, colour, and some structures, etc in terrestrial animals. It influences their pigmentation in several ways. Skin colour may be indirectly affected by light through the mediation of eyes or other receptors Vertebrates possess an adaptation to ultraviolet light in which the outer layers of the skin are thickened and develop keratin, pigments, and corns, etc., all of which help to absorb the sunshine and reduce the penetration of light of all wavelengths. Lizards of the genus Crotaphytus expand the dark pigment of their skin to absorb sunlight until they reach an optimum temperature, after which they contract the dark pigment and expand the light-coloured pigments for light reflection. The absence of light results in the loss of colour in cave animals:

According to Rasquin (1947) blind cave amphibians and fishes develop pigments in the skin after exposure to normal daylight.

The role of protective colouration and mimicry is well known in terrestrial animals, including insects. In birds the light affects reproductive activities through colouration, protective resemblance, and migration. The latter in birds is evidently a secondary effect of the influence of light. The light factor also affects the orientation behaviour and the normal day-to-day activities of many terrestrial animals. The bee does not fly directly toward or away from the sun but orients at an angle to the changing azimuth of the sun. The bee's compound eye has been shown to be sensitive to the angle of polarization of sky light and the bee apparently uses this information in combination with other orienting forces to determine the proper line of flight when the sun is obscured. Locuts also stop their flight when the sun is hidden by the clouds. The attraction to light is so strong in insects such as moths that they fly into the source of light that may prove lethal to them. According to Kramer (1952) certain species of birds may use the direction of sun light as a means of orientation and are able to follow the change in the sun's position during the day.

However, the positive and negative orientations are not the only forms of behaviour influenced by light. But the level and type of activity may also be affected. Decreasing light intensity draws nocturnal moths, owls, bats, and many small mammals out of their hiding places, and the twilight before sunrise drives them back again. The fact that numerous species of insects interrupt the, development of the young stages in summer, despite the apparently favourable high temperatures and abundance of food, and enter a resting phase (diapause) can definitely be attributed to the influence of light. The latter sets in motion certain hormonal regulatory mechanisms evolved, in association with the climatic or other peculiarities of the animal's environment and thus during the resulting diapause the seasons unfavourable to development and reproduction are passed over.

Though diapause may be elicited directly by light, but in some cases it is "preprogrammed", in the sense that light conditions to which early developmental stages (such as larvae still within the egg) are exposed include diapause indirectly later on in the pupal stage. For example, if caterpillars of Apatele rumicis are exposed to short periods of day light, less than 15 hours, all the pupae that develop enter diapause. But if the light period is increased to more than 16 hrs, the pupae omit the resting stage. Among other structures the light affects eyes or vision most. There is a correlation between the types of eyes and the nature of light that they perceive.

Among terrestrial animals with well developed eyes, some of the eyes are so located on opposite sides of the head as to view the entire surroundings of the, animal, whereas other animals have their eyes facing in one direction so that both eyes may focus upon one object. Nocturnal animals usually have large eyes for gathering faint light, or small eyes where other means of guidance are available. The burrowing habit and the deficiency of light in the environment similarly leads to the reduction of eyes in many burrowing animals.

Effects of light on aquatic organisms

Light is certainly one of the major factors controlling the distribution, movement (speed and direction of swimming), colouration, and the development of certain structures in aquatic organisms. Primary food production in the marine environment is virtually confined to the illuminated surface layers of the sea where there is sufficient light to support plant life. The depth of this photosynthetic or euphotic zone varies from about 40 to 100 metres. Plants are restricted to the euphatic zone due to their dependence upon light for energy. As animals drive their food directly or indirectly from plants, they are also most numerous in or near the surface layers.

Below the, euphotic zone, there is little or no light in dysphatic (100-200 m) and aphotic zones (below 200 m). Hence plants ace almost absent and animals are almost entirely depended upon food sinking to them from above. However, many organisms do not remain consistently at one *level* but perform vertical movements which are related to changes of light. Another effect of light on animals and plants is exhibited by a 24 hour cycle in their activities or diurnal periodicity. The most fundamental diurnal rhythm is that of photosynthesis itself which fluctuates due to daily change of light. Many plants exhibit other more specialized reactions because of change of light, such as the opening and closing of flowers and folding of leaves.

Diurnal changes of distribution are shown by a great variety, organisms of both plankton and nekton, including medusae, siphonophores, ctenophores, chaetognaths, pteropods, copepods, cladocerans amphipods, mysids, pelagic decapods, and some cephalopoda and fish. The moon light and the lunar cycle also control many animals activities. For example, marine alga, Dictyota produces its gametes at the time of full-moon spring tide. Similarly, in some marine polychaete worms spawning takes place at a particular period of the moon.

The colouration of marine organisms is also related to the illumination of their surroundings. The fish of shallow water are usually protectively coloured, being dark on the upper surface and whitish underneath. Deep water creatures which often come to the surface may have a reflecting surface or be almost transparent. In the total *dark* ness of the sea there are numerous nonpigmented forms. The common Occurrence of bioluminescence in the sea is probably the result of much lower intensities of light in the sea. According to Harvey (1952) bioluminescence may be used for One or more of the following functions: recognition, illumination, lure warning, prey-catching, etc.

In marine environment there may be a correlation between the of colouration bioluminescence, and the occurrence of eyes. The latter are well developed among animals of the surface layers of the ocean. At deeper levels the eyes of the fish become enlarged or 'telescopic' and at greater depths they are often degenerate of absent. However, some deep sea fish possess both well developed

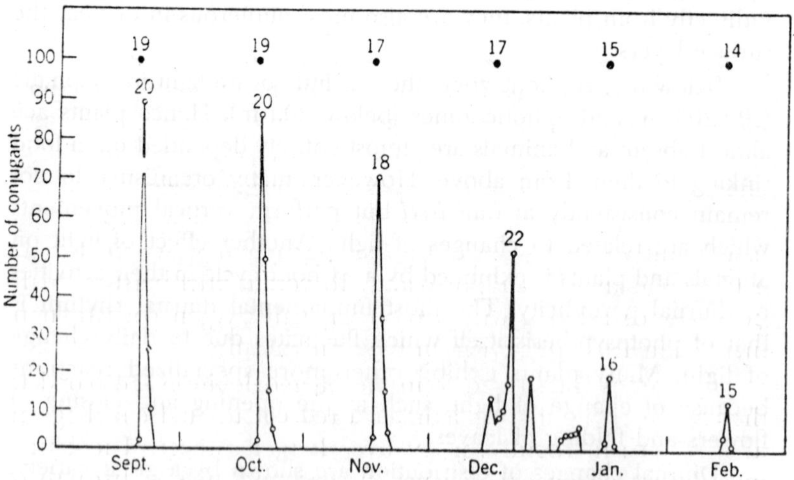

Fig. 2.27. Average daily number of conjugants of the ciliate Conchophthirius lamellidens on the gills of the fresh-water mussel Lamellidens marginalis.

eyes and luminescent organs. But bioluminescence is definitely an adaptation to the absence of light in the sea.

Oxygen Supply

Organisms need oxygen to break down organic matter and thus obtain energy. When no oxygen is available other temporarily when there is a sudden marked increase in energy consumption (this probably happens only in animals), or it can be a maintained condition; there are organisms capable of living in a milieu extremely low in oxygen or permanently lacking it. Such a situation arises only underwater or in the soil; actual terrestrial organisms never encounter it. On the land–that is, in the air–there is always an adequate supply of oxygen. It is limited to a certain extent in the high mountains. Here birds and mammals find it very difficult to achieve peak performance.

Land animals seem to have had hardly any success in evolving haemoglobin with greater oxygen affinity. This fact is remarkable, in that aquatic animals have been able to take this evolutionary route; modifications utilizing the Bohr effect and the Root effect have made it possible for the oxygen to be released to their organs as required despite the high oxygen affinity of their haemoglobin. Adaptation to the high mountains mainly involves an increase in the amount of haemoglobin in the blood, so that more oxygen is transported. To some extent this increase in blood haemoglobin content is a simple acclimatization acquired after a prolonged stay at altitude, but in some cases genetic factors play a role. At the moment it is impossible to decide whether these measures–a slightly increased oxygen affinity of the haemoglobin and a greatly increased haemoglobin concentration–really represent the only response of animals to the low oxygen supply at great heights. Relevant findings are few, although an intensive programme of physiological research is currently underway. Data are particularly scarce for all ectothermic animals and for plants. To permit brief bursts of high-performance activity, with short-term high oxygen consumption, many animals store oxygen in their musculature.

Oxygen is transferred from the haemoglobin in the blood to the muscle tissue of these animals a red colour, and enabling the animals to run rapidly or fly over long distances. The breast musculature of pigeons, falcons, curlews and gulls is, because of its myoglobin, as red as the muscles of hares, deer and antelopes. By contrast, animals that move at top speed only briefly, and then hide, do not store oxygen in their musculature; such muscles (those of rabbits, for example, and the pectoral muscles of most fowl) are white. They consist entirely of contractile fibres, and are capable of maximal activity only for a short time. Oxygen is also stored' in the myoglobin of whale muscles, which are a deep red; whales dive without first filling their lungs with air and so must rely on the oxygen in the myoglobin while they are under water. There are additional ways of supplying energy where it is needed during such "active oxygen deficiency," via metabolic pathways other than the normal glycolysis followed by citric acid cycle and respiratory chain. These alternative pathways are present in bacteria and thus do not represent recent evolutionary acquisitions.

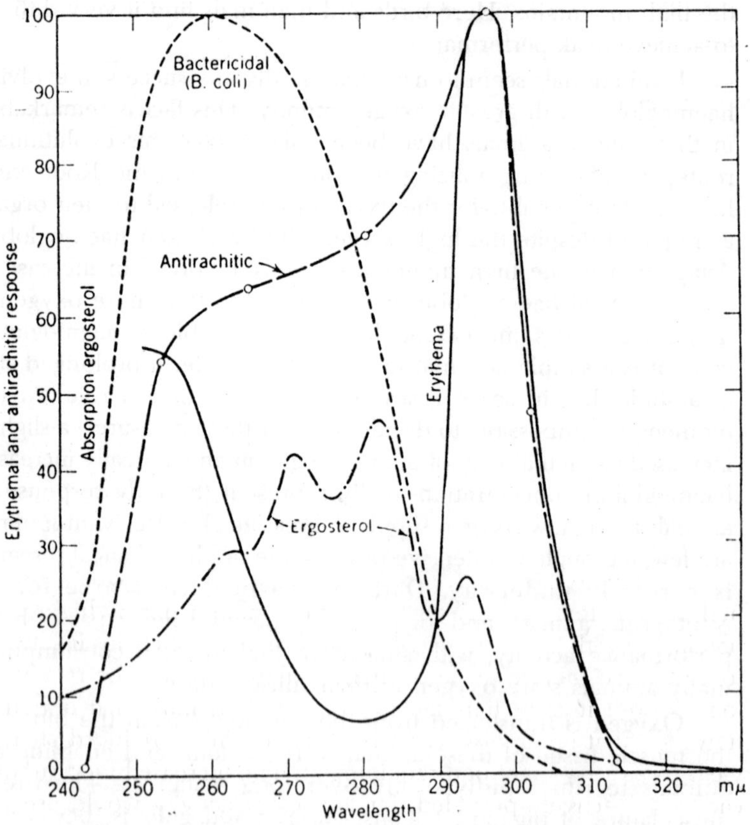

Fig. 2.28. Relation between wavelength in the ultraviolet region and the bactericidal, erythemic, and antirachitic effect of ultraviolet radiation, together with the absorption of ergosterol.

We must assume that they have been carried along during evolution as a "genetic load," until it suddenly again became selectively advantageous to use them (Thauer, 1977). Chief among them in warm-blooded animals is lactate fermentation; here glycolysis does not feed into the citric-acid cycle but branches off with the conversion of pyruvate to lactate and ends there. The yield of energy is only 2ATP per mole of glucose, as compared with 38 ATP for the classical pathway. This alternative may be more widespread than was previously thought. It has been demonstrated, for example, in marine worms. During the escape movements of cephalopods and molluscs octopine fermentation occurs, another strictly temporary, low yield alternative. Octopine, like lactate, is subsequently reconverted under normal conditions and returned to the metabolic pool.

If an oxygen deficiency should arise in the surrounding medium many organisms can switch, for prolonged periods or even permanently, to other means of obtaining energy. Familiar examples are the fermentation of acetic acid or alcohol in microorganisms and that of succinate by worms parasitic in the intestinal canal. These forms appear to have discarded the "normal" (and phylogenetically younger) pathway by which energy is derived. But recently it has become increasingly evident that many aquatic animals are very flexible, whereas they ordinarily operate with the citric-acid cycle and the respiratory chain, they can use fermentation pathways under short or long-term environmental oxygen deficiency. This is true of Arenicola, many marine and freshwater clams, Tubificidae and earthworms-and no doubt of many other organisms. The alternative pathway as a rule involves conversion of succinate to propionate; a small fraction of the malate is converted via pyruvate to lactate or to acetate.

The energy yield of succinate fermentation is probably 8 ATP, higher than that of lactate fermentation though not as high as is obtainable with the citric-acid cycle and respiratory chain. It suffices to guarantee the existence of many aquatic animals for relatively long periods in a medium lacking oxygen. One must bear in mind, however, that in all these cases there is not only a reduced energy yield but a very high food requirement. The increased food supply cannot be actively provided, for any extra activity would use up more energy. The amount of food immediately available must therefore be extraordinarily high if the alternative pathways described are to mean more to the animals than just survival with loss of energy (for a summary see Zebe, 1977). In the ocean, oxygen deficiency occurs primarily in the tidal zone during ebb tide; clams and mussels close their shells, thus shutting off the flow of fresh oxygenated water past their gills as effectively as it is shut off from the tubes occupied by crustaceans and polychaetes. In fresh water and in enclosed bodies of sea water (the Baltic and Black Seas) the conditions bringing about oxygen deficiency are of particular current importance of humanity. The oxygen in the water is derived from two sources.

First, it can diffuse into the water from the air. This is a slow process, and it supplies oxygen to underlying layers only if the surface water moves into the depths. In lakes such circulation is brought about only by high winds; because such strong wind is rare during the warm summer months, there is hardly any stirring of the water at the time when the temperature is high–just when the metabolic rates of all the organisms, and thus their oxygen consumption, are especially high. The second supply of oxygen comes from aquatic plants, especially algae floating in open water, in the plankton. These are of course restricted to the upper, illuminated water levels. If the body of water contains abundant nutrients for the plants, a particularly large mass of planktonic algae develops.

The consequence is increased turbidity of the water, so that light penetrates less deeply and photosynthesis is concentrated in the uppermost layers. When the water is heavily fertilized, then, it is precisely the most endangered levels that are deprived of oxygen. The danger is increased in that during such a "bloom" dying planktonic algae sink to the depths in vast numbers. There they are decomposed by bacteria–a process that naturally requires oxygen. The oxygen at the bottom of the river or lake can be used up completely, so that all the animals living there die. This is the reason why eutrophication of inland waters is such a severe threat; fertilization here has effects quite different from those on land. In a lake poor in nutrients light penetrates deeply; planktonic algae can carry on photosynthesis and produce oxygen at depths of more than 50 m. The end result is that fish production in a nutrient-poor lake is higher than in a lake rich in nutrients.

Of course, there are lakes that have a high nutrient concentration even without human intervention. Reichholf (1975) showed that flocks of ducks overwintering on nutrient-rich reservoirs along the Inn River graze off all the water plants, removing from the water organic material which otherwise would sink to the bottom and rot there–a process that would use up oxygen. The ducks themselves, being air breathers, withdraw no oxygen from the lake. The important point is that the ducks must not be disturbed, so that they will remain in large numbers. They will then prevent loss of oxygen from the water and thus postpone the reversion of the lake to land.

Fire

Fire is a regularly recurring factor in many natural ecosystems. Spontaneous combustion and lightning are its most common causes. Fires at regular intervals are a feature of the dryer parts of the tundra, throughout the taiga, in all savanna and steppe regions, and in all Mediterranean plant communities in the broadest sense–including the chapparal of California and the pine forests of Florida as well as the sclerophyll woodland around the Mediterranean Sea. Some eucalyptus forests in Australia are also exposed routinely to fires of natural origin. The detailed investigation Zackrisson (1977) made of the burn, scars on very old trees showed that in the northern European taiga, before humans intervened to protect it, the forests were swept by about two fires per century. Similar figures have been obtained for the Canadian tundra and the tundra of Alaska. Fires are still more frequent in savannas, steppes, and the Mediterranean regions.

To simplify matters, one can say that in general pines (the genus Pinus), oaks (Quercus) and all the Ericaceae are typical "fire plants". Because fire in a regular, event in their natural habitats, they have become highly modified in adaptation to such conditions. Their thick bark gives excellent protection, against fire, allowing them to survive without difficulty. If large regions of bark should be destroyed by the fire, regeneration hardly ever presents a problem. The seeds of pines and of the heather Calluna a germinate particularly well after being subjected to heat stress. Indeed, the cones of many pine species release the, seeds only after they have been warmed to 70°-80°C. Germination thus Occurs after competitors have been eliminated; the seedlings of pines and of calluna are very sensitive to competition.

The lichens Lecidea anthracophila and Lecidea friesii grow only on charcoal and are therefore strictly dependent on forest fires for their existence. Similar adaptations to fire are exhibited by the animals inhabiting these regions. Beetles of a number of species seek out very warm wood in which to lay their eggs. One metallic wood borer has actually been found to have infrared sensors that enable it to find freshly burned over areas. The health that germinates and grows rapidly after a fire provides very many birds and mammals (hares, especially) with far better food than old or cut heath. For centuries the heath, in Scotland has been burned regularly to keep the population of willow grouse and alpine hare large; the effect on black grouse populations is similar. The new growth of heath after a fire contains more nutrients and appears to be more palatable to animals than either old heath or the new growth following mowing.

In the pine savannas of the southeastern United States the parts of the plants that grew out after either fire or cutting were found to contain distinctly more N, P, K, Ca, and Mg, and the amounts of N, Ca, and Mg were greater after fire than after cutting. Not until 4-6 months had elapsed did the mineral content of these plant parts return to; normal (Christensen, 1977). Other species are extraordinarily sensitive to fire–spruce; for example, and most deciduous trees such as beech and linden. Pure spruce forests or beech forests catch fire only after strong winds have blown sufficient combustible material to the ground. It is generally valid to *say* that fire is not a natural event in most deciduous forests (apart from oak woods, and Mediterranean sclerophyll regions).

Fire plays a leading role in determining the composition of plant communities. The northern European-Siberian taiga belt, in which spruce and pine predominate in different proportions, is a product of fire. Without repeated burning the forests here, where it is not too dry, would consist exclusively of spruce. The recurrent fires do severe damage to the spruce trees, and it is only for this reason that the pines can maintain themselves. Fire resistance could be regarded as a direct mechanism for competition; pines shed a great deal of burnable material onto the ground, where it accumulates in loose piles. The fire sweeps rapidly through this fuel, injuring-the-spruce as it goes. By contrast, a dense stand of spruce is practically invulnerable to fire. The short needles become tightly packed in the ground and so are almost impossible to set alight. Once it covers a certain minimal area, then, a spruce forest is to a great extent protected from burning. The situation is similar

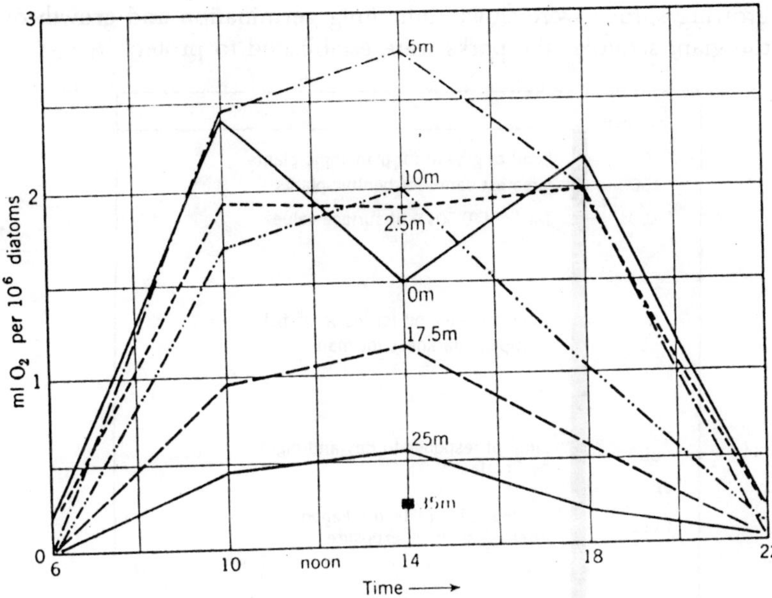

Fig. 2.29. Photosynthesis of a species of phytoplankton (Coscinodiscus) at the indicated depths during the course of the day off Stoke Point, England.

in the oak savanna of North America, where the oaks (like the cork oaks in the Mediterranean region) develop specific kinds of bark that enable them to survive even intense fire. When the acorns germinate, they are free of competition. The well-known, oscillation of hare (and thus of lynx) populations may perhaps be associated with fire cycles. Man has fought fire wherever he could. As a result, the frequency of fires in all regions, inhabited by humans soon declined sharply.

The result was a drastic change in the composition of the local plant communities. In northern Europe the range of spruce distribution expanded, and in the North, American oak savanna brush displaced the original oaks. Bushes also invaded the savannas of Africa. Specifically fire adapted animals became rare. To offset such changes, a system of controlled burning was introduced some time ago in North and South America, Africa, and recently in Europe. Because of this conservation measure the oak savannas of North America have regenerated in many places. Fire has also been used to keep the spruce population down in, parts of northern Europe and in the national parks of North America, where the growing spruce were slowly inhibiting germination and growth of the giant sequoias the parks were established to protect.

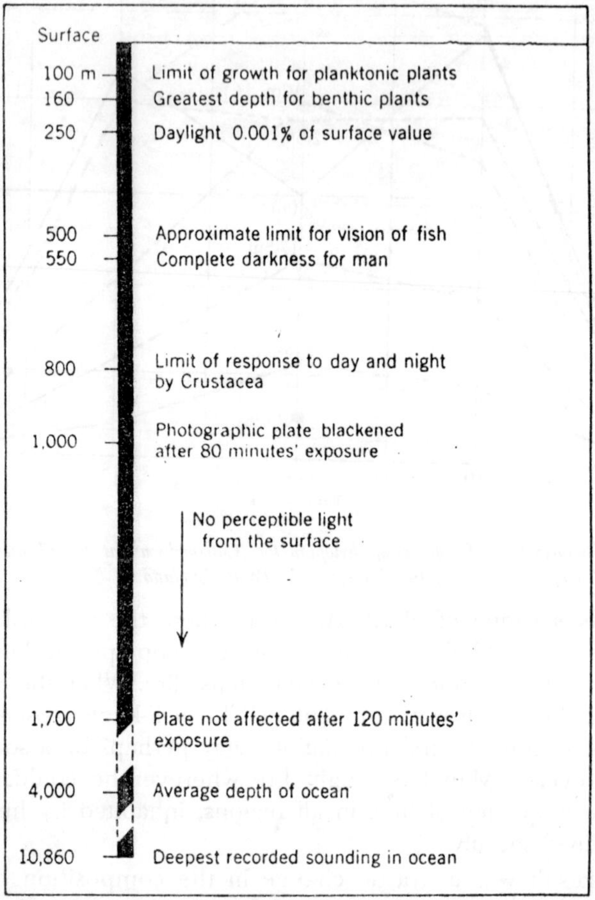

Fig. 2.30. Summary of limitations by light factor in aquatic environment. Values are for clearest water.

In the Florida Everglades, too, fire has been and is being widely employed to maintain the natural vegetation. Today we know how to light very cool fires, with an effect like that of mowing, under certain moisture conditions; we can also lay very hot fires that burn off part of the raw-humus layer and speed up the breakdown to real humus. We know the difference between hot fires that go with the wind and the relatively cool fires against the wind. The insect world suffers to varying degrees from the different kinds of fire, but in general the effects are less severe than was expected.

Nutrition

Unfortunately, it is practically impossible to give a description of plant nutrition that will satisfy the physiologist. Only minute, quantities of the essential nutrients are freely available in the soil and water. Even fertilization causes little change in this situation, for the added nutrients are immediately adsorbed on soil particles or react with other components of the soil and water. It requires activity of the plant (through the roots of which hydrogen ions and organic acids can be excreted) to make these nutrients accessible. Other nutrient ions are immediately incorporated by microorganisms and are thus out of reach of the plants, but when a microorganism dies–if it does so in close proximity to a root–the ions liberated become temporarily available. The advantage of this situation to the plant is that although the supply of nutrients is small at any moment it is quite uniform in the long term. The best example is given by the primeval forests of the Amazon basin, which thrive on soil that is practically free of nutrients, with ground water containing no nutrients at all. These forests have been compared with a firm that holds no funds in reserve, but keeps its entire capital in the business.

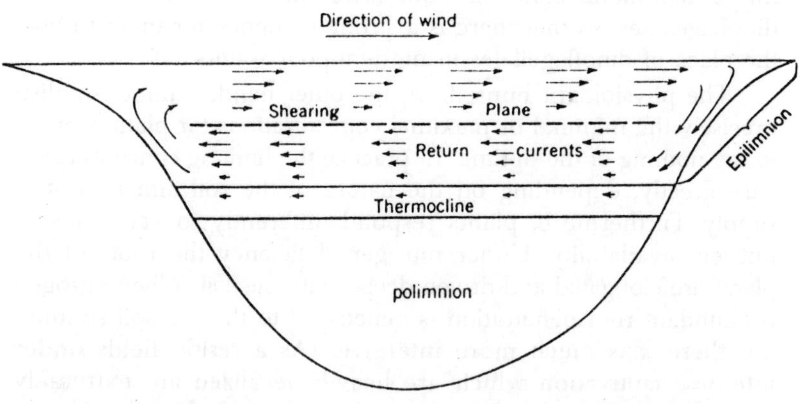

Fig. 2.31. Wind-driven currents in a lake with a pronounced thermocline. Circulation is limited to the epilimnion.

The same can be said of aquatic environments. When a body of water is fertilized to increase the yield of fish, what-happens to the fertilizer? It rapidly disappears from the water, sometimes in

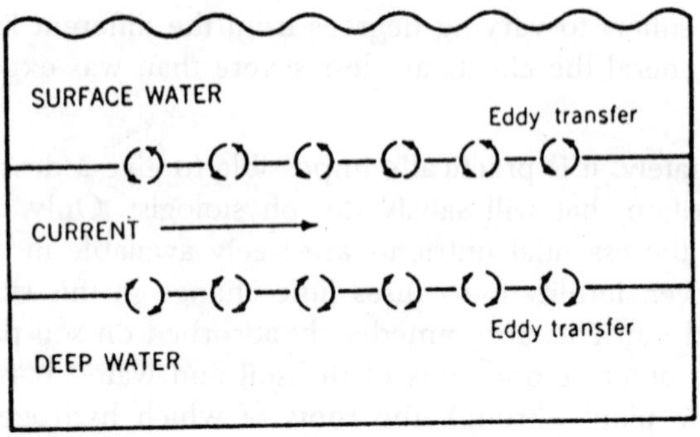

Fig. 2.32. Diagrammatic vertical section of a body of water, indicating vertical transfer by eddies across the current boundaries. Horizontal eddy conduction similarly takes place at the lateral margins of the current.

less than a week. In part this is due to the plants, which utilize it immediately, and in part to mud particles, to which the fertilizing substances adhere and from which releases slow. The latter fraction must be regarded as temporarily lost. This withdrawal of substances makes it impossible even in water to find the true relationship between the amount of minerals supplied and the resultant plant growth. Such experiments are feasible only in hydroculture, but then the results cannot readily be applied to the field situation. Only a rough idea can; be obtained–for example, that diatoms thrive at mineral concentrations lower than those necessary for dinoflagellates, so that there is a strong tendency for them to take the place of dinoflagellates in nutrient-poor waters.

The physiologist himself, on the other hand, cannot establish precisely the minimal or maximal concentrations for plant growth, to say nothing of the optima. In practice the limiting concentrations vary greatly, depending on the nature of the soil and its water-supply. Furthermore, plants respond differently to variations in nutrient availability. Under nitrogen deficiency the roots of the plants are elongated and driven deeper into the soil. When nitrogen is abundant root penetration is penetrated to the top soil stratum, but there it is much more intensive. (As a result, fields under intensive cultivation which are heavily fertilized are extremely vulnerable to brief period or drought. Shallow rooting is also the cause of the greater danger on erosion in intensively cultivated regions such, as Icelandic, pastures. In an ecological context, the uptake of nutrients involves, microorganisms to a bilge extent. This involvement is especially clear in the case or the many land plants associated with symbiotic microorganisms or fungi that live in or near their roots. Here hardly anything can be said about the requirements of the individual species; the system fundamentally comprises; two species.

In considering the nutrition of animals, we must distinguish between the qualitative and the quantitative aspects. Let us first turn to the qualitative nutrient requirements of animals. Usually food specialization is evident in the structure of an animal. Carnivores all have shorter digestive tracts than herbivores, even when they are members of the same species. The intestine of a plant-eating tadpole is very long, whereas that of the adult, carnivorous frog is very short. The modification of the forelimbs of arthropods to form a great variety of mouthparts–biting, licking, piercing-sucking–is a well-known phenomenon. Specialization can lead to extreme selectivity; the snake Dasypeltis scaber subsists entirely on the eggs of birds, and some nematodes eat only diatoms. There are a particularly large number of specialists among the insects, Some of which are restricted to one or a few species of plant. Such obligate feeding is based predominantly on secondary substances in the plants.

But it has been established that only a few animals really require these secondary plant substances. In actuality the alkaloids, terpenes, and phenols were developed in evolution as defense mechanisms against grazing. Through coevolution specific animals developed resistance to these defense substances, and at the same time they came to use the substances as a means of recognizing their specific plant. If the sense organs on the mouthparts of the tobacco-moth caterpillar are destroyed it no longer feeds exclusively on tobacco, as before, but will eat and thrive on a great variety of other plants. These specialists have "made a virtue of necessity". Not only do they tolerate the poisonous, secondary plant substances; most can store them in their bodies and as a result become unpalatable, inedible or even toxic to predators. In fact, not only herbivores product themselves in this way. Marine slugs that feed on coelenterates with nematocysts not only are unharmed by these stinging organs but carry them in their own body walls, ill firing condition; these "kleptocnids" make life difficult for the slugs predators. The animals that have coevolved to tolerate the defense substances of plants are not physiologically dependent on them, although they benefit from a food supply in accessible to other herbivores under the competitive conditions in the field, But not all secondary plant substances are dispensable. It is a familiar fact that animals have lost the ability to synthesize a great many biologically important materials, which they must then obtain in their food.

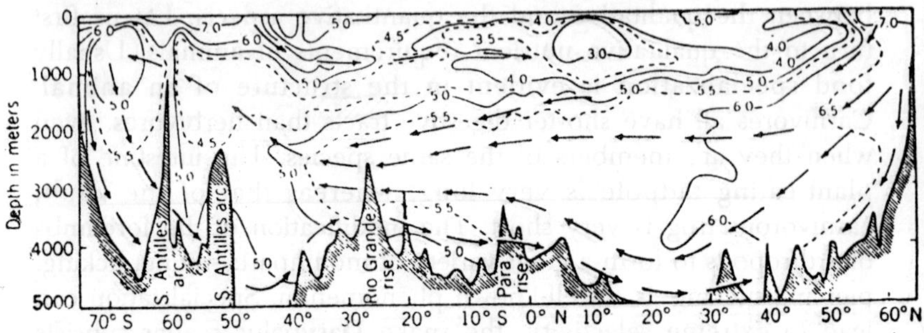

Fig. 2.33. Cross section of the Atlantic Ocean, showing the distribution of oxygen in cubic centimeters per liter. The north-south and vertical components of the oceanic currents are indicated by the arrows.

A few remarkable cases, of interest in the ecological context, deserve mention. Brancles (Balanus balanoides) can use a great variety of living or dead foods, which they filter out of the water, for growth. But to reach sexual maturity they need the planktonic diatom Sceletonema costatum. Without this food they can attain a normal size but cannot reproduce. Similarly, the marine isopod Idotea requires green algae plus the attached diatoms to get through the first stages of growth, whereas later the green algae alone suffice. The butterfly Iphiclides podalirius enters diapause in the fall, when the days become shorter. The mechanism that elicits diapause functions better when decreasing day length is accompanied by the availability of autumn leaves as food which ordinarily occurs in nature. But autumn leaves can also nave a diapause-triggering effect if provided when the days are growing longer. The males of many danaid species were found to be utterly unattractive to the females under laboratory conditions. It turned out that the male imagines are incapable of synthesizing the pheromones that stimulate the females unless given the opportunity to suck on dried Boraginaceae (Heliotropium), from which they obtain the precursors (the pheromones.

Normally the food consumed by an adult butterfly or moth is considered it negligible element among the insects ecological

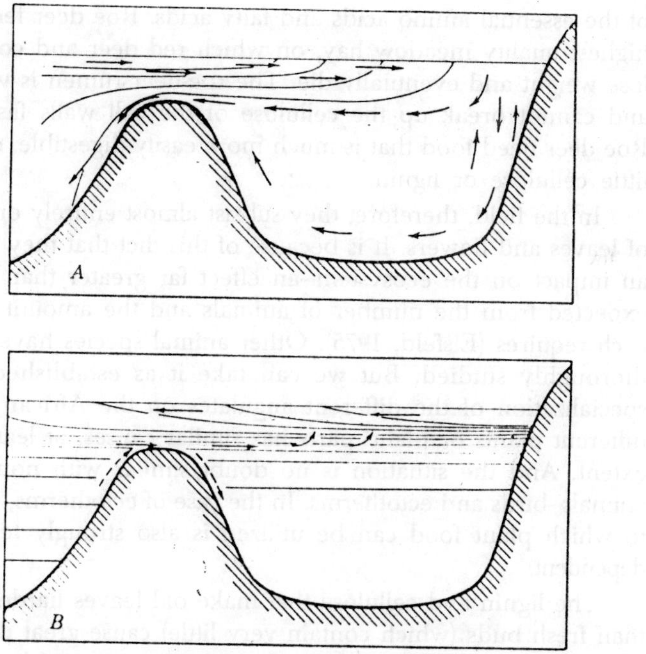

Fig. 2.34. A–Basin with local formation of heavier water and deep outflow across the sill, as in the Mediterranean Sea, the Red Sea, and the Inner Gulf of California. B–Basin with surface outflow of lighter water and occasional inflow of denser water across the sill, as in the Black Sea, Baltic Sea, and many fjords.

requirements. But it is quite significant apart from its energy contribution. This example demonstrates that a plant that is not food for the larvae and is no alive during the flight phase of the imagines (the males suck on the dried stems) can be crucial to the existence of the species. The subtle requirements of animals for particular food constituents imply adaptations of sense organs and nervous system that are now receiving greater attention from neurobiologists. Toads, for instance, when presented with a horizontal stripe moving along its long axis (a sight resembling an earthworm), respond with prey-capture behaviour; they exhibit no interest when the stripe is moving in the perpendicular direction. Grasshoppers have a specific grass receptor.

Even with respect to the basic nutrients–carbohydrate, fat and protein animals make qualitative demands on their food. These substances must be available in a form the animal can utilize and in the right proportions, and they must contain the right amounts of the essential amino acids and fatty acids. Roe deer fed with the highest quality meadow hay, on which red deer and cows thrive, lose weight and eventually die. The roe-deer rumen is very small, and cannot break up the cellulose of the cell walls fast enough. Roe deer need food that is much more easily digestible, containing little cellulose or lignin.

In the field, therefore, they subsist almost entirely on the buds of leaves and flowers. It is because of this diet that they have such an impact on the ecosystem–an effect far greater than would be expected from the number of animals and the amount of energy each requires (Eisfeld, 1975). Other animal species have been less thoroughly studied. But we can take it as established that the specialization of the different angulates on the African steppe to different plants or plant' parts has similar causes, at least to some extent. And the situation is no doubt similar with nonmamalian animals–birds and ectotherms. In the case of ectotherms, the degree to which plant food can be utilized is also strongly temperature dependent.

The lignin and cellulose that make old leaves harder to digest than fresh buds (which contain very little) cause great difficulty to the ruminant digestive system. It is not a matter of breaking the substances down, but of doing so as rapidly as possible; otherwise more energy is used up than is gained. When an organism selects its food on the basis of digestibility it uses much energy in looking for the food but does not need any to digest it and very little-ballast material is consumed. The alga-eating chironomids are less selective and consume far. More than they utilizes: many algal cells leave the gut of such an insect undamaged and fully capable of further growth. Special significance attaches to the composition of food. If this is not appropriate, animals *(e.g.,* some that suck plant juices) may be assisted by symbionts; in other cases, considerably more food is eaten than the energy requirements of the: organism demand.

The surplus components in the food are absorbed in the intestine but then eliminated. This phenomenon has become famous in the case of the aphids, which excrete: "feces" in which sugar is highly concentrated and can be used by other animals such as bees. Aphids are forced to excrete so much sugar because their food contains such great amounts in comparison with the low content of protein or amino acids. Similar relationships can be found with regard to many nutrients. Herbivorous mammals living in regions with a long winter or a long dry season must make do for extensive periods with food of very low quality, consisting almost entirely or lignin and cellulose. In the animals' very warm intestinal canals this food is broken down with the help of symbionts, so that it provides sufficient energy but not enough protein or amino acids. When mammals with several-chambered stomachs (kangaroos, camels and llamas, ruminants) are restricted to such a diet, can recycle nitrogen. The urea formed in the liver is transported to the stomach, taken into the rumen and there converted by microorganisms to amino acids and protein that can be used by the animal.

As a result the urine contains hardly any nitrogen. This sort of nitrogen, processing has been shown to occur in kangaroos, camels and most ruminant groups. The animals also conserve water in this way; it is not necessary to produce urine in order to get rid of' nitrogenous products of metabolism. Naturally this procedure is not used On a permanent basis, for it does not permit growth. But it does permit an animal to maintain the status quo for fairly long periods of time. Nagy and Milton (1979b) showed that the howler monkey (Alouatta palliata) is adequately supplied with minerals only if its food is very diverse. The figs and young leaves of Ficus insipida that constitute he most important food of these animals during the dry season do not contain enough copper, sodium, or phosphorus; these minerals must be obtained from other food.

Many tropical animals leave the crowns of the trees to supplement their diet with minerals from the soil. Howler monkeys do not descend to the ground; evidently they can find the appropriate nutrients in the treetops. The alternative approach to specific nutrient scarcity–eating enough to supply the scarcest substance and eliminating the excess of the others, as the aphids do–does not seem to be realized in other animals. The regulation of amount eaten appears to be based entirely on energy requirement. Naturally animals choose the qualitatively best food available; but if this contains too little protein either recycling occurs or the animals show deficiency symptoms. Hyperphagy in, insects has not yet been completely explained. When given, food deficient in protein crickets eat distinctly more than they do of better-balanced food. But this hyperphagy cannot compensate for the protein deficiency, because it begins only when the protein content of the food is extremely low. However, some leaf beetles are said to be able to compensate for the low nitrogen content of leaves by hyperphagy. This complex of questions deserves considerably more attention in the future, because of the abundance of trace elements present in widely varying amounts in potential food sources.

The ecological effect of animals in a system could, under certain Circumstances, be increased many times by hyperphagy–but no more than this can be said with the limited data now available. Experiments along these lines can be laborious and frustrating. Hahn and Aehnelt (1972) were able to demonstrate that the fertility of male domestic rabbits and cattle was greatly reduced when the animals were fed on hay from a heavily fertilized meadow in which only one grass species was growing. Animals fed on varied sorts of hay from unfertilized meadows bad considerably more reproductive success. The search for ions responsible for this difference in the various foods gave consistently negative results. This is the problem that confronts us everywhere: minute differences can have marked effects.

One can speculate further, that a massive increase in the size of the herbivore population leads to increasing unformity of the plant community. Is this uniformity in food one of the factors responsible for the collapse of a herbivore population. Recently we have begun to appreciate the significance of the discovery that many marine animals can nourish themselves directly on dissolved substances (amino acids, carbohydrates). Such a capacity seems obvious hi the relatively large pogonophorans, which lack a gut, but it also plays a considerable (though not decisive) role in many other animals. Sea anemones (Actiniaria, Cnidaria), for example; bear on the body surface an absorbing epithelium with microvilli, and molluscs and polychaetes take up dissolved substances from their surroundings. Calculations by Schlichter (1975) have shown that the quantity of such dissolved substances is quite sufficient to supply the entire nutrient requirement of these animals. In fresh water and on land this principle is probably of no importance.

One reason is that such direct absorption is more difficult in view of 'the osmotic gradients involved; furthermore, the numerous bacteria in fresh water remove from it essentially all of the free amino acids and carbohydrates (most of which are derived from the metabolism of planktonic algae and lower animals). The freshwater clam Pisidium does take up glucose even when it is present only at the concentrations found in the natural environment. But the amounts so accumulated meet less than 0.05 percent of the animal's energy requirement–in marked contrast to the marine situation, in which 100 percent of the required energy can be obtained from dissolved substances.

Nor can the freshwater sponge Ephydatia meet its requirements; it takes up dissolved proteins by micropinocytosis, a method quite different from the enzymatic uptake by marine animals. Statements about the biomass of animals in a habitat, then, are not in themselves informative. At least the mean body size of the species must be given. The relatively higher food consumption and the relatively higher metabolic rate of small organisms naturally finds expression in other fundamental ecological parameters. Small organisms, for example, grow more rapidly than larger animals (this is also indicated in Kleiber's table); the productivity per unit time is greater and the rate of reproduction is higher.

The consequence of all this is that small organisms are more likely to starve, and do so more quickly, than large ones. Shrews must eat at intervals of at most two or three hours; they cannot follow a light/dark cycle of rest and activity. Field mice, too, must leave their burrows at least once every 2-3 h, although their chief activity period is at night. An ecological consequence of this distributed activity is that both nocturnal predators (owls) and those that hunt by day (buzzards) can feed on the same prey. Small animals are also better suited than large animals to take advantage of the sudden occurrence of favourable conditions by reproducing very rapidly. Large animals are not so flexible, but they are better able to survive unfavourable periods without serious losses. We shall return to these questions in considering predator-prey relationships.

Animals	1 cow	300 rabbits
Total body weight	600 kg	600 kg
Daily food consumption	7,5 kg hay	30 kg hay
Daily heat loss	20000 kcal	80000 kcal
Daily weight increase	0,9 kg	3,6 kg
Weight increase per t hay	108 kg	108 kg
Meadow with 3 t hay suffices in theory for	1 year	90 days

Fig. 2.35. A cow weighs as much as 300 rabbits. The ecological effects of the two, however, are quite different.

The next step is a detailed analysis of the food requirements of an, animal in the laboratory or in captivity. First, the validity of the general formula can be tested; it can be expanded by determination of the additional food required when the animal is regularly in motion. With small animals (for example, insects or clams), the quantitative food requirement estimated in the field can be made relatively precise. But correction factors must be introduced to allow for the complicated effects of temperature; at higher temperatures food is more readily utilized, metabolic rate rises, and the amount of food necessary increases (development is more rapid). Such effects present no problem in the case of warm-blooded animals; on the other hand, with warm-blooded animals it is more difficult than with most ectotherms to transfer the data obtained in captivity to field conditions. For many animals–from filter-feeding clams to fish, frogs, birds, and mammals–there is great uncertainty as to bow the two situations compare.

Some plant-eating insects have been usefully studied in the field with analysis of both the plant substance utilized nod the excreta produced. Under favourable circumstances feces production by mammals and birds can be determined in the field, and such data can be used to infer consumption if the digestibility of the food is known. This approach has been moderately successful with wild geese in Iceland and reindeer on Spitsbergen. A promising new method of estimating energy requirement in the field has been developed by Nagy and Milton (1979, a, b). They captured howler monkeys (Alouatta palliata) and injected them with weakly radioactive (tritium-labelled) water. The blood thus acquired a certain initial level of radioactivity. Because water is continually produced by metabolism and then eliminated, the radioactivity in the blood steadily declines, providing a measure of metabolic rate. The results implied that the active howler monkey under field conditions uses up about twice as an such energy as when at rest. But there are still many gaps in our knowledge. Consider, for example, the uncertainty that prevails in the current estimates of consumption by mammals.

A widely used procedure is simply to multiply the basal metabolism by and take that as the food consumption of herbivorous mammals. Recently, with no particular justification, people have begun to regard a factor of 1.5 as sufficient. In the case of roe deer there are indications that a factor of 4 is more realistic. But probably there is no generally applicable factor. Rate of locomotion, need by no means; be linearly related to energy consumption, Birds have an optimal flight speed well above the minimum; slower flight costs more energy. The relatively quick "five-legged" shuffle of a kangaroo looking for food on the ground costs considerably more energy than the smooth forward leaps it uses for long-distance travel. It will probably be impossible to find any general solution to these problems.

The particular feeding strategy of a species has a marked effect on its energy consumption in its habitat. A "sit-and-wait" animal needs little more than its basal turnover, as experiments of wolf spiders have indicated (Ford, 1977 a, b). And the extra energy required by a web-building spider is only that needed for the construction of the web Studying a pair of snowy owls with young in a zoo. Ceska (1974) found the same food requirement as measured for snowy owls with an equal number of young in Alaska When the lemming population is in the exploding phase, at least, the owls do hardly any physical work when hunting. Things are quite different for the roe deer, which' accepts only the most fresh and delicate buds and must therefore hunt for each separate bite. An animal feeding by this method is in constant motion. The fact that metabolism is regulated by size also implies that no simple factor is applicable. Movement increases the metabolic rate of a small animal much more than that of a large animal. Finally, short term analyses of an animal's food requirements are extremely uninformative. Even adult animals undergo great changes in food requirements during the year, although the supply may be constant. This is true of both birds and mammals.

The situation is doubly complicated in that the quality of the food in the natural habitat changes during the year. And the feeding strategy and energy consumption of a given species is not always the same; in a dry summer the grasshopper studied by von Gyllenberg (Chorthippus parallel us) had to travel; further than before to find favourable food, so that Gyllenberg (1969) had to expand his population model by adding an "activity block." The activity of roe bucks and does during the mating season expends considerable effort and they lose weight correspondingly; later this loss is made up by increased food consumption. Of course, the nutrient content of food-whether its, chemical composition or its caloric value–is also subject to change. When a wild mouse is caught and fed in the laboratory the energy content of its body rises distinctly after only a few days because of the accumulation of fat; laboratory mice have a higher caloric value (per unit weight) than wild mice.

The composition and caloric value of plant parts change in the course of a year and even during a day. As important as the problem of quantitative animal nutrition is to ecology, then, it is still far from a solution. Very rough estimates can be made on the basis of approximate, calculations of basal metabolism, if the feeding strategy of the animal is known so that the required locomotor activity can be taken into account. But for any more precise estimate one would have to know the energy "budget of each individual from birth to death, and this information is available only for certain plant-eating insects. In the International Biological Programme studies of this sort have been done and a number of appropriate formulas and units established; on this basis, an ecosystem analysis can be undertaken. There are hardly any detailed analyses of the degree to which animals satisfy their needs. The few studies of wild mammals, birds, springtails and crickets, though, indicate that "the normal animal is a hungry animal" Times of abundance probably occur everywhere, but only rarely.

The normal free living wolf is undernourished, just like the normal springtail. The choice of suitable food for larvae is determined not only by quality, but by quantity as well. Parasitic wasps are famous in this regard, for in general they check very closely whether their victim has already been invaded by parasites or not. When Trichogramma finds a host egg already parasitized, it adds no eggs of its own. Most other insects behave in the same way. The number of eggs laid by the beetle Oryzaephilus depends on the mass of food on which they are deposited and which the larvae will eat; only about 1.16 as many eggs are laid as the mass could feed. Flies and gnats with larvae that live in a rotting substrate (Drosophila, Limosina, Pseudosmittia) also measure out their eggs with relative precision. If there are already very many eggs in the substrate, no more are laid. The eventual disposition of the energy acquired by feeding is of fundamental ecological interest.

Some of the food is digested and incorporated into the body, some is used up in respiration, and some is excreted as feces. In simplified terms, one can define assimilation as production plus respiration: A=P+R. In times of a world nutritional crisis it is of course not a matter of indifference what capacities animals have for assimilation. Simplifying the situation greatly, we can say that warm-blooded animals assimilate 80 percent to 90 percent of the energy in the food they eat. Similar values are reached by ectothemic predators, which feed on easily digestible, high-protein flesh.

Ectothermic herbivores, by contrast, in general can assimilate only 20 percent to 40 percent or the energy they take in. Departures from this rules occur, for example, when warmblooded herbivores receive food extremely hard to utilize; then their assimilation efficiency can fall as low as 30 percent. Of the assimilated energy, some must be used to maintain the normal vital processes, and some of the remainder goes into production of new matter (growth and reproduction). It costs a great deal of energy to maintain a high body temperature, so that the ratio of consumption to production in warmblooded animals is incomparably worse than that in ectotherms. For this reason it has occasionally been suggested that instead of raising warm-blooded domestic animals for meat, ectothermic herbivores should be provided with favourable food and used as a source of protein. But this suggestion is not realistic. The composition of the protein in ectothermic animals is not as suitable for our needs as that of warm-blooded animals. If we relied on ectothermic animals for protein we would need more than we would if the protein came from a warm-blooded animal. Instead of the effect we were hoping for, we would achieve the reverse (Chart 2).

Data on ecological efficiencies should never be taken too literally. When food is abundant, owls dissect their prey and eat only certain parts; the American badger (Taxidea tax us) has a digestion efficiency of 85.6 percent when food is scarce and only 72.2 percent in times of plenty. A general rule of thumb is that an animal uses 1-10 percent of the food eaten to synthesize new matter for its own body. Up to 90 percent to 99 percent of the food assimilated is burned in metabolism. The end products water and carbon dioxide are eliminated. Other components of the food (protein contains nitrogen) are given off as excreta. It is on the picture of energy flow through an organism so obtained that subsequent calculations and discussions in the realm of ecosystem research are based.

Materials that can neither be burned nor be eliminated accumulate in the body. Examples of such substances are DOT.

PCBs, lead, and mercury. When one animal is eaten by another and that by a third, there is a progressive accumulation of these substances in the food chain. This is the physiological basis of the danger inherent in modern environmental poisons (cf. Ehrlich et al., 1975). There is an additional consideration. Because predators are always fewer in number than their prey, the predator population is more likely to be affected by any biocide application than the prey it controls.

Chart–2. Data from Schwerdtfeger

Species	*A/C*	*P/C*	*R/C*	*P/A*	*R/A*
Rotatoria					
Brachionus plicatilis	19	11	8	57	43
Annelida					
Polychaete Nereis virens	85	44	41	52	48
Enchytraeidae	54	36	18	67	33
Archnida					
Harvestman Mitopus morio	46	20	26	55	45
Orb-web spider Araneus quadratus	85	57	28	67	33
Wolf spider Pardosa lugubris	82	24	58	30	70
Crustacea					
Water flea Gammarus pulex	37	24	13	65	35
Crab Menippe mercenaria	96	68	28	71	29
Isopods, several species	25	4	21	16	84
Sowbug Oniscus asellus	29	6	23	21	79
Insecta					
Mayfly Stenonema pulchellum	53	15	38	28	72
Stonefly Preonarcys scotti	11	4	7	43	57
Damselfly Pyrrhosoma nymphula	90	53	37	58	42
Grasshopper Orchelimum fificinium	28	10	18	37	63
Grasshopper Melanoplus sp.	33	5	28	16	84
Grasshopper Chorthippus parallelus	40	17	23	41	59
Lepidoptera, Chimabacche flagella	24	11	13	47	53
Lepidoptera, Ennomos quercinaria	32	20	12	60	40
Lepidoptera, Hyphantria cunea	29	17	12	57	43
Bug leptopterna dolabrata	33	18	15	55	45
Cicada Neophilaenus lineatus	33	18	15	52	48
Mollusca					
Clam Scrobicularia plana	61	13	48	22	78
Snail Littorina irrorata	45	7	38	14	86

Pisces					
Gudgeon Gobio gobio	80	7	73	8	92
Roach Rutilus rutilus	80	7	73	9	91
Bleak Alburnus rutilus	80	7	73	7	93
Perch Perca fluviatilis	76	20	56	27	73
Aves					
Bunting Passerculus sandwichensis	90.0	1.0	89.0	1.1	98.9
Mammalia					
Waterbuck Adenota kob thomasi	84.4	1.1	83.3	1.3	98.7
Elephant Loxodonta africana	32.6	0.5	32.1	1.3	98.5
Ground squirrel Citellus sp.	68.0	2.0	66.0	2.9	97.1
Mouse Peromyscus polionotus	90.5	1.6	88.9	1.8	98.2

The consequence of any generalized poisoning will thus, from the outset, probably be a mass multiplication of the pest, for the predators that had previously kept its numbers down will have been destroyed. Finally, the much higher number of prey individuals has yet another effect. The probability that offshoots resistant to our poison will appear rises, the greater the number of individuals. Such poisoning, then, brings the danger of eradicating the predator and making the prey immune. The spatial distribution of food is highly significant. A dense monoculture–whether animal or plant–is easter for a predator or grazer to exploit than a very irregular distribution, which requires an animal to spend considerable time and energy in finding a suitable food source. There is a certain minimum food density below which the effort per unit time to obtain enough to eat is prohibitive. Here, again, we have the problem of cost vs benefit. If the water flea Daphnia pulex is offered the diatom Scenedesmus acutus as food, it filters the diatom from the water and eats it. A 1-mm-long water flea in water at 10°C can survive if 0.04 mg carbon is available per liter of water (1 mg carbon corresponds to about 20 mg animal matter). Water fleas 3 mm long at 25°C need much more to maintain themselves–1.9 mg carbon per liter. Copepods filter-feed more effectively and can live even in nutrient-poor lakes, in the winter and at depth; in this regard they are superior to water fleas.

Quantitative data of this kind are scarce; there are only a few other examples. In some places redshanks feed primarily on the crab Corophium, which is found in enormous numbers in mud flats. If the density of the crabs is about 200 animals/m^2 or less, the redshank switches to polychaete worms which provide more food for a given amount of time spent searching. The sand-dwelling shrimp Crangon feeds chiefly on polychaetes (Nereis, Nephthys). But experiments by Gerlach (1969) showed that this laborious, hunting strategy costs more energy than it brings in, so that the weight of the shrimp slowly declines. Badgers (Meles) in southern England feed almost exclusively on earthworms. (Lumbricus), catching them on the surface of the ground by night. There are vast numbers of these worms everywhere in the soil, but they do not come to the surface everywhere, and not enough are available to the badgers to satisfy their needs.

Chart–3. Symbols and abbreviations used in production ecology, as recommended by the International Biological Programmme. Diagram of the flow of matter and energy through an ecological unit (organism or population)

Food ignored ← Available food

↓

Part not utilized NU ← Food removed MR=C+NU

↓

Rejected FU ← Consumption C=A+FU

↓

Assimilated part A=C-FU=P+R → Respired part R

P_g ↓ P_r ↓

Production $P=A-R=P_g+P_r=E+\Delta B$

↓ Eliminated part E ↓ Change in biomass ΔB

MR : Food removed from the system
NU : Unutilized part of the food removed
C : Consumed part of the food removed
F : Feces
U : Urine and other excreta
A : Assimilated part of the food
D : Digested part of the food
P : Production

R : Respired part of the food
E : Eliminated part of the food (molted, shed)
P_g : Production associated with growth
P_r : Production associated with reproduction
ΔB : Change in biomass

These are terms in a number of equations, for example:

MR=NU+C C=P+R
A=P+R FU P=A–R=C–(R+FU)

Ratios (usually given in %) indicate the ecological efficiency: for example:

MR/P C/P C/A
C/R R/P etc.

These conventions should be adopted in publications on animal ecology, so that the data are comparable; when ecological efficiency is concerned it should always be made clear which of the possible efficiencies is meant.

Therefore no relationship can be established between the density of the earthworm population and that of the badgers. The sole decisive factor is the availability of the food, which depends on specific geomorphological features (Kruuk, 1978). In the case of colony-breeding sea birds, a relationship has been found between reproductive biology and the distance that normally must be travelled to find food. Terns usually lay three eggs and feed the young until they can fly. They can manage to only if there is a good feeding site without 5 km of the nest, where they can fish without losing time. Among the auk species one can find a progressive series of brood-feeding behaviour. The black guillemot is the only bird in this family to lay two eggs; it breeds individually near the coast and usually finds its food by diving in the water just off the coast. The fish it catches are brought directly to the young which very soon leave the nest and go along on the fishing expeditions.

The other species breed in very large colonies and each bird lays only one egg. The parents usually fish relatively far from the coast; they feed their young until they have grown to about one-quarter of the adult size. Then the young leave the nest and follow their parents out to sea. The puffin flies out to the high seas in its search for food. Its young live in holes in the ground and are thus particularly well protected; they are fed for a very long time. This sort of behaviour is taken to the extreme by the tube-nosed albatrosses and their relatives, These birds also lay only one eggs; they hunt for food on the high seas, out of sight of the coast. Therefore the young can be fed at most once a day, and those of many species are fed only every second day. The adult birds fly at great speed; their feeding grounds in general lie between 30 and 200 km from the nest, and it is quite possible for them to extend to even greater distances. Among the larger forms, then the period during which the young are fed lasts more than three months. By that time the young bird has grown to a weight greater than that of the adult.

The young of most species at this stage leave the brood-hole at night and wander independently out to sea, where they develop their flying skills. Some other species (for example, the arctic fulmar. Fulmaris glacialis) are fledged while still in the nest, with marked loss of weight, for the adults have already stopped coming to the nest rocks and feeding them. The young bird is fully fledged when it leaves the nesting site. The large procellariids and albatrosses can breed only every second year; the breeding ground is occupied each year, however, by different pairs of birds in alternation. The price these birds pay for the opportunity to exploit the food offered by the high seas is a reduction in rate of reproduction (only one offspring per reproductive season) and an extremely long breeding period. They can afford this cost because their life span is so long.

3

Aquatic Ecosystems

About three quarters of the earth is covered by oceans, lakes, ponds, livers, or streams. Not only is water all around us, but it is underground, too. When it rains, part of the water runs off into streams, but much of it seeps into the soil. As it does, particles absorb and hold some water near the surface. Usually it is this soil water that plants use. Some of the water trickles down through subsoil until it is eventually stopped by an impenetrable barrier of rocks. Unable to drain further, groundwater accumulates on top of the impervious rock. The upper surface of groundwater is the water table, which more or less parallels the surface of the land. Water fills every pore and crevice between the water table and the underlying rock. Swamps, springs, and other bodies of water develop wherever the water table is high enough to intersect the surface. Groundwater, rainwater and runoff water all participate in the water cycle driven by the sun.

Solar energy evaporates tremendous amounts of water every day: This evaporation draws water from the soil, from plant leaves through transpiration, and from each body of water, large or small. Once in the atmosphere the water falls again to earth as rain or snow. Thus there is continuous recycling of moisture between earth and atmosphere and back again. Falling rain and snow are powerful erosive forces. Most landscape even those of arid countries, are carved by water action. Water freezing in tiny crevices splits mighty boulders, and glaciers grind wide, smooth valleys between knife-sharp ridges. Violent streams dig narrow canyons that in time become, flat plains through which a much-tamed waterway meanders

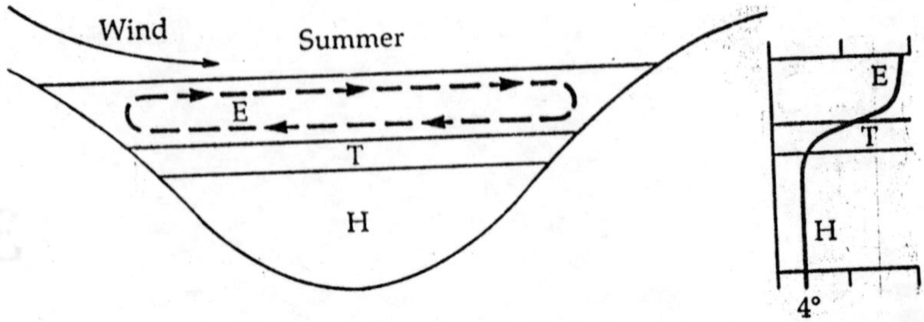

Fig. 3.1. Cross-section of lake stratification and profile of temperature with depth. E–epilimnion; T–thermocline; H–hypolimnion. Temperature scale on the right starts at 0°C and increases to the right.

10 the sea. Without doubt the sun that powers the water cycle and the precipitation that condenses from water vapour are among the mightiest forces on earth. Oceans, lakes, and rivers me only a part of this great system.

The Oceans: The Biggest Part of the World

Water returning, to the oceans from the land carries huge amounts of sediment. The buildup of these sediments, together with erosion of shores by pounding waves, has created the continental shelves that border the coasts. In general, a sharp drop, the continental slope, marks the edge of the continental shelf and the boundary of the deep oceanic basins. The shelf, slope, and in many places the ocean bottom are covered with a fine mud or ooze consisting of silt, minerals precipitated from sea water, and the microscopic shells of dead marine animals. Much of the ocean floor is a broad abyssal plain. Here and there the floor is, studded with underwater mountains, often of volcanic origin. Where, these seamounts just above the water they form small, isolated islands, such as those of Hawaii. Also interrupting the flatness of the abyssal plain are deep trenches and mountainous submarine ridges. The Japanese islands are the exposed tops of one such ridge.

Drifting Continents and Changing Oceans

Recently it has been widely agreed that the continents were once jointed together, and that the present oceans were found after the continents drifted apart. This is the theory of plate tectonics, or as it is more popularly known, continental drift. There is a significant body of evidence in support of this theory. Equally important, it explains many previously puzzling features of the earth's surface. Close inspection of a world map reveals that some

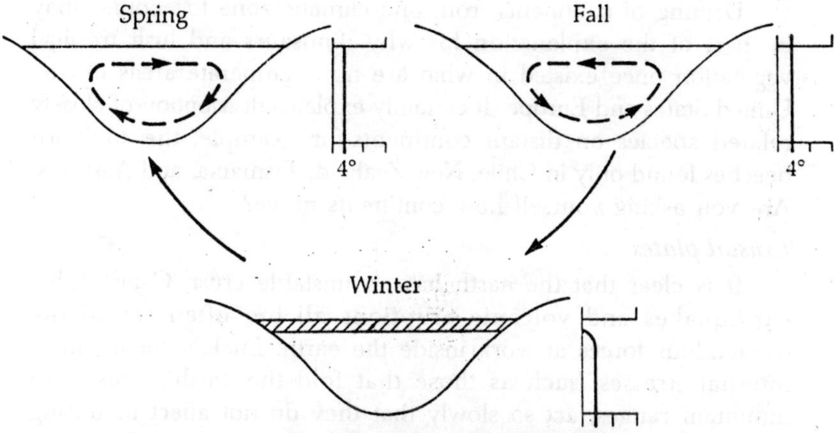

Fig. 3.2. Annual cycle of a lake.

continental coastlines have complementary shapes. Like pieces of a jigsaw puzzle, they seem to fit together. Furthermore, the present ocean floors appear to be relatively young. Examination of cores drilled from bottom sediments suggests that the Atlantic Ocean formed only about 200 million years ago. That was probably when the Americas began to separate from Africa and Europe. The separation was a slow Process, which still continues.

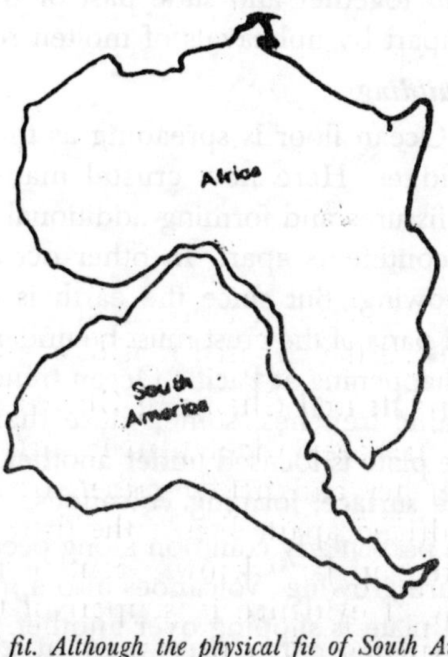

Fig. 3.3. Continental fit. Although the physical fit of South America and Africa is quite remarkable if one considers only the surface features as shown, the result is even more striking if the undersea continental margins at the edge of the continental slopes are taken into consideration.

Drifting of continents from one climatic zone to another may be part of the explanation for why dinosaurs and lush tropical vegetation once existed in what are now temperate areas of' the United States and Europe. It certainly explains distribution of closely related species on distant continents-for example, the southern beeches found only in Chile, New Zealand, Tasmania, and Australia. Are you asking yourself how continents move?

Crustal plates

It is clear that the earth has an unstable crust. Catastrophic earthquakes and volcanic eruptions all too often reveal the tremendous forces at work inside the earth. Luckily for us, most internal stresses, such as those that fold the earth's crust into mountain ranges, act so slowly that they do not affect us during our lifetime. Yet these forces move: continents. Present information indicates that the earth's crust is not an unbroken skin.

Instead it consists of at least six–and probably more–fairly rigid blocks or plates. These plates do not correspond exactly to any of our customary geographic boundaries but often include both continental masses and ocean bottom. Being composed of rather light rock, these crustal plates float on top of the heavier, more plastic material beneath. And like ice cubes floating in water, adjacent plates rub together and slide past or over each other as they are pushed apart by upheavals of molten rock from below.

Tectonics means building

The Atlantic Ocean floor is spreading as the result of activity at mid-oceanic ridges. Here new crustal material seems to be seeping out from fissures and forming additional sea floor. This, in turn, pushes the continents apart. In other oceans similar crustal plates are also growing. But since the earth is not expanding in circumference, old parts of the crust must be undergoing destruction. This seems to be happening at Pacific Ocean trenches, among other places. Such oceanic trenches, some, more than six miles deep, develop when one plate is focused under another. In places, molten crust comes to the surface, forming volcanoes.

Volcanoes are particularly common along ocean ridges at places where the plates are growing. Volcanoes also appear near trenches where one crustal plate is slipping over another. In such locations, strings of volcanoes create island arcs. Not surprisingly, earthquake zones are associated with these island arcs as well as being common along other plate margins. Collisions between plates are also responsible for folded mountains. The union of India (which once was an island) with Asia is believed responsible for the crumpled crust we know as the Himalayas.

Crustal movements are an ongoing characteristic of the earth, extending far back in time. Before the Old and New Worlds were united in one supercontinent, there was a time when they were separated by a body of water called the Protoatlantic Ocean. Considering the behaviour of the crustal plates, no marine environment is changeless, although some are exceptionally stable compared with most habitats.

Ocean Habitats

Salts are constantly added to the ocean in river water and rain water runoff. Depending on drainage, currents, and temperature, the saltiness of the oceans varies at different locations. The average salinity is about 3.5 percent, or about 35 parts by weight of salt to each 1000 parts of water. Most of the salt in seawater is ordinary table salt (sodium chloride), but sulphate, magnesium, calcium, and potassium salts are also abundant. In fact, seawater contains traces of just about every element. These accumulated salts make seawater much denser than fresh water. This is the reason it is easier to swim or float in :the ocean than in a river or a lake. It also means that salt water has a much lower freezing point than fresh water. In addition to salt, seawater also contains dissolved atmospheric gases. But seawater has less oxygen than fresh water, since salt decreases the solubility of oxygen in water.

Ocean layers and life

The sun heats and illuminates the ocean, although few of the sun's rays penetrate very deep into the water. As a result, water temperature, illumination, and salinity (which is temperature dependent) are related to depth. Of course, all these characteristics influence the kind of organisms to be found at a given location, but light is of special importance. Because photosynthetic plants need light in order to live, they are found only in shallow water and in the surface layer of the open ocean. Almost all photosynthesis occurs in the upper 80 meters (260 feet) designated as the euphotic (good light) lone. Underneath the euphotic zone is a region of ever-deepening twilight. No sunlight penetrates below 600 meters (1967 feet). As a result, most of the ocean is in perpetual darkness and can be inhabited only by animals and decomposer bacteria and fungi.

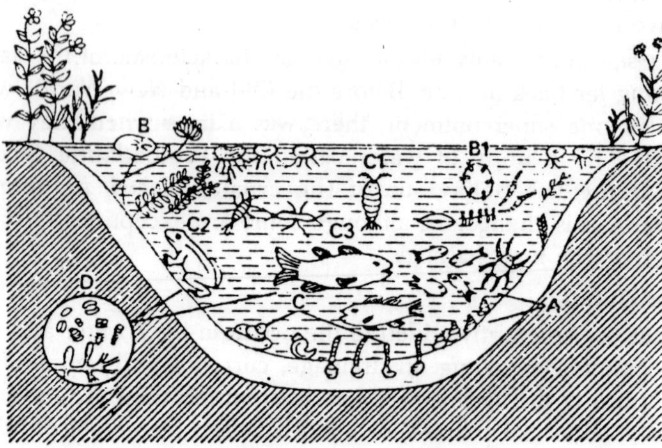

Fig. 3.4. A pond ecosystem.

Life at the top

Microscopic floating algae, known collectively as phytoplankton are the main producers in the oceans. Diatoms and dinoflagellates arc the most prominent phytoplankton. Their collective biomass far outweighs that of the more obvious algae known as "sea weeds." Most seaweeds are attached to rocks near the coast. Phytoplankton are never evenly distributed. Not only are they limited to the top layers where there is light, but within that layer their density follows distribution of minerals. Organisms that die in deep water usually drift to the bottom. Similarly, faces settle out, and the valuable nutrients they contain are lost from surface waters.

Currents and diffusion only slowly return precious nitrogen and phosphorus from the bottom to the euphotic zone. Shallow water over the continental shelves is usually fairly well mixed and contains more nutrients than surface water over the open ocean. Here the deep water is quite undisturbed, and minerals are concentrated far below the euphotic zone. During the winter, because sunlight striking temperate waters is less intense than at other seasons, photosynthesis is limited to the top few meters of water. With the coming of spring the phytoplankton multiply rapidly in response to increasing illumination.

Exploding plankton populations are termed plankton blooms, because the masses of algae often colour the water bright red, brown, or green. Although clearly visible, marine plankton blooms are never as dramatic as those of freshwater lakes. This difference in productivity may result from mineral deficiencies in illuminated marine waters. Microscopic animals, often referred to as zooplankton, feed on phytoplankton. Copepods and krill, tiny relatives of crabs and shrimp, are among the most abundant zooplankton. Others are numbered among the unicellular protozoa and include foraminifera and radiolarians.

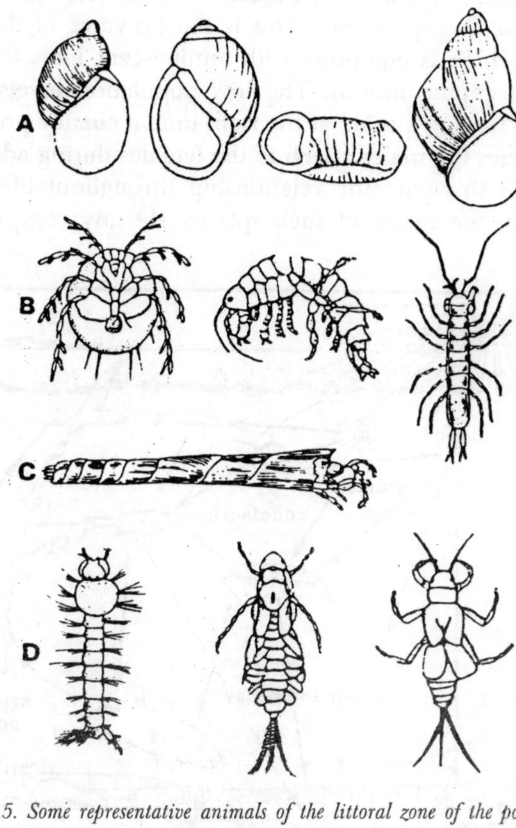

Fig. 3.5. Some representative animals of the littoral zone of the pond and lake.

Typically, zooplankton bear numerous projections that increase their surface area and help, suspend them in the water, Most zooplankton both swim and float. For the most part, relatively small fish, such as anchovies, herrings, and sardines prey on zooplankton.

There arc exceptions, however. In the cold but fertile water around Antarctica the penguins, fish, squid, and even baleen whales feed directly on krill. The amount of zooplankton consumed by a large animal can be astounding. A single blue whale can eat three tons of krill a day.

The briny deep

Beneath the productive upper layer of the ocean lies as much as six miles of water. Throughout most of this space, life is sparse. Food from the top decomposes quite slowly. Fish in deep, dark waters are mostly mouth. Good meals are rare here and few items are rejected as being too big. This is the province of the famous angler fish, which is equipped with iuminescent lures that attract prey into its gaping mouth. The low population levels in deep water makes finding a mate at the right time a chancy proposition. In some species the males attach to the females during adolescence and maintain this parasitic relationship throughout life. As you might expect, the males of such species are tiny compared with their mates.

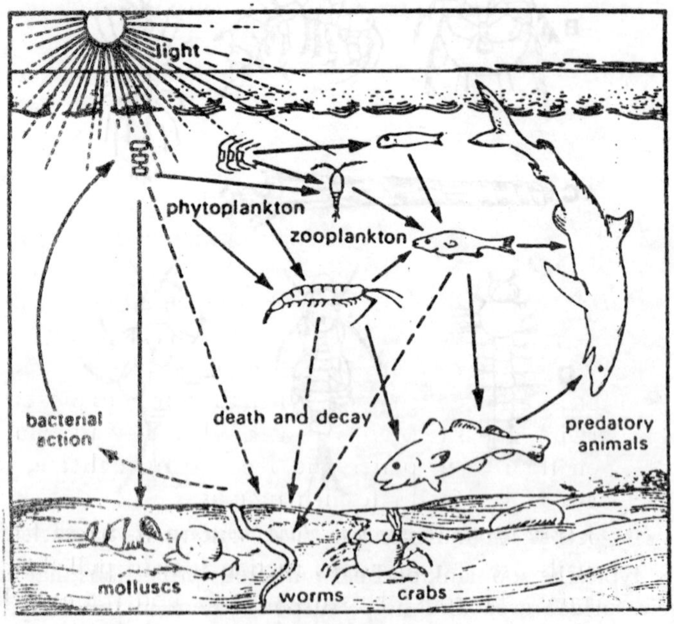

Fig. 3.6. Oceanic ecosystem.

Rocks and Sand: Marsh and Muck

In contrast to the relatively uniform conditions of the open sea, a variety of habitats occur where water meets land. Each of these environments is unique; each is intriguing in its own way. Perhaps rocky shores are the most interesting. Here we find a diverse community attuned to the rhythmically changing environment.

Living between the tides

In the band that spans high and low tide, the first priorities are to stay put, to keep wet, and to avoid being crushed. Waves pound the shores so vigorously that delicate organisms can be destroyed or carried away. Alternately submerged and exposed, intertidal species (those that live between lowest low tide and highest high tide) must also withstand cold, heat, and a wide range of salinity. Summer sun can cook tissues, and evaporation of seawater deposits a crust of salt on every surface. Only a few months later, winter tides pull the protective layer of water away and leave organisms exposed to freezing cold or to torrents of fresh water. As adaptations to this extreme environment, intertidal organisms either cling tenaciously or dart about to find shelter. Their bodies are protected by shells, tough body walls, leathery surfaces; or muscus.

Since tidal exposure varies every day, there are distinct life zones along the shore. The following applies to the central California coast, but similar areas occur on rocky shores from Mexico to Alaska and along the coast of Maine and the Maritime Provinces, as well as on other continents. A dark band of cyanobacteria (blue-green algae) marks the highest zone where there is much marine life. Here numerous periwinkles scavenge. A distinct band of white barnacles grows below the black zone. Barnacles lie on their backs, stuck to the rocks. When the tide goes out, barnacles close their limy shells to conserve water. When they are submerged, they open their shell plates and feed, using delicate, jointed appendages to kick food into their mouths.

Limpets are abundantly distributed among the barnacles. These marine snails use a tremendous suction foot to pull their shells tightly against the rock. This traps water under the shells and prevents dehydration during low tide. A crowded strip of mussels stands just below the mean-sea-level mark that coincides with the bottom of the barnacle region. The clamlike mussels spin proteinaceous fibers that attach their blue-black shells to the rocks. Tidal pools harbour rock crabs, hermit crabs, and large green sea anemones. Farther out, on the underside of large rocks and steep ledges hide sponges, sea cucumbers, sea squirts, chitons and abalone. Shrimp and spiny lobsters lurk in the lowest tide pools.

No one has made a complete census of animals in the intertidal zone, but biologists familiar with the California coast estimate that there may be 3000 species there. The showy life of rocky shores appears missing from sandy beaches, but even though beaches appear barren, animals are present. It's just that most of them are bidden from view. Burrowing, digging, and tunnelling creatures, among them clams, mole crabs, tube worms, and olive shells, bury themselves in the sand. Low tides expose some species periodically; others are permanently protected within their burrows. Typical residents below the low-tide line include whelks, swimming crabs, sand, dollars, and hermit crabs.

Wetlands and estuaries

Bays and river mouths, where fresh and salt water meet, are unusually fertile environments. The most valuable of such estuaries are broad, shallow basins created by silt deposits. Here mud flats and tidal marshes line channels of open water. Most of the silt and organic matter dumped from the river becomes trapped in its estuary. Slow currents, a large surface for evaporation, and presence of rooted vegetation all help make estuaries places where nutrients tarry. Salt marsh grasses, eel grasses. and other rooted plants take advantage of rich water and soft bottoms.

Algal scums cover mud flats, larger plants, and every solid surface, and dense phytoplankton blooms colour the water. There is so much food that only a tiny part of the photosynthetic product finds its way directly into the mouths of herbivoures. Instead, most plants die and, in the absence of swift currents, simply sink to the bottom and rot. The abundant organic matter supports a teaming broth of bacteria and fungi that use all available oxygen from the muddy bottom.

As a result, other anaerobic microorganisms thrive there and release hydrogen sulphide. This noxious gas blackens marshland mucks and imparts a characteristic stench. Their tremendous photosynthetic productivity and extensive decomposer food chain permit estuaries to literally teem with life. Microscopic decomposers are eaten by filter-feeding zooplankton, as well as by larger filter feeders, such as worms, clams, and oysters. Estuaries serve as nurseries for numerous coastal fish. Menhaden, striped mullet, summer flounder, king whiting, croakers, striped, bass, smelt, and sturgeon are all commercially important species that rely on estuaries sometime during their lives. The shrimp fisheries also depend on this resource, because young shrimp require the estuary habitat.

Oysters, blue crabs, and several commercially valuable elam species are permanent estuarian residents. Furthermore, saltwater marshes are the natural home of most waterfowl. As valuable as they are in their native state, estuaries are often prized more for their land. To take advantage of a protected harbor, most seaports are built on estuaries. As early ports grew, the surrounding marshes and mud flats were usually drained, and regions of shallow water were dredged or filled to provide areas for docks, industries, airports, and even residences. In Washington, D.C., only the Tidal Basin remains to remind us that the Jefferson Memorial, Lincoln Memorial, and Washington Monument were built on land "reclaimed" from a marsh. Once leached free of their load of salt, drained tidal marshes become fertile farmland.

Over the centuries, carefully engineered liking and pumping schemes created much of the Netherlands. But present generations seem little aware of how much humans are responsible for familiar shorelines and of how important estuaries are in the schemes of aquatic life. Even now, many of our least-modified estuaries are threatened with "development" into resorts.

Lakes: Quiet Changes

In many ways lakes resemble oceans. Of course, lakes are much smaller, and they usually contain fresh water. Diatoms and desmids are the most abundant phytoplankton of lakes. And though the shoreline and broad-leaved floating plants contribute much more to the productivity of fresh waters than marine seaweeds do to the oceans, zooplankton and fish are abundant in both habitats probably the most distinctive animal components of the freshwater food chain are the many insects, especially in their juvenile, stages. Variations in climate, fertility, shape, and bottom characteristics cause lakes to be quite different from one another. But ultimately all these factors translate into nutrient availability. Unlike the oceans, temperature zone lakes are stirred regularly. As a result, there is efficient nutrient cycling. This is what accounts for the much greater productivity of many lakes compared with any ocean.

Seasonal Turnover

Swimmers know that lake water is warmer at the surface and colder at depths. Thus in the summer, lakes are stratified with lighter, warmer water at the top and colder, denser water toward the bottom. As winter approaches, falling air temperatures and brisk winds slowly cool temperature zone lakes. The fact that fresh water is most dense at 4°C (39°F) permits cooling lakes to be thoroughly stirred. When surface waters cool to 4°C, they sink to the bottom and displace the warmer, less dense water there. This turnover brings nutrients from the bottom and spreads them throughout the lake. At some point the entire lake is at 4°C and lacks any density stratification. Absence of stratification allows the lake to be further stirred by the wind. Only after the whole lake reaches 4°C may the temperature of the surface dip lower. When this occurs, the colder, less dense water floats and a layer of ice may eventually form.

In the spring, the surface water gradually warms to 4°C. Then, as during the fall-cooling, there is a period when all the water is 4°C and all of the same density. At this time winds blowing across the surface may set up currents that travel toward the shore and turn downward and move across the bottom. As you can see, there are two turnovers a year, one in the spring and the other in the fall. Both stir bottom sediments, and both also replenish the oxygen supply at the bottom. With available nutrients and increasing sunlight and warmth, plank ton blooms colour lake water beginning in the spring. Changes in temperature and nutrient levels cause one phytoplankton species to bloom and die, only to be followed by a population burst of another. The exact succession of blooms is determined by the requirements of the various species.

Nutrients, Pollution, and Aging

Lakes have a relatively short life span. It may take thousands or millions of years, but most lakes that we know are in the process of disappearing. At the time of formation, a young lake is usually low in nutrients and sparsely inhabited. As time goes by, the lake is enriched by accumulation of river sediments land nutrients leached by rainwater from the surrounding land. Further fertilization occurs as dead plants fall into the water and decay. This normal enrichment of a lake is known as eutrophication (addition of good food). Eutrophication permits more and more organisms of increasingly different kinds to live in the lake. Clearly, the biological nature of a lake gradually changes as the lake ages. With time, lakes undergo a process analogous to succession on land. Eventually eutrophication may end with disappearance of the lake altogether. The Fate of Lakes. As sediments build up, lakes become more shallow and vegetation encroaches. Filled-in lake margins develop into marshes or swamps.

By definition a marsh is a wet grassland and swamps are wet forests. Like their saltwater counterparts, freshwater marshes support diverse communities adapted to slight variations in water levels. So do swamps. And like marine wetlands, those of fresh water are gradually disappearing. Conversion of the rich muck into farmland is the goal of many government-funded drainage and stream channelization projects. In cold climates lakes may eventually turn into spongy, wet bogs. Because of the cold, accumulated plant material decomposes poorly, and it is gradually transformed into peat. Humic acids derived from the peat further inhibit decomposition and give the water a characteristic brown tinge. Peat is often dug up, dried, and burned as fuel.

Excess eutrophication

Although natural eutrophication is, part of the aging process of every lake and tends to favour mineral accumulation, human activities that accelerate the process can lead to early death of a lake. Fertilizer, washed into lakes from nearby farms and the addition of minerals in household or industrial sewage can create an overabundance of nutrients in the water. Under these conditions algal blooms deplete oxygen in the water. Fish and other animals may simply choke to death. Studies of the Great Lakes trace their "eutrophication" to excess "phosphorus. Experimental addition of phosphate: to small lakes in the same region has produced responses similar to pollution in the Great Lakes. Neither nitrogen nor carbon alone has a similar effect. Apparently simply removing phosphate from waste water, principally by banning phosphates in house hold detergents, would markedly improve conditions in many bodies of water.

Rivers and Streams: Running Waters

The continuous one-way flow of running waters distinguishes them from lakes. Indeed, many of the special characteristics of these waters are related to their turbulence. Because currents constantly bring new water into contact with a given organism running water is effectively richer than still water in which a static, "tired" layer surrounds each individual. And to be sure, turbulence mixes air into running water and thus usually ensures a plentiful supply of oxygen.

The Food Web of an Open System

Rivers and streams inevitably reflect the land they drain. Stream water arises mainly from groundwater, and surface runoff provides most of the remainder. Water from both sources carries mineral nutrients and human additives, such as pesticides. But most of the organic matter that supports the food chains of running waters also washes in from the surrounding land. Or simply falls in. Probably tree leaves supply more energy and more carbon than any other single source. Additional sources of carbon and energy as adult insects that flounder into the water, earthworms, and other small soil animals. Except in quiet streams that closely resemble lakes or ponds, there are, few phytoplankton in rivers. Thus running waters are unusual ecosystems in that primary production accounts for but a small portion of the energy needed. Most nutrients enter rivers and streams from the outside.

Adaptations of Currents: Don't Get Carried Away!

Only careful strategies prevent the current from carrying stream organisms to different and less suitable habitats. For example, diatoms of such habitats have stalks. Larvae of Blackflies and many other insects anchor themselves with holdfasts. Yet other insects spin sticky threads about themselves or burrow into the mud. Many stream animals are streamlined. They have compact and flattened bodies and usually live under stones, where the current is reduced and where they are less visible to predators.

Fish can expend a great deal of energy swimming against the current–so they don't. Fish survive by finding shelter. As anyone who fishes knows, the best place to catch trout is in a deep bottom hole or along the downstream side of large rocks or other obstacles. Generally, the larger the shelter, the larger the fish. The adaptations of stream organisms truly illustrate the selective forces of this environment. But of course, every environment is selective.

A Special Way of Life in Water

To an animal, the most fundamental difference between living on land and living in water is the scarcity or oxygen. Oxygen is so poorly soluble in water that it is hardly surprising that no warm-blooded animals can live submerged for more than short periods. Aquatic mammals, such as whales, and aquatic birds, such as penguins, all breathe air and thus obtàin enough oxygen to sustain the high metabolic rate associated with a constantly warm body. And surely the complexity of gills is related to the need for extensive surfaces for gas exchange with water. Most often gills are located where water can be readily pumped past them to maintain a constant supply of oxygen. As we will see in the next chapter, the deadly effects of aquatic pollution can be traced to exhaustion of the oxygen supply. Aquatic ecosystems are also unusual in that most producers are microscopic.

Since so many people disregard anything they can't see, this may be why the structure of underwater communities escapes the understanding of many citizens. Microscopic producers dominate in water, because they have a large surface area compared with their volume. This ensures efficient absorption of nutrients. Furthermore, vascular tissue is unnecessary in this habitat, and large plants are subject to damage by waves or current. Finally, the high specific heat of water gives aquatic environments a temperature stability unknown on land. Water is slow to warm and slow to cool. Large bodies of water, especially, fluctuate little in temperature. Not only are aquatic organisms freed from the usual problems of temperature regulation, but density changes associated with minor temperature shift account for lake turnover and also affect ocean currents. Remembering these facts can help us maintain the integrity of aquatic ecosystems. This is an important step toward ensuring stability of the biosphere.

4

Terrestrial Ecosystems

To understand the biosphere, we must examine major biomes and ask not just how communities differ but why. Although no single factor defines an environment, environmental extremes may dramatically limit options. Consider, for example, the earth's poles. These ice-covered regions that the penguins and polar bears call home are the least hospitable areas of the earth. They are cold most of the year because of the pattern of sunlight, water currents, and wind. The brief summers are never warm enough to melt the accumulated ice packs. As a result, rooted plant life is impossible. In the north the food web is based on the producers of the Arctic Ocean. The presence of a continent at the South Pole results in an environment with little life. A well-developed food chain is present only on the edge of the continent. In the arctic north, the land of permanent ice is edged by a tundra that develops where more moderate day lengths and increased solar radiation leave the soil bare and soft for a few short months. Conditions permit terrestrial food chains to develop.

Large, nomadic animals, such as reindeer and caribou, graze on the low, hardy vegetation but must range widely to find enough food. Predatory wolves and foxes follow the great herds or hunt lemmings and other small rodents that burrow in the ground. Although tundra soil is frozen much of the time, sunlight thaws the surface each spring. Suddenly plants burst into flower as the desolation of the dark, snowy winter is broken by the long, warm days of arctic summer. Black flies, mosquitoes, and other insects emerge in great swarms. Migratory waterfowl, shore birds, and song

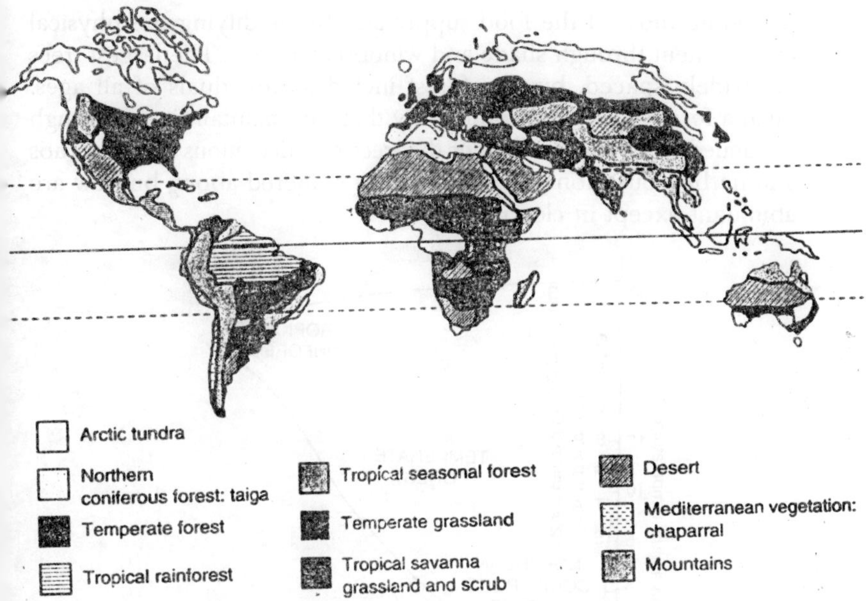

Fig. 4.1. The terrestrial biomes of the world.

birds arrive from thousands of miles away to nest and feed on hordes of insects from temporary ponds and thawing rivers. Further south, cold is not so pervading; here the land form and rainfall become important variables. As a result, the earth is blanketed with biomes like a giant patchwork quilt. Forests, grasslands, and deserts are scattered about. No two are alike, and of course, none remains untouched by human hands. Some we have only exploited. Others we have manipulated beyond all recognition. The story is at the same time fascinating, inspiring, and sad.

Temperate Deciduous Forests

European colonists found most of the eastern seaboard of North America covered with deciduous trees–trees that shed their broad leaves in the fall. Similar forests once blanketed most of Europe, the temperate coasts of Asia, and smaller areas of Australia and South America.

The Plant Community

Mature deciduous forests often contain dozens of tree species, but in any given area one or two kinds of oak, hickory, chestnut, maple, basswood, buckeye, or beech are common enough to give the forest their name. Big trees dominate the community by providing most of the food supply and by modifying the physical environment through shade and windbreak action. The largest trees are widely spaced, but the forest includes individuals of all ages. Such a forest is a stable community that can maintain itself through continued reproduction. Under the trees of a deciduous forest, shrubs and herbaceous (nonwoody) plants are scattered about, but few are abundant except in clearings.

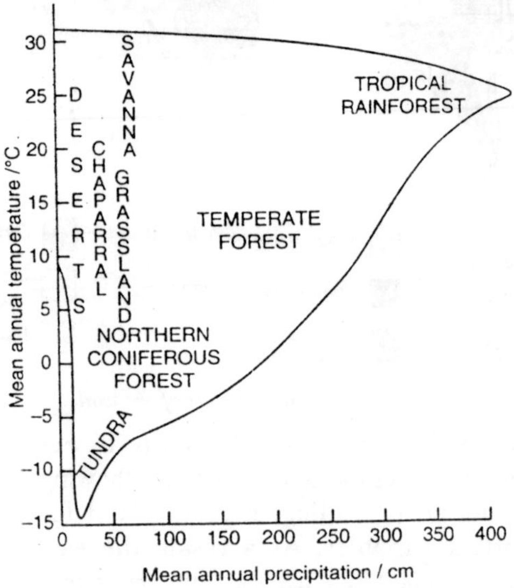

Fig. 4.2. The relationship between the eight major terrestrial biomes and mean annual temperature and mean annual precipitation.

The density of the leafy canopy overhead and the time each year that the canopy is present regulate growth of understory (below the canopy) plants. Many of these are small herbs or "spring flowers" that grow rapidly as days lengthen. They reproduce and carry out much of their photosynthesis before the trees leaves open and reduce the sunlight available on the forest floor. Soils of some deciduous temperate forests are brown and rich-looking, but most of the available minerals cycle through the plants each year. Only a small portion of the minerals absorbed are retained in the wood of the tree. The remainder returns to the ground as twig and leaf litter. Not all the returned materials can be reused the following year, because decomposition takes a long time. Of the minerals released from the litter, some invariably leach away in rainwater.

Humus from decomposing leaves and twigs helps hold both water and minerals.

Table 4.1. Average Numbers of Organisms/2.6 km^2 (1 Square Mile) of Temperate North American Deciduous Forest During the Summer.

Plants	
Trees 7.5 cm (3 in.) or more in diameter	75,000
Tree seedlings	78,600
Shrubs	281,000
Soft-stemmed plants	345,000,000
Animals	
Invertebrates (Insects, snails, centipedes, millipedes, earthworms, etc.)	2,688,000,000
Pairs of small nesting birds	768
Large predatory birds (owls and hawks)	2 to 5
Mice	240,000
Gray squirrels	1,500
Flying squirrels	1,500
White-tailed deer	40
Wild turkeys	20
Gray fox	3
Black bear	0.5
Mountain lions	0.2

Temperate deciduous forests are highly productive, and they support a large number of organisms, as suggested by Table. Such forests are also quite diverse. An often overlooked part of the community lives in the litter on the forest floor. Much of the forest enters this decomposer food chain yearly. Litter decomposition involves the activity of multitudes of tiny animals and microorganisms. Insects such as springtails and other near-microscopic arthropods eat dead leaves. These primary consumers absorb less than 10 percent of the nutrients in the litter. However, as a result of their digestive process, the remains that become feces are easily attacked by bacteria and fungi. Animals not only consume and alter litter; they also mix it into the soil. Earthworms are among the most active litter mixers. The carnivores that eat the litter feeders also stir litter and mix it into the soil as they endlessly seek their prey.

Adaptation of Climate

Temperate deciduous forest have moderate temperatures and rainfall (about 70-100 cm or 28-40 in./yr) but are characteristically subject to forest. Shedding leaves in the fall protects the trees against cold damage. Otherwise the large leaf surface that favours transpiration and photosynthesis during the warm season would allow heavy evaporation and severe water loss during the winter. You will recall that the absorption of water by roots relies on diffusion, as well as on energy-consuming activities of root cells. Both are slowed by cold, and of course, frozen soil yields no water. Cold weather can dehydrate plants more than hot weather, especially when it is combined with strong winds. Furthermore, leaves are difficult to protect against freezing.

Deciduous trees have evolved an effective strategy to prevent damage from winter cold. First they concentrate sugars and other organic compounds from their leaves into roots and stems. Here these large molecules lower the freezing point of cell fluids and prevent formation of ice crystals that could rapture cells. Then the trees simply shed their leaves. Leaf fall is a preprogrammed process coded into the genetic structure of the plant and only slightly affected by the weather any particular year. In response to a decrease in auxins, an abscission (cutting off) layer forms where the leaf stalk joins the stem, and the cement between the cells in that area softens.

In the absence of strong intercellular cements, the short cells of the abscission layer separate, permitting the leaf to drop off or blow away in a gentle breeze. The colour changes that foretell leaf drop in deciduous forests result from cessation of chlorophyll formation. Because chlorophyll breaks down spontaneously in light, leaves remain green only through continued synthesis of chlorophyll. With the coming of fall, chlorophyll manufacture ceases, and this green pigment gradually bleaches in sunlight. In the absence of chlorophyll we see yellow or reddish pigments, mainly carotenoids that aid in light absorption for photosynthesis. Another source of red colour results from interruption of phloem by the developing abscission layer. This causes sugars to accumulate in the leaves, where some are converted into reddish compounds.

Community Succession in Deciduous Forests

When the vegetation in an area is destroyed, it is usually replaced. Often the new vegetation is different from what was destroyed. And in most cases, this new vegetation is gradually replaced with another and then another. The sequence for any particular area is always the same. Eventually the vegetation becomes stabilized. This permanent, stable vegetation is said to be the climax vegetation for that particular region. The replacement of one vegetation, and then another, in a predictable sequence is known as succession. The fact that successional stages and climax vegetations differ from place to place results from differences both in the environment and in the history of the vegetation present. Obviously, a particular species can appear in a successional vegetation only if there is a seed source.

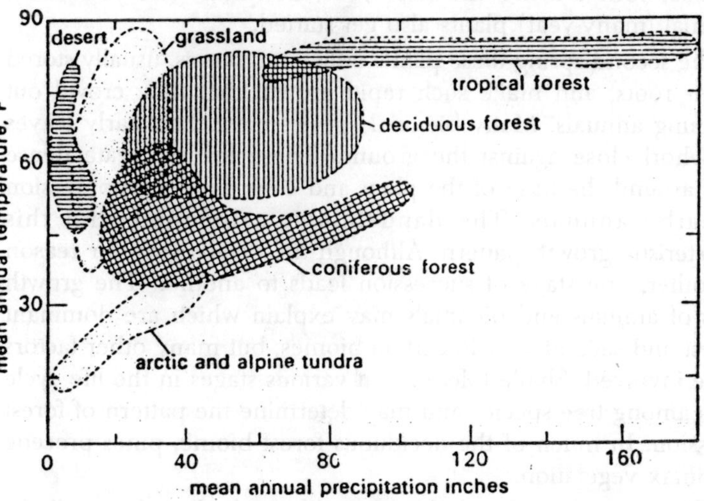

Fig. 4.3. A co-relation between mean precipitation and annual temperature in different terrestrial ecosystems.

The seed may be present in the ground or provided by patches of successional vegetation. Because climax vegetation is susceptible to destruction by fire, flood, or some other catastrophe, no area is long without successional growth. Most human societies continually disrupt natural vegetation. Consequently, the vegetation people see daily is successional rather than climax vegetation. Examples may be found in roadsides, fence rows, neglected yards and fields, and cut-over land. When first visited by Europeans, much of the Americas supported climax vegetations. Because the forests seemed untouched by humans, they came to be known as virgin forests. Usually what the layperson calls a virgin forest the biologist recognizes as the local climax vegetation. The common term second growth most often refers to a successional forest.

The factors controlling successional patterns differ from place to place, and they are often poorly understood. Most studies have involved land that was cleared for agricultural use and then, later, abandoned. Whenever bare ground is left undisturbed, rapidly growing annuals (plants that set seed the first year and then die) may appear. Although there is a characteristic first-year vegetation in each plant community, in many places these plants are replaced by introduced annuals that everyone calls "weeds." This is one example of the fact that the history of an area affects the pattern of succession. During the first year, biennial (two-year) and perennial (many-year) plants also get started.

The second spring these plants draw on reserves, usually stored in their roots, and make such rapid growth that they crowd out the young annuals. Many biennial species have their early leaves in a whorl close against the ground. These leaves create dense shade around the base of the plant and contribute to suppression of nearby annuals. The dandelion is a biennial with this characteristic growth pattern. Although the details vary for reason or another, one stage of succession leads to another. The growth habits of annuals and biennials may explain which are dominant the first and second year in certain biomes, but many other factors can be involved. Shade tolerance at various stages in the life cycle differs among tree species and may determine the pattern of forest succession. In much of the deciduous forest biome, pines precede the climax vegetation.

Most pines are particularly vulnerable to shading. Lower limbs die as the pine forest grows dense, and pine seedlings never survive in deep forest shade. The monks, maples, beeches, and other members of the climax forest are more shade-tolerant. Small trees become tell and thin in heavy shade but survive and eventually grow taller than the pines. Soon shade from the deciduous species kills the pines. The seedlings of the deciduous trees also survive in the shade of larger ones, so there are always small trees in the understory. When a tree is blown over, others quickly replace it in the canopy. Thus shade tolerance helps explain why the deciduous trees replace the pines and also why the deciduous trees are the climax vegetation. For details of succession in a well-studied part of the deciduous forest biome. The series of vegetational stages that follow destruction of a climax vegetation, as we have just described, is often called secondary succession. Note that secondary succession begins when soil is available. Primary succession is a series of events whereby plants colonize an area of barren rock, such as fresh lava. Presumably this process has occurred at least once wherever there is now vegetation.

In many instances primary successions have been associated with the evolution of species and have required millions of years. However, we see examples of the early stages of primary succession today on rock outcrops. The original colonizing plants are usually lichens that produce acids. The acids dissolve certain minerals in the rocks. This process roughens the surface, so the lichens can attach better, and it also makes minerals available for lichen growth. Over the years, mineral grains released by the acids and decaying lichen components create a small amount of soil. Other plants, particularly mosses, may grow in the lichen soil and so constitute the second stage of primary succession. Later events depend on the environment and the species present, just as we noted with secondary succession.

What Humans Have Done with Deciduous Woodlands

Most of Europe was once covered with climax deciduous forests except at high elevations and in places where standing water created marshes. The forest canopy was so dense that, it has been said, a squirrel could travel from central Russia to the Atlantic coast without ever touching ground. On much of the continent the dominant species were oaks, but in some regions elms, beeches, and birches were prominent.

Fire and axe

Many of these forest succumbed to fire or the flint axe. At first people cut trees mainly to obtain building material or fuel, but sometimes trees were felled merely to clear the land. Initial clearing were often abandoned and recovered by forest, as has been shown by the study of pollen buried in the mud of nearby lakes and bogs. Decreases in pollen of oak ivy (a native understory vine) mark the first clearing of the forest. This stage is followed by high levels of weed pollen, which indicate a period of cultivation. Later peaks in hazel pollen show when these shrub by trees, took over fields that must have been abandoned. Vast tracts of climax forest still flourished in Roman times. Throughout that period the forest edge expanded and contracted with wars and other changes in human fortunes.

Eventually, the relentless demand for plowed land and forest products doomed the woodlands. As commerce and trade grew, more and more timber was needed for ship building. Glass and soap manufacture required wood ashes as raw materials. The smelting of tin, lead, copper, and iron depended on charcoal made from wood. By the eighteenth century these industries had scalped the forests of England, and the British had turned to digging coal. The woodlands of western Europe disappeared a short time later. Such drastic alternation of the landscape naturally had far reaching effects. Rains began to wash the soil from the naked land. It is no coincidence that several British seaports filled with slit during the twelfth and thirteenth centuries and were abandoned when their waters were no longer deep enough to float seagoing vessels.

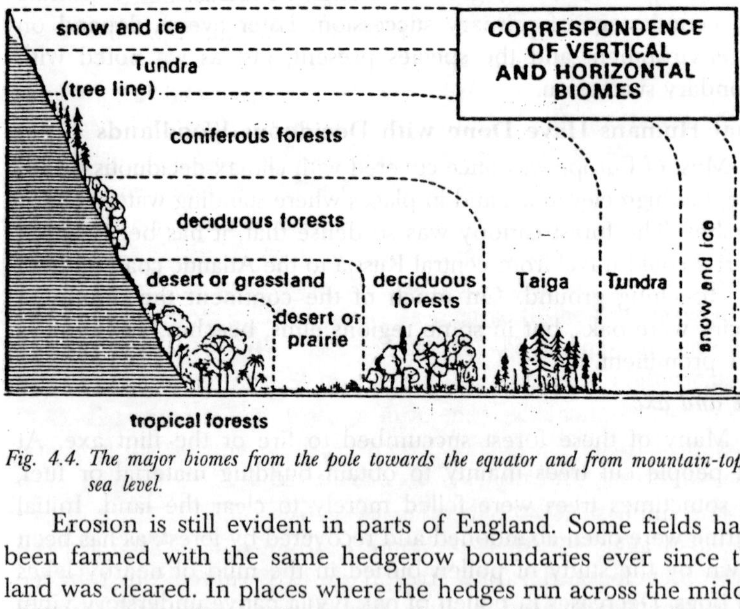

Fig. 4.4. The major biomes from the pole towards the equator and from mountain-top to sea level.

Erosion is still evident in parts of England. Some fields have been farmed with the same hedgerow boundaries ever since the land was cleared. In places where the hedges run across the middle of slopes, the soil is as deep as six feet on the upward side of the hedgerow but so thin on the lower side that rocks lie exposed. We have no idea how much more soil eroded away entirely and washed into the ocean. Domestic animals prevented regeneration of most British woodlands. A great sheep industry thrived in England during the Middle Ages; grazing was so heavy that the seedlings had no opportunity to grow out of reach of livestock.

As sheep-rearing decreased, rabbits, introduced from the Continent, increased dramatically. The rabbits stripped bark from the young trees and did so much damage that few seedlings survived. Conversion to Grassland and Moor. In many deforested areas of Britain the climate, soil, and grazing animals interacted to create a habitat favourable to coarse grasses. Manure from the sheep and rabbits returned minerals to the soil, established an efficient nutrient cycle, and permitted the formation of a crumbly soil that supported a heavy turf. Not all the deforested British lands went to grass. On some soils the loss of trees raised the underground water level enough to create bogs. There is a simple explanation for this drastic change. Each day during the leafy season, large trees transpire tremendous volumes of water.

Deforestation reduces this drain on the groundwater. Therefore cutting certain forest can lead to their replacement with bogs or wet, peaty moors that support low shrubs. Fate of the Forests. Wherever civilization flourishes, forests are destroyed. The history of other forests differs little from that of European forests. The Chinese began clearing ground for crops at least four thousand years ago. The Mediterranean forests were already seriously damaged by Plato's times. As a rule, few forests remain where civilization has long thrived. Although we must infer much of the history of Eurasian vegetation, we know a great deal about what happened in North America. Along the East Coast, the abundance of land and the need for a lightweight crop the could be traded in Europe promoted slash-and-burn tobacco farming. Not only did cutting and burning the trees clear the land, but it provided ashes that contained sufficient minerals to support a few crops of tobacco. However, these minerals were soon lost from American soil.

The mineral-rich tobacco leaves were shipped to Europe and connected again to ashes, this time in foreign pipes. The broken nutrient cycles and rapid erosion quickly ruined the tobacco plantations. Whenever that happened, the owners merely moved westward and cut another area. In this way the forests around Chesapeake Bay were destroyed before the American Revolution. So much soil washed into streams and rivers and settled into the bay that the shoreline we know is quite different from the one the early settlers found. Wherever the tobacco fields, were abandoned, the vegetation began to regenerate. Probably some fields in Virginia and nearby states have been cleared three or four times over as many centuries. A few bear climax vegetation today, but because of the chestnut blight, we have not a single forest just like those the first planters cut down. Chestnut blight is a fungus introduced into New York early in this century on disease-resistant Chinese chestnuts. Soon the blight swept through the eastern part of the United States leaving great stands of dead trees. Some old chestnuts still have live roots that sand up new shoots, but there are on new trees, because the blight is lethal to sprouts.

Coniferous Forests

The conifers are gymnosperms that bear seeds in cones instead of within fruits as do the flowering plants. Familiar conifers include pines, spruces, firs, cedars, and redwoods.

Boreal Forest

A wide band of conifers girdles Eurasia and North America north of the deciduous forests and grasslands. Most of these conifers retain their leaves the year around and shed one set only after the next is in place. Persistence of the leaves permits conifers to take immediate advantage of good weather, since they are able to begin photosynthesis without waiting to develop new leaves. This is a distinct advantage to ward the poles, where the warm season is short. These conifers also tolerate low temperatures. One reason is that they have stiff, wax-covered leaves known as "needles." When cold winds blow through the forests, the thick layer of wax retards water loss through evaporation.

Of course, the leaves do dehydrate, and they would wilt if they were not rigid. Thus the stiffness of the needles, due to tissues with exceedingly thick cell walls, prevents damage that would otherwise be caused by wilting. The leaf structure of conifers clearly suits them to subarctic regions, where winter winds sweep across the landscape. Conifers communities have fewer species than do deciduous forests. One reason is that the ever-present shade of the year-round canopy restricts photosynthesis by understory plants. As a result, the ground is nearly bare except for a deep carpet of slowly decomposing needles. Slow decomposition coupled with leaching of the soil by acids from the needle litter results in low fertility.

Conifers generally have root fungi, known as mycorrhizae, which aid in accumulation of minerals. Trees without these symbiotic

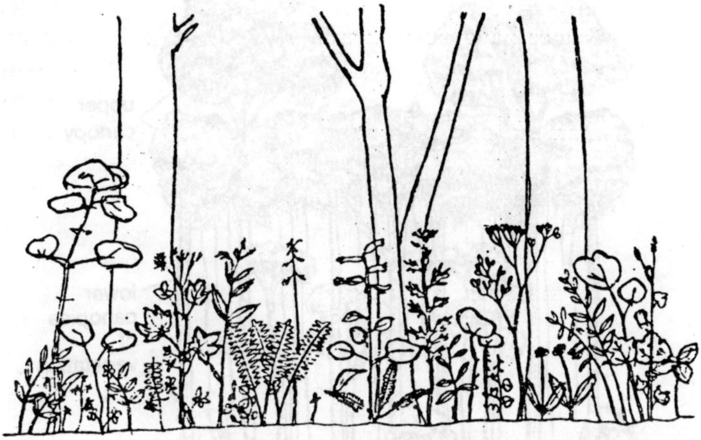

Fig. 4.5. Flora of deciduous forest ecosystem.

partners are at a disadvantage. A boreal forest usually consists of one species of high latitudes. The major differences are in the length of the day and the intensity of sunlight. Where mountains are at temperate latitudes, organisms experience a day length typical of the temperate zone. But because mountains stand above much of the earth's atmosphere, the sun's radiation is more intense here than at lower attitudes. (This results from the fact that the atmosphere absorbs part of the radiation that reaches it.) Yet mountains cool quickly at night because of loss of radiant heat through the thin atmosphere. This and exposure to winds make high altitudes colder than those in adjacent lowlands. The cool weather of high altitudes creats environments resembling those towards the poles.

Southern Conifers: A Fire Climax

The same adaptations that fit conifers to winter dryness in cold climates suit many to hotter areas, especially where the soil is sandy and the water supply is unstable. In the southeastern United States, pines colonize abandoned fields and form major forests. Perhaps their initial success depends on mycorrhizae, since these soils are usually poor. Left undisturbed, many such pine forests are replaced by oak, hickory, or magnolia. Apparently fire set by lightning or by humans maintains these pine forests, since they tolerate fire, whereas the broad-leafed trees do not. The pines have a thin trunk bark through which new buds sprout if existing limbs are destroyed.

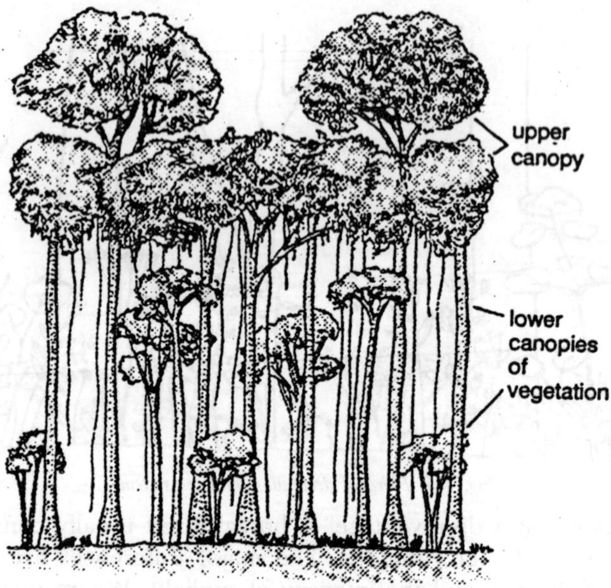

Fig. 4.6. A tropical rain forest, showing the upper and lower canopies of vegetation that effectively reduce the amount of light reaching the forest floor.

A thick tuft of needles protects the terminal bud of the southern longleaf pine and makes it extremely resistant to fire. As a matter of fact, longleaf pine seedlings seldom survive in the absence of fire, for they are sensitive to crowding and shading. Only when other vegetation is burned back can the little pines get enough light to grow. If southern pine forests are burned every few years, the fires do little damage to wildlife, because the animals can escape by running through of flying over the low fire line.

In fact, occasional burnings of patches of pine forest increase the population of bob-white quail and wild turkeys. These birds find food and shelter in the successional stages that follow that fires. We are not suggesting that fire is always beneficial even in the forests. In contrast to the low, cool surface fires of frequently burned regions, fires in pine forests that have not burned recently blaze up high and hot. Over many years without fire, a thick blanket of needles and fallen limbs accumulates on the ground. Once this litter is on fire, it provides so much fuel that the heat is sufficient to ignite the tree-tops and produce great crown fires. Such fires destroy the plant community and much of the wildlife.

Grasslands

Most temperate grasslands have long since succumbed to the cow and the plow. Only a few retain the unmodified vegetation of the plains. prairies, steppes, velds, or pampas. But whether they bear, native vegetation or support the highly selected grasses we know as grains, these grasslands supply the bulk of human food. Cattle and sheep graze on their grasses; cattle, hogs, and chickens fatten on corn grown on former grasslands; and most of our cereals, especially wheat, are produced on soils that once supported native grasses.

Biology of Grasses

All the grasses belong to one large family and share distinctive characteristics that fit them to their environment and make them of particular use to humans. Many people are surprised to learn that grasses are flowering plants. Open grasslands are windy, and it is the wind that pollinates grass flowers. Among these wind-pollinated plants, natural selection has favoured those that have conserved resources by producing small, petal-less flowers. Therefore, grasses and some other wind-pollinated species lack the showy petals that other plants use to attract animal pollinators. A single head of oats or other grass contains numerous flower. Each tiny flower can form a single fruit called a grain, or kernel. The ovarian tissue of the fruit fuses with the single seed inside in such a way that they become one continuous structure.

Adapted to withstand grazing

As grasslands evolved, their abundant thin leaves offered a bonanza to herbivorous animals. As a result, there evolved a large group of grass-eating mammals, including the ancestors of today's cattle, sheep, and horses. The selective pressures of these grazers promoted evolution of traits that resist grazing damage. For one thing, grasses store nutrients in their roots. As a result, loss of stems does not substantially reduce their reserves. Another factor is the presence of silica, the substance we know as glass, in grass cell walls. Of course, grazers have evolved excellent grinding teeth. But grasses take their toll. Tooth wear is so severe that may old herbivores eventually starve to death. The growth pattern of grasses has several distinctive characteristics that many also adapt them to withstand grazing. Grass leaves originate singly at prominent nodes along the stem. The lower part of each leaf forms a sheath that surrounds the stem for considerable distance before bending sharply outward as a flattened blade.

The leaves retain a meristematic growth region at the base of the sheath and another at the base of the blade. If the blade is cropped back, these meristems resume growth and lengthen the leaf to compensate for lost photosynthetic tissue. Thus grass leaves grow from the base rather than from the tip. You can see this growth pattern in any lawn. Until time of flower, the grass stem remains short, and the leaves grow upward beyond the tip of the stem. Thus the stem tip is protected from damage as long as possible. Only when it is time to reproduce does the stem elongate between several nodes and the tip stretch skyward. Here the flowers develop where they are exposed to the winds that pollinate them.

Binding the soil

Grasses bear abundant roots. Most are adventitious roots, which arise not from the primary root formed by the embryo but from the lower nodes of the stem. The sod-forming perennial grasses have extensive stems lying just on top of the soil or in upper soil layer; these serve both to hold the soil and to spread the plant. From rhizomes (underground stems) arise adventitious roots and occasional branches that penetrate to the surface, giving rise to what appear to be new plants. Quackgrass, a garden pest across the northern half of the United States, often reproduces through rhizomes.

Plowing or cultivating disrupts and scatters the quackgrass rhizomes. Each piece with an intact node may form a new plant. Often grasses have stolons, stems that lie on the ground and root at intervals. Each rooting node may give rise to aerial branches. Rhizomes and fibrous roots constitute over half the mass of a grass plant. Together they form a network penetrating throughout the soil and binding it into a nearly inseparable plant-soil complex known as sod. Neither wind nor water can erode a healthy sod. Growing roots and rhizomes break the soil repeatedly and contribute to the characteristic crumbly texture of grassland soils. Many of the roots of perennials die each year and are replaced by new roots. Decay of dead roots supplies humus and leaves species that aerate the soil.

Decaying roots return minerals to the soil, where they are immediately reclaimed by other roots. This process is part of the rich and efficient nutrient cycle of grasslands. Dead stems and leaves provide an extensive ground cover in most grasslands. Unmoved grasses accumulate as much as 10,000 kg of humus per hectare (about 9000 lb/acre) each year. It takes three of four years for litter components to decompose. Hence the litter forms a deep layer that shades the ground and lessens evaporation due to the wind. The litter also holds rainwater and aids its penetration into the soil. Because the water soaks in, there is no surface runoff and therefore little erosion.

Except in the most moist grasslands, the rainwater seldom soaks deep enough to join the groundwater. Instead, the extensive root system picks up most rainfall while it lies in the upper layer of the soil. Then the xylem carries the water immediately to the leaves, where it is transpired back into the atmosphere. Because little water percolates down far into grassland soils, minerals are not leached away; instead, they remain near the roots. Retention of minerals contributes to the richness of grassland soils. This and the crumbly texture produced by the roots make grassland soils excellent for agriculture.

Grassland Communities

The factors that limit forests and permit grasslands to develop are hard to determine. Some biologists maintain that grasslands develop only where there is insufficient rainfall to support trees that could shade out the grasses. On the other hand, there is both historical and experimental evidence that many grassland edges, such as those between the American prairies and the eastern deciduous forest, were maintained by fire. Where forests and grasslands meet, burning usually favours grasses over trees. Variations in climate and accidents of grazing have resulted in complex and dynamically balanced grassland communities. Contrary to popular belief, the American prairies were never a uniform sea of grass ranging from the deciduous forests of the East westward to the Rocky Mountains. Mixed in with the grasses there were other herbs, especially members of the aster family and legumes, such as the lupines.

Nitrogen fixation by symbiotic bacteria in the nodules of the legume roots contributes to the fertility of the soil and the vigour of the community. The roots of various species extended to different levels and were best developed in specific regions of the soil. Concentration of roots of one species at a particular level reduces competition between species. But because each level is populated by roots of some type, resources are fully exploited. The grassland landscape is even more varied along the moist borders of streams, river, marshes, lakes, and ponds. Where inland waterway pass through grasslands, groves of trees and shrubs thrive, along with other organisms ordinarily associated with woodland communities.

Grassland animals

A large and diverse animal community lives in every grassland. The deep soil that covers rocks and the absence of woody vegetation limit above ground shelter, except where grasslands merge with woodland or along rivers and streams. Thus most small animals depend on the soil for protection. Numerous little mammals burrow beneath the sod. Prairie dogs, rodents related to squirrels, are symbolic of the short grasslands or Great Plains of North America. Stockmen exterminated most of the prairie dogs, believing that these animals competed with livestock for forage. Undoubtedly the selective grazing of prairie dogs on grasses favoured the growth of other plants.

Pronghorn antelopes, which were once almost as numerous as bison and are now few in numbers, relied heavily on the broad-leaved vegetation of prairie dog "towns". Elimination of prairie dogs also resulted in near extinction of one of its predators, the black-footed ferret. This handsome, weasellike animal hunted in the prairie dog burrows. Because vast numbers of prairie dogs are necessary to support even a tiny breeding population of ferrets, this species is now extremely rare–or perhaps extinct except for a few in captivity. Prairie dog burrows and those of gophers, pocket mice, kangaroo rats, and ground squirrels provide shelter for other animals, including grass-hopper-mice and snakes.

At least one bird, the burrowing owl, relies on old rodent burrows for nesting sites. So do cottontail rabbits. Beetles and camel crickets also use burrows; the dung, fungi, and hoarded vegetation afford these insects a ready food supply. Ants are abundant in grassland soils; some species build huge mounds surrounded by a zone stripped clean of all visible life except for the ants themselves. Animals that dig underground benefit the grasses, because they aerate the soil and mix in humus from the surface. Even those that are underground transients contribute fecal material where it is directly available to the roots. Some animals, such as grasshoppers, are found only above-ground. Nevertheless, they depend on the soil for shelter.

The female grasshopper uses the tip of her abdomen to penetrate deep into the soil and bury her eggs. The adults perish in the rigour of winter, but the species continues because the developing embryos of the next generation lie safely protected in the soil. A number of medium-sized animals that live in the surface vegetation face danger from predators. Rapid movement through thick grasses is very difficult, and vision is limited; consequently, it is easy for predators to stalk their prey. Escape must be fast and sure; otherwise it comes too late. Consider how grasshoppers, jackrabbits, and jumping mice meet this challenge. Huge aerial leaps permit such animals to clear the top of the vegetation, get an unimpeded view for a moment, and then drop some distance away without leaving a trail.

No hiding place

In wide-open habitats large animals find little or on shelter. Under these circumstances "predator control" becomes as group activity, and the animals aggregate together. The pronghorns, bison, and muskoxen of North America, the kangaroos of Australia, the saga antelopes, wild horses, and asses of the steppes of Russia, the gnu and zebra and even the ostriches of Africa show herd instincts. In a group there are many eyes to watch for predators, and alarm spreads rapidly. Frightened pronghorns raise their tails and display a large white patch on their rumps.

The flash of white rumps and white flaglike tail on a running pronghorn starts every other pronghorn in sight on its way! Mixed herds, such as are common in Africa, cooperate in watching for danger. Ostriches are taller than most mammals they herd with and are usually the first to sound alarm. Large grassland animals are exceedingly fleet; they have long length that cover the ground quickly. The faster runners in the world inhabit grasslands.

The American pronghorn, our faster native animal, can do 60 miles an hour. At this rate the pronghorns leave a solitary predator or a pack of wolves so far behind that no amount of strategy or cunning can lead to another encounter until the pronghorns have had time to feed and recuperate from the original confrontation. Of course, natural selection has also promoted evolution of fast running grassland predators. In Africa, the cheetah has been clocked at 65 miles an hour. It is known to accelerate from a stand-still to 45 miles an hour in a few seconds. Grassland herbivores rely on the group not only for detection of predators but also defense.

When threatend, bison from protective head-outward circles. The young are secure in the center of the circle. Herding is itself a defensive measure. Predators are confused by large numbers of running animals. Thus if the individual a predator is following loses itself within the herd, the predator stops in confusion. For this reason, most predators kill only aged, diseased, deformed, and very young animals that are unable to keep up with the herd. These same animals are especially vulnerable, because they are also weak.

Grazing and Overgrazing

Each grassland is a balanced ecosystem of producers and consumers. The native herbivores have evolved with the plants and are adapted to the vegetation and to one another. Likewise, grasslands are adapted to withstand grazing by certain animals. Unfortunately, domestic animals often have destructive effects on ecosystems in which they do not evolve. The abuse of natural grasslands by domestic animals arises from two factors. First, the domesticated grazers, with few exceptions, are from other parts of the world. Hence local grasses are not adapted to circumvent their feeding patterns.

Domestic sheep, in particular, can graze more closely than native North American herbivores can. Most grasses are vulnerable to extensive loss of stems and leaves. Second, overgrazing by domestic animals is common. The usual population controls that limit wild grazers are nearly if not totally absent from populations of domestic animals. Humans do their best to protect their animals from severe weather, predators, or temporary food shortage. When they slaughter animals for food, people carefully select surplus young adults, particularly excess mates. Very young animals, which have a period of rapid growth ahead of them, pregnant females, and females of good breeding age are seldom killed.

In almost all cultures, people exploit their understanding of animal growth and reproduction to maximize the number of animals and obtain as much food as possible. Unfortunately, most societies lack a similar understanding of the ability of grasslands to support animals for long period. Perhaps this difference in understanding animals compared with grasslands results from the fact that the life cycle of the animals is comparatively short. Each human has the opportunity to see many generations of cattle grow up and reproduce. In comparison, the effects of overgrazing may appear only over decades and sometimes require several human generations to become acute.

Damage from overgrazing

Repeated close cropping of grass blades reduces the photosynthetic potential of the plant and hence the energy that can be stored in the roots to support the following seasons' growth. In response to grazing, most grasses undertake a compensatory growth. The leaves elongate at their meristems, and new branch stems arise near the ground. Both results from the loss of auxin-producing tissue at the tips of leaves and stems. Under excessive grazing some grasses continue to make new growth until their underground stores are exhausted. Heavy grazing prevents the stems from growing out, blooming, and setting seed. Sometime the supply of new plants is cut off completely.

Close cropping of leaves and stems also inhibits normal growth of roots that must occur to replace old, dying roots, to ensure absorption of water and minerals, and to bind the soil against erosion. If all leaves are gone, the grass crown (base of the stem) lies exposed to physical damage from freezing or trampling. Finally, overgrazing harms hams the soil directly. Reduced vegetation and surface litter increase evaporation and reduce water absorption. Much of the rain simply runs off along the surface, eroding the soil. Overgrazing has accelerated erosion of the deep gullies and arroyos of the arid Southwest. Heavy grazing also changes the soil texture. Continuous trampling compacts barren soil and thereby destroys spaces between the particles. This reduces ability of the soil to hold air and water.

Range management

Except for the constant pressure for short-term economic gains, range management is quite simple. Grasslands must be permitted to set seed. The manager can achieve this goal by dividing the land into plots and rotating the grazing so that each plot remains ungrazed every few seasons until after the seed has ripened and dipersed. If overgrazing on the other plots is to be prevented, more land must be available for a herd of a given size, or the total number of animals must be reduced. Another problem arises from the fact that grazing animals never use available forage uniformly. They always prefer some areas, especially those near water holes. Sometimes judicious placement of salt blocks can balance the distribution of grazing.

Natural checks

If range use must be so carefully managed, how did the thundering herds of bison live in equilibrium with the native grasslands of North America? The answer is quite simple: there really were not many bison in relation to the amount of grassland. The highest estimates of bison populations suggest that there was only one animal per 20 acres of range. Although the bison travelled in herds, raising clouds of dust with their hoofs and trampling the vegetation badly in places, most plants went many seasons without a single bison passing their way. Some pond and stream margins did suffer repeated damage, and these probably supported successional vegetation. But in general the grasslands were lightly used. Natural checks limited the bison population. Although only grizzlies from the western mountains could have killed healthy adults, the young bison and ill ones were prey for wolves and perhaps coyotes. Weather may have been another limiting factor.

According to Indian tradition, a large herd of bison disappeared from Illinois during a terrible blizzard late in the eighteenth century. Always some bison became trapped in river muds, and some surely died from disease or parasites. But the best animals usually survived, and numbers were in balance with the food supply. However, there can be little doubt that there were times when drought brought starvation to the bison and damage to the grasslands.

Drought, Dust, and Deserts

The moderately dry climate of grasslands carries the constant risk that fluctuation in rainfall will create drought. Twenty-three separate droughts have been recorded in the Russian steppes alone during the past century and a half. But the grassland community survives, because the plants are adapted to the ravages of drought. Overgrazing reduces the resistance of grasses to drought, and plowing the sod can lead to erosion that will nearly destroy the entire ecosystem. Let us examine the history of American grasslands and the effect of humans and drought on them.

The Dust bowl

The late 1920s and early 1930s were marked with increased rainfall that permitted tall grasses to flourish on the American prairie. The seven-year drought that struck in 1933 drastically altered the species composition of the unplowed prairies, most of which had been subjected to heavy grazing. Dry-adapted grasses quickly replaced those requiring more moisture. The big bluestem grass that could grow nearly eight feet high almost disappeared; only a deep root system and underground food reserves permitted some plants to survive. Even more drastic changes occurred in the normally drier Great Plains to the west. Grazed areas were badly damaged, but ranchers were reluctant to reduce their herds.

One study in western Kansas showed less than half of the ground covered with vegetation in 1935 and only five percent in 1936. Great dust storms were both the cause and the result of the decreased vegetation. High temperatures and low rainfall prevented crop growth on plowed fields. Dry winds raised clousd of dust from these fields and from the poorly covered grassland; when the dust fell from the air, it drifted like a horrible dark snow. A deposit of only an inch was often sufficient to smoother short grasses and produce another field of unstable soil. Stressful as the great drought was, no species is known to have been completely eradicated.

All survived in favourable habitats or as ungerminated seeds. Such survival would be expected, for only a small portion of the seeds of wild plants will sprout upon encountering favourable conditions. Not will all the remainder sprout in any one season. So after several years in which all germinating grass seedlings shrivel from lack of water and die, there will still be live seed left in the ground. When a normal year finally comes, a few seeds will sprout and begin to replenish their kind. Although no species were lost in the drought, the effect on the soil was another story. The rains that followed the drought fell on loose, naked soil that eroded into steep gullies.

Turning grasslands into desert

The drought of the 1930s was certainly not the first one Americans had known, nor was that the first time dry weather and overgrazing had damaged the grasslands and hurt the ranchers. Increasing herds of sheep and cattle during the latter part of the nineteenth century had culminated in starvation of stock and disastrous economic losses when drought struck in the 1890s. Following the Spanish pattern of migratory sheep grazing, some regions of the West have been overgrazed for several centuries.

By moving the animals continuously, mainly to higher elevations in the spring and back downwards in the fall, ranchers can maintain large herds on very poor land. Arid western lands often show the effect of this practice. The hills are terraced by continuous rows of sheep paths. Except on the highest peaks of the Sierra Nevada and the Rockies, the vegetation has been altered, perhaps permanently, by such practices. In some regions natural grasslands have been replaced by weedy desert vegetation.

In western North America sagebrush now dominates many of what were once cool, short grasslands. To the south, creosote bush, cacti, and mosquito have replaced other grasses. It is important to recognize that these "new deserts" lack the diversity and complexity of old "true deserts." But once overgrazing has produced bare sports, the mesquite thrives. It sends roots out literally, as well as deep into the soil, sometimes as far as 8 meters (25 feet).

Overgrazed grasses with poor root systems cannot compete for water with mesquite. Soon the grasses die back further, and the mesquite expands. Hungry cattle feed on the fleshy mesquite frui pods, but mature seeds pass unharmed through the cattle. Thus cattle scatter mesquite seeds about, especially in the trampled areas where they congregate. Fertilized by the cattle dung, the mesquite seeds are well planted in the barren but hospitable soil. American cattlemen have no monopoly on the conversion of grasslands to deserts. It has long been said that the Sahara Deserts marches farther south each year. During the early 1970s drought coupled with high human and cattle populations led to loss of herds and massive starvation among the people of the sub-Saharan or Sahel grasslands Some climatologists trace this disaster to major weather trends, bu drought, overgrazing, and disaster are not strangers to these or most other arid grasslands.

Deserts

The story of the desert has one theme–water. The distribution of water in time and space determines the nature of fragile desert ecosystems. An average rainfall of less than 25 cm (10 in.) a year will produce a desert; but each desert is different, just as are each forest and each grassland. Relative humidity, temperature extremes, the underlying rock and drainage–all contribute to the uniqueness of each desert. Deserts are never totally dry; there is always some water. Although dew can be of considerable importance in the water budget of certain organisms, most desert life relies larger on sporadic rainfall. Particles of desert soii hold surface layers of water, as do other soils. Any excess water flows downward to collect over impenetrable layers of rock. Groundwater lying near the surface produces springs and oases.

Substantial layers of groundwater underlie many deserts; some are believed to represent the accumulations of thousands or even millions of years. Perhaps this "fossil water" can be traced to rainfall during periods when these lands that are now deserts had more humid climates. Or groundwater from regions with higher rainfall (such as adjacent mountains) may feed into the rock layers under deserts. In any case, water can be drawn from desert wells faster than it is replaced. Only rarely does vegetation carpet the desert. Consequently rains create flash floods the dig deep gullies and carve rugged landforms in the desert. Some desert soils developed where we find them; the wind has carried others hundred of miles from their origin. Desert soils are surprisingly fertile; many support substantial vegetation after a good rain or when irrigated.

Desert Plants: Many Solutions of One Problem

Plants have evolved seemingly endless adaptations to desert conditions. Individuals divide the water supply so efficiently that they tap almost every drop. Most desert plants exhibit special modifications to gather and conserve water, but many merely become dormant during dry periods. For example, lichen on desert rocks are active only intermittently. When dehydrated by hot desert days, these lichens suspend most life functions and for a time tolerate extremely high temperatures (to 80°C or 176°F). But after being dampened by night dews, these same plants revive and resume photosynthesis at dawn. When the Rains Come. Although no higher plants withstand the severe dryness that lichens undergo, many live through drought as seeds, bulbs, or other low-moisture structures. Such plants usually have shallow roots that grow laterally through the top soil for considerable distance.

The wide distribution of roots permits them to adsorb substantial amounts of water from a light shower. In more moist deserts, seeds, of annuals germinate every year, and the plants bloom predictably, usually within a few weeks in early spring. But in many deserts, bloom is irregular. Certain species appear only every 10 to 20 years. They survive because their seeds remain alive but inactive in the soil. Seeds of some desert plants germinate any time there is sufficient moisture, but others rely on one or more of such additional cues as light and temperature. The short-styled sedge of the Negev in Israel responds to rain in another way. Although the old leaves lie dead on the surface of the ground, their base remain alive and start to grow right after a rain. The desert becomes green immediately with new sedge leaves, but each supports dead material at its tip.

Every drop counts

Root systems of perennial desert plants vary markedly. Some rely on extensive near-surface roots, especially when growing on hills or ridges far above the groundwater. Others, particularly plants near dry stream beds, produce long taproots that reach far into the soil and rock. Plants of stony deserts may send a single root under each nearby rock. This way they can absorb water that condenses from vapour during the cool desert nights. Some cacti sprout new roots within a few hours of a heavy rain. Once the water is absorbed and stored in the thick fleshy stem, the new roots shrivel and dry up. Around many desert shrubs, such as the creosote bush and sagebrush, lie areas barren of vegetation. These species secrete toxins that accumulate in the soil and inhibit other species. Thus these shrubs can use all nearby moisture without competition.

Many succulent (fleshy) desert plants, such as sedums and aloes of Africa and agavas of North America, store water in thick, permanent leaves. But whether the water storage tissue is modified leaf or a stem, as in cacti, it is usually green with photosynthetic pigments. In bright, unshaded deserts, sunlight penetrates to the deepest cells. Packing the photosynthetic tissues in stems or thick leaves permits sufficient photosynthesis and reduces the surface area from which water can evaporate. Few desert perennials have large leaves. Some, like the gray sagebrush, grow bigger leaves in the moist season and progressively smaller ones with the coming of dry weather. The last leaves formed are mere scales.

Other species, such as crown of thorns of Africa, the ocotillo of the U.S. Southwest and the boojum tree of Mexico, send out ordinary leaves during the moist season but drop them completely as soon as water is in short supply. Special mechanisms doubtless regulate stomata in many desert plants. For example, agaves open their stomata only at night; during this time they absorb carbon dioxide and store it for use the following day. This is the reverse of the common pattern; plants of moist environments open their stomata during the day and close them at night. Much dry-adapted vegetation bears spines or hairs. The adaptive value of these structures has been the subject of speculation.

Obviously spines deter some animals from feeding on the water-rich tissues. Spines also shade and insulate the plant. When hairs are dense, they interfere with air movements near the surface. This reduces evaporation and holds an insulating layer of air that can slow transfer of heat to the plant as the environment warms. Perhaps these spines and hairs also provide a surface on which water vapour condenses into dew. Despite all these adaptations, desert plants can suffer severe water deficits during midday temperature peaks. Such daytime deficits are made up during the night but they may actually be beneficial as physiological acclimatization to extreme heat. It is well established that enzymes are less sensitive to denaturation by heat in dry than in most environments. Perhaps midday dehydration permits desert plants to tolerate temperatures that would "cook" other vegetation.

Animal Adaptations to the Desert: A Table of Water Conservation

Desert animals obtain their limited water rations from dew, occasionally from drinking water, but very largely from the water released during cellular respiration. In fact, this may provide the entire water supply of kangaroo rats. (You will recall that the respiration of glucose yields both carbon dioxide and water.) Birds that eat insects may also obtain sufficient water from their food. But those with a diet of dry seeds fly many miles daily to reach drinking water. Mammals, too, search for water. Some, such as the ibex and gazelle, will dig deep into the bottom of dry springs or temporary stream beds to reach groundwater. Whatever their water source, desert animals have evolved strategies to conserve water.

Many, including some reptiles and a multitude of small rodents, are nocturnal. They spend their days in cool underground burrows, where they waste little water in evaporative cooling. Although desert birds are active during the day, they seek shelter in the shade of rocks, cliffs, or plants. The camels, donkeys, sheep, and goats that people raise in arid lands grow thick woolen coats that insulate these animals from the desert sun. Both donkeys and sheep keep their body temperatures in check through evaporation associated with painting.

Consequently, they must tolerate tremendous dehydration, because they use so much water in evaporative cooling. In fact, a well-dried-out donkey can guzzle a quarter of his body weight in water within only a few minutes. Instead of using water in panting, goats and camels tolerate a rise of several degrees in their body temperatures. They accumulate heat during the day and dissipate the stored heat during the cool desert night. Most desert rodents excrete as little water as possible with their wastes. Perhaps you have noticed the milky urine of a pet hamster. These native of the Syrian deserts reabsorb to much water from its kidney tubules that the minerals precipitate and give its urine a milky appearance. A large surface area-to-volume ratio could make dehydration a serious risk for desert insects that are active during the day. Many such species have evolved water-conservation adaptations that parallel those of desert plants. A thick waxy cuticle covers the bodies of many desert insects, just as desert plants have an extremely thick waxy cuticle on their leaves. All insects have multiple openings for gas exchange, not unlike the stomata of plants. But the gas exchange openings of some deserts are recessed in pits, just as are the stomata of desert plants.

CHAPARRAL

Variations in terrestrial environments seem endless. Grasslands naturally merge into deserts, and a little abuse will make deserts out of dry grasslands. Similarly, deserts grade into forests. Indeed, dry-adapted trees, such as pistachios and acacias, characterize certain deserts. Some dry, warm environments support thorn forests. The shrubby, leather-leafed plants that now cover much of the North Africa and Southern Europe are similar to the chaparral of the south-western United States. As far as can be determined, chaparral is the climax vegetation along the coast of Central and Southern California, where summers are dry but where winter can bring up to 50 cm (20 in.) of rain. In this Mediterranean-type climate the chamise, manzanita, and other shrubs bear leathery leaves the year around.

Severe dehydration during the dry season is largely avoided by simply closing the stomata. Nevertheless, the plants suffer water shortages during the summer. Extensive root systems penetrate as far as 8 meters (25 feet) into the soil. Fire sweeps the chaparral regularly, destroying all vegetation above the ground. The roots of many plants survive, and new sprouts appear quickly, attesting to the availability of water in the soil. In other species, heat activities seed germination.

The chaparral community is apparently a fire climax. Most of the minerals of this ecosystem are converted into ashes during periodic fires. Minerals leach from the ashes when it rains. This loss keeps the soil poor. Almost all chaparral plants have mycorrhizal associates that gather minerals, and many have nitrogen-fixing symbionts. Chaparral soils are subject to heavy erosion. Homes built in chaparral areas many survive a fire, only to be washed out in a sea of mud and water when rain falls on the denuded soil.

LAND AND PEOPLE

Terrestrial ecosystems vary in their resilience, that is, in their ability to withstand abuse and to grow back when the abuse ceases. If protected from unusual erosion and loss of seed sources, most temperate forests can support successional stages and reestablish themselves in a few centuries. Grasslands seem more fragile, perhaps because it is so difficult to prevent erosion in semiarid areas. Although we have less data concerning the resilience of tropical forests and tundra, they apparently even less stable ecosystems. In any case, it is sobering to realize how drastically people have altered terrestrial environments. The degree of destruction in most areas seems roughly proportional to the length of time that agriculture has been practiced and to the state of industrial development.

During the past 10 years, timber shortages in temperate areas have stimulated heavy logging of tropical forests. Also, well-intended but misguided attempts to increase food production have promoted clearing of tropical forest lands. Meanwhile, development of such arctic resources as petroleum is creating traffic that gouges up the tundra. All these assaults on terrestrial ecosystems trace to increasing human populations. More people require more food and more lumber and more raw materials. In evaluating the impact of growing human numbers, we must include overall effects on the biosphere. Since the continents are a major portion of this great, worldwide ecosystem, reduced resilience in terrestrial ecosystems must contribute to a generally less stable biosphere. This reason alone is sufficient to cause concern. If people are to survive, we must take care of our precious terrestrial ecosystems.

5

Ecological Niches

In this chapter a close look at species packing in parasite communities will be undertaken in an attempt to resolve the divergence of opinion. Attention is limited to the literature on helminth parasites of vertebrates because it illustrates the fun range of types of coexistence seen in any taxon and it contains some of the most celebrated examples of coexistence of parasites, some incredibly rich communities within one organ system, excellent experimental work reveling the fundamental and realized niches (Hutchinson, 1957) of parasites and is deserving of much more attention by ecologists. In the review by Schoener (1974) on resource partitioning in ecological communities covering 80 studies, none on helminths is included. Some studies are summarized in Table 5.1 with a statement on the predominant type of coexistence in the community. The kind of evidence provided in the studies is also listed with a check indicating quantitative data or experimental manipulation. The general pattern of coexistence is clear (Table 5.1).

The vast majority of cases indicates a predominantly non-interactive niche occupation (no substantial shift from fundamental to realized niche in presence of another species), either in nonoverlapping niches or in strongly overlapping niches. In the latter case all species occur in the alimentary canal and are usually members of very rich helminth faunas. Although studies have not gone far in determining how these species coexist while overlapping so broadly along the length of the alimentary canal, it is clear that other niche dimensions must be considered. Schad (1963a, b) found in the genus Tachygonetria in the tortoise gut that some species occur near the gut mucosa while other can be found throughout the gut lumen. Some species feed on large particle sizes, others specialize on bacteria, and others imbibe purely liquid food. Given the great physicochemical diversity of the gut, many other parameters may be involved in the niche diversification in these gut parasites (See MacKenzie and Gibson, 1970; Williams, McVicar, and Ralph, 1970). Thus the group of studies showing predominantly overlapping but noninteractive niche occupation illustrate overlap only on the most obvious dimension of gut length, which has been the most commonly studied. Since niche divergence on other gradients is evident, this group of studies should be classed with the other noninteractive examples making 18 studies out of the total or this type.

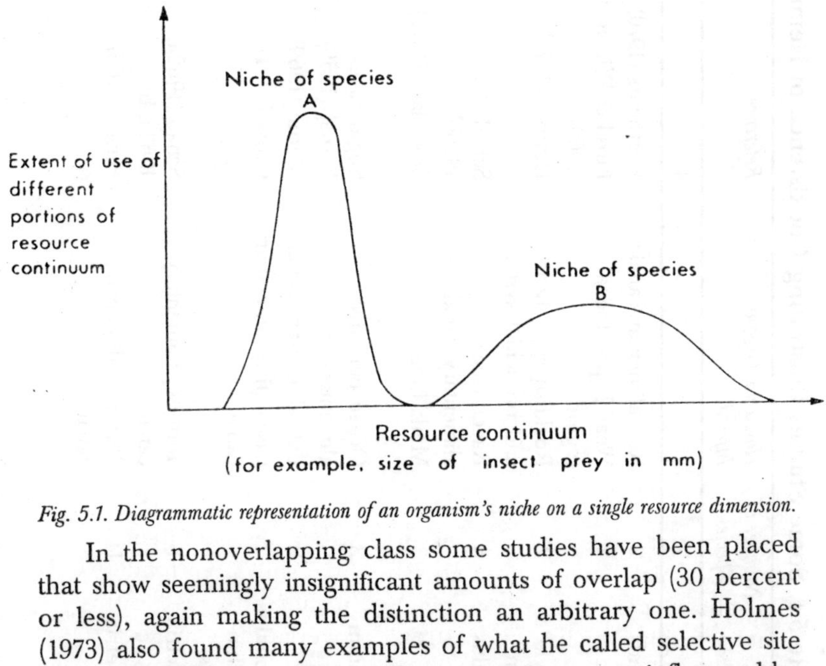

Fig. 5.1. Diagrammatic representation of an organism's niche on a single resource dimension.

In the nonoverlapping class some studies have been placed that show seemingly insignificant amounts of overlap (30 percent or less), again making the distinction an arbitrary one. Holmes (1973) also found many examples of what he called selective site segregation of niches where niche occupation was uninfluenced by the presence of other parasite species. As in Table 5.1 he supplied many examples of niche segregation of this type and only few where coexistence involved reduced realized niches in the presence of competitors, *i.e.*, interactive site segregation. He concluded that

Table 5.1. Some Studies Illustrating Coexistence of Heiminth Parasites in Vertebrates

	Genus of Parasite	*Number of Species*	*Host and Organs Infeted*	*Reference*	*Quant. data*	*Expermtl. Manip.*	*Type of coexistence*	
	1	2	3	4	5	6	7	
1.	Strongyloids	2	Rat, alimentar canal	Wertheim (1970)	√	√	Non overlapping	Non interactive
2.	Calycotyle	2	Weasel, proctodeal region	Euzel & Williams (1960)	–	–	”	”
3.	Aporocotyle & Psettarium	2	Rockfish spp., heart and bronchial system	Holmes (1971)	√	–	”	”
4.	Kalicephalus	3	Racer (snake) alimentary canal	Schad (1956, 1926b)	–	–	”	”
5.	Filaroides & others	3	Mink, lung	Stockdale (1970)	–	–	”	”
6.	Apocreadium	3	Ocean tally (fish) alimentary canal	Sogandares Bernal (1959)	–	–	”	”
7.	Castroia	2	Bat, alimentary canal	Martin (1969)	√	–	”	”
8.	Proteocephalus & Neoechin-orhynchus*	2	Cisco (fish) alimentary canal	Cross (1934)	√	–	”	”
9.	Tachygonetria	8	Tortoisc, alimentary canal	Schad (1962a, 1963a, b)	√	–	Over-lapping	Non interactive
10.	Tachygonetria & others	14	Tortoisc, alimentary canal	Petter (1966)	√	–	”	”

11. Cylicocyclus & other strongylids	32	Horse, alimentrary canal	Foster (1936)	√	–	”	”
12. Hymenolepis & others*	43	Scaup, alimentary canal	Hair & Holmes (1975)	√	–	”	”
13. Crepidostomum & others*	6	Trout, most in alimentary canal	Thomas (1964)	√	–	”	”
14. Cucullanus & others*	11	Flounder, alimentary canal	MacKenzic & Gibson (1970)	√	–	”	”
15. Capillaria & others	9	Ray, alimentary canal	Williams, McVicar & Ralph (1970)	–	–	”	”
16. Lepidapedon others	15	Cod, alimentary canal	Williams, McVicar & Ralph (1970)	–	–	”	”
17. Hymenoliepis & Moniliformes*	2	Hamster, alimentary canal	Holmes (1962b)	√	√	”	”
18. Echinorhynchus*	>1	Fish spp., alimentary canal	Chubb (1964)	–	–	”	”
19. Itygonimus & Omphalometra	3+	Mole, alimentary canal	Frankland (1959)	√	–	”	”
20. Hymenoplepis & Moniliformes*	2	Rat, alimentary canal	Homel (1961, 1962a)	√	√	Overlapping, niche	Interactive, shifts
21. Proteocephalus & Neoechin-orhynchus*	2	Stickleback, alimentary canal	Chappell (1969)	√	–	”	”
22. Dactylogyrus	4	Carp, gills	Paperna (1964)	√	√	”	” exclusive

*Indicates presents of an acanthocephalan in the community.

the majority of parasite communities are mature because the residents have evolved discrete niches without competition. That is, communities have reached the evolutionary phase in Wilson's (1969) concept of community development and have presumably passed the earlier noninteractive, interactive, and assortative phases. The fact that there are two phases in community development with predominantly non interactive niche occupation, and these at the two extremes of community age, should make us pause before reaching conclusions from the observation that many parasite communities are predominantly noninteractive.

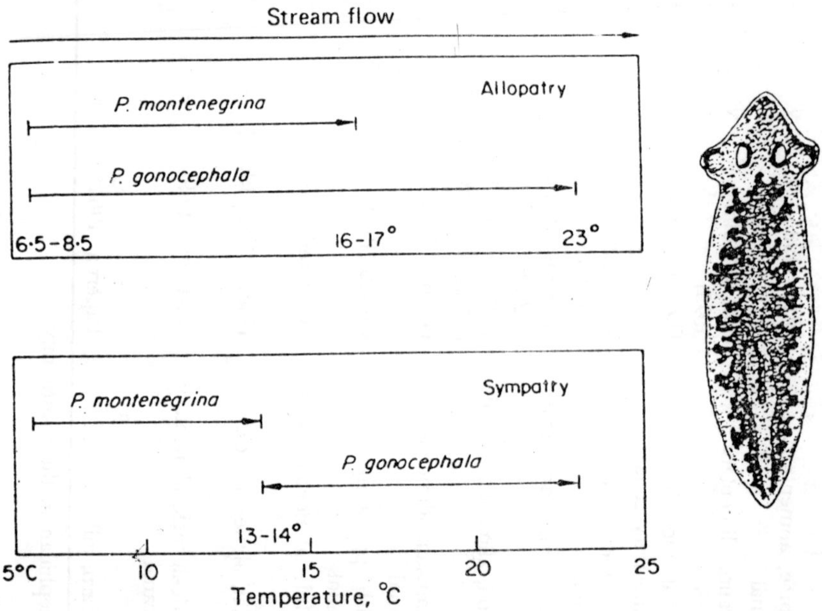

Fig. 5.2. Distributions of two Platyhelminth flatworms, Planaria montenegrina and P. gonocephala along temperature gradients in streams when they occur separately (above) and together (below). Each species is restricted to a smaller range of thermal conditions when in the presence of the other.

The concepts stressed in this monograph favour the conclusion that parasite communities are in an early stage of development. Noninteractive niche exploitation is common because resources remain unutilized since species have not evolved to use them. Communities are largely in the noninteractive and interactive phases of development and seldom have the assortative and evolutionary phases been reached. Each species is specialized to a narrow range of resources because of coevolutionary demands other than competition. A closer look at the helminth community literature, which follows in the next two sections of this chapter, justifies the latter conclusion in my opinion. Of course, competition can be observed among helminth gut parasites (e,g., see studies 20 and 21 in Table 5.1), but I contend that it affects a minority of species.

The influence of immune responses may also prove to be important, but, again, evidence is wanting that large numbers of species interact via the host's immune system. Crompton (1973) reviewed the literature on 252 species of helminth in the alimentary canal of vertebrates, but he cited no clear examples of interspecific interaction through immunological responses and only four examples of niche shifts through interspecific competition (including Holmes, 1961; and Chappell, 1969 in Table 5.1). But eight examples were provided for the extension of niche breadth in response to intraspecific competition as parasite population increased in a host, suggesting that interspecific constraints on niche expansion were absent. Particularly in the gut, but perhaps elsewhere also, resources may be replenished by a host so rapidly that they are not limiting except at the most extreme parasite densities (when the host is likely to die with the consequent death of the parasites).

Noninteractive Coexistence

There is little evidence to suggest that competition has been an organizing force in the studies indicating nonoverlapping, noninteractive coexistence of parasites (studies 1 to 7 in Table 5.1). In mink pulmonary tissues, Stockdale (1970) reports three species of metastrongyle nematode each with a very distinct niche Perostrongylus pridhami in alveolar ducts and terminal and respiratory bronchioles; Filaroides marits in peribronchial connective tissue; Crenosoma hermani in bronchi. Stockdale notes that pulmonary tissues provide a diverse array of resources available to nematodes: alveoli and alveolar ducts, terminal and respiratory bronchioles, bronchi, lamina propria of the bronchial bifurcation, peribronchial connective tissue and pulmonary arteries. In a range of animals all these sites have been colonized but in the mink little more than three of the six sites are utilized.

The nematodes are loosely packed, almost twice as many species could coexist in the pulmonary community, and it is most unlikely that competition could cause such complete divergence of ecological niches. Two species of blood flukes in rockfishes had fundamental niches that showed a similarity of 7 percent and when together this similarity was reduced by only 3 percent (see data in Holmes, 1971). Indeed, the species could not overlap significantly because they are adapted in fundamentally different ways to lodge in different parts of the vascular system. Aporocotyle macfarlani is a squat fluke that wedges into the walls of blood vessels, largely in the gill arches.

Psettarium sebastodorum is a longer, narrower fluke that loops its body and is so held largely in the chambers of the heart. These species have a grossly different body form that preadapted them for coexistence. Divergence of niches once both species had colonized that host was probably miniuscule. Similar examples are described by Llewellyn (1956), Williams (1960), and Uglem and Beck (1972). In the snake Coluber constrictor three species of Kalicephalus can be found: K. inermis in the anterior part of the

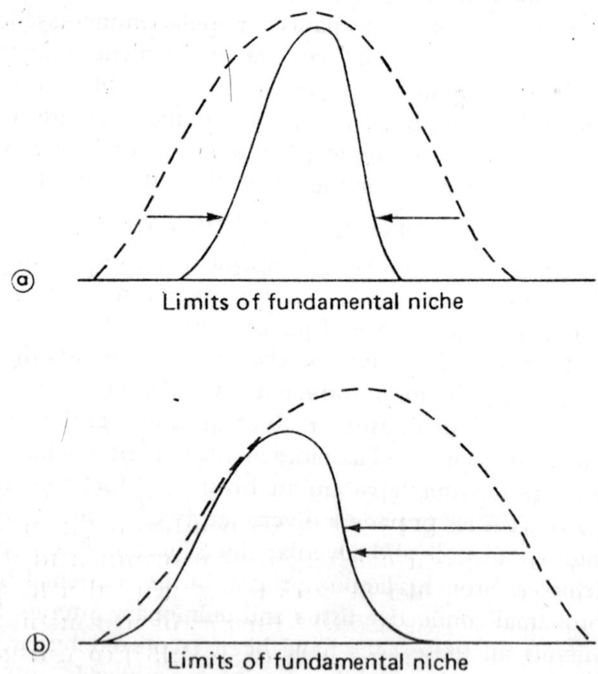

Fig. 5.3. Effects of interspecific competition in the realised niche. In (a) uniform competition over all parts of the resource continuum results in an even shrinkage of the niche. In (b) intense competition on only one side of the niche causes both restriction of the niche and displacement.

esophagus, K. costatus in the duodenum, and K. rectiphilus in the rectum (Schad, 1956, 1962b). Species packing is so loose it could not be created by competitive interaction but rather by the demands for specialization, leaving resources available for other species to colonize. Three species of Apocreadium occupy largely nonoverlapping niches: A. balistis the first third of the intestine, A. uroproctoferum the second third, and A. coili the rectum. Had competition been a force in organization we should expect more overlap even to the extent that some species have Hutchinsonian niche differences of only 20 to 30 percent (see Hutchinson, 1959).

In the literature describing species with largely overlapping niches the evidence for competition commonly acting as an organizing force is also very unimpressive. Schad (1963b) claimed that the pinworm communities in tortoises showed non overlapping niche occupation with the inference of tight species packing. But this claim was based on correspondence of relative abundances to MacArthur's (1957) broken stick model that can be derived from very different assumptions and is thus discredited as a means of distinguishing organizational influences in a community (Pielou, 1969; Poole. 1974). Based on Schad's (1962a, 1963a, b) studies, we could reasonably expect in the tortoise gut a group of species specialized to each of at least four sections of the gut, each group segregated into those in the lumen and those mostly against the mucosa, and within these subgroups species feeding on large and small particles and liquids.

Even at this moderate scale of specialization 24 species could be expected in the gut whereas Schad, Kuntz, and Wells (1960) found only 10 species of Tachygonetria. Holmes (1973) chose to emphasize the complementary distributions of Atractis dactyluris with Mehdiella uncinata and Tachygonetria dentata found by Petter (1966) in tortoise guts. But of the 14 species of common nematode studied in detail by Petter, A. dactyluris was the only species possibly involved in competitive interactions, and it became dominant only in older tortoises when many other factors could be changing in the gut. Of the 43 species of helminth in the scaup alimentary canal. Hair and Holmes (1975:153) concluded that the community is "composed of a chance combination of ecological specialists." Where they claimed to find interspecific competition, only a pair of species was involved and the mean overlap between them was only 14 percent, but in some cases large numbers of each species coexisted over 20 to 30 percent of the intestine suggesting that competition was not the mechanism leading to the small mean overlap value.

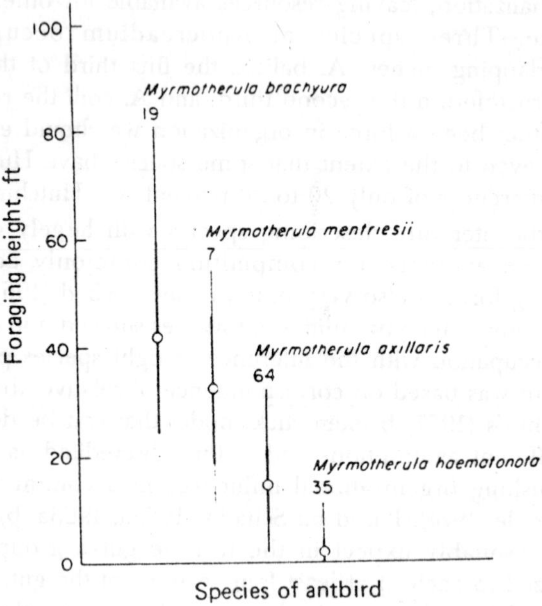

Fig. 5.4. Foraging heights of four species of sympatric antbirds (Myrmotherula). Means are represented by open circle, standard deviations by thickened vertical bars.

In one gut Hymenolepis abortive was abundant but completely overlapped by *H. spinocirrosa.* For the 13 parasite species in one scaup gut (data for which were given in detail), there was a significant correlation between population size and the number of segments of the gut occupied (r^2=0.26, $p < 0.10$) indicating niche expansion under population pressure unimpeded by other species. The abundant species were not competitive dominants as there was no sign of complementary abundances in the 10 scaup guts examined, even between *H. abortiva* and *H. spinocirrosa* (which showed a positive but nonsignificant correlation in abundance). Hair (1975) concluded that the majority of helminth species in the scaup responded individualistically to gradients in the gut, which I interpret as indicating a lack or positive or negative relationships between parasite species. Chubb (1964) noted no interactions between Echinorhynchus clavula and any other parasitic helminths in the guts of fish.

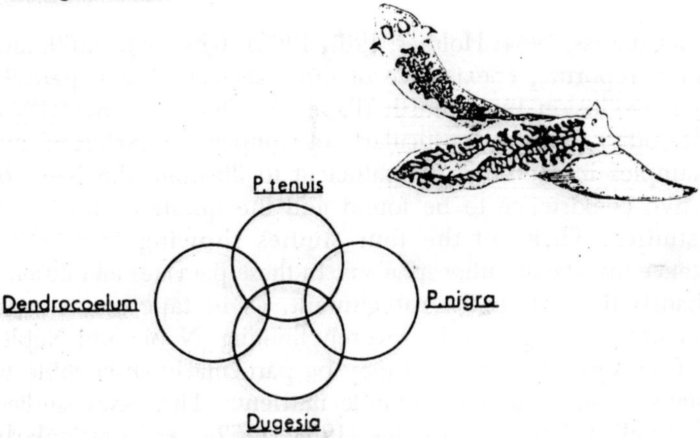

Fig. 5.5. Niche overlap amongst Lake-dwelling flatworms.

Thomas (1964) also concluded after it very extensive study of brown trout parasites that competition between parasites occurs only rarely. In pairwise partial correlations between six helminths over four years and twice a year (165 pairwise comparisons), he found many more positive than negative significant correlations. In a similar set of correlations involving only the four helminths in the alimentary canal whose distributions in the gut largely overlapped, but with up to three seasonal samples per year (264 pairwise comparison) he found only 14 cases that were significantly and negatively correlated (whereas 13 such cases should be expected by chance) and 12 cases showed significant and positive correlations. Although the three intestinal trematods of the mole show some differences in distribution along the gut, the probability of colonization for each species was so low that they hardly ever interacted (Frankland, 1959). Less than one percent of the moles studied had more than one species of the three, and Frankland noted that the distribution of each species, even within an area where it occurs, is very patchy and local. He also found that the presence of these flukes had no effect on the other intestinal helminth parasites of the host.

Interactive Coexistence

Interactive niche occupation is defined as the substantial shift by one species from fundamental to realized niche in the presence of another species. The cases in 5.1 involving inter active coexistence concern relatively few species: three studies on pairs of species (Cross, 1934: Holmes, 1961, 1962a; Chappell, 1969) and one study reporting coexistence of three species of grill parasite but their exclusion by a fourth (Paperna, 1964). Holmes (1973) provides other example, particularly of competitive exclusion, but the examples in the table are sufficient to illustrate the types of interactive coexistence to be found and the questions raised by such studies. Three of the four studies showing interactive coexistence involve acanthocephalans. In these parasites all nutrients are absorbed through the integument, as in tapeworms, and carbohydrate shortage can be severely limiting (Noble and Noble, 1976). Therefore, these species may be particularly susceptible to exploitative competition for soluble nutrients. The cases studied by Chappell 1969) and Holmes (1961, 1962a) are particularly convincing examples of competition.

In the former case the fundamental niches of the two parasites in stickleback showed a percentage similarity (PS) of 46 percent, but when occurring together their realized niches had PS= 17 percent. In the rat, the two parasites had fundamental niches that were 64 percent similar, but in concurrent infections the realized niches were only 10 percent similar (Holmes, 1961, 1962a). The excellent experimental studies of Holmes also raise some important questions. The same pair of parasites in rat showed strong niche shifts, particularly by Hymenolepis diminuta; but in hamster there was no indication of competition (Holmes, 1962b).

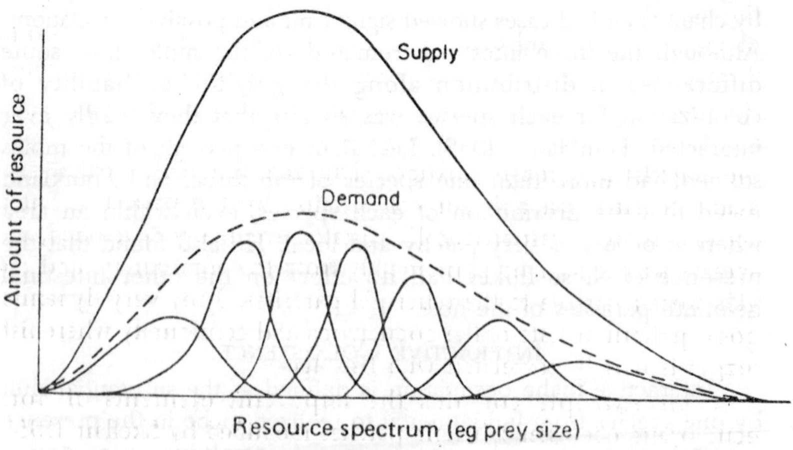

Fig. 5.6. Niche relationships among potentially competing species are often visualised and modelled with Bell-shaped resource utilisation curves.

Proportional similarities of distributions along the gut were 45 percent in single infestations and 46 percent in concurrent attacks. Also in rat H. diminuta had the broader niche on the gut-length dimension while in hamster M. dubius had the broader niche. It becomes necessary in order to understand coexistence of parasites and their community organization fully to study the full range of coexisting hosts since parasite species and species pairs behave differently in each host. Also where host tissues are damaged and/ or immune responses triggered, infection may be transient and a rapid succession of species may be expected. Niche occupation and community development in time may be extremely dynamic. The, dynamic nature to be expected in some parasite communities is fully realized in the gill parasitic community of carp studied by Paperna (1964). The community is composed of four species of monogenean trematode in the genus Dactylogyrus: D. anchoratus, D. extensus, D. minutus, and D. vastator.

Dactylogyrus extensus and D. anchoratus were the better colonizers because they could be infective throughout the year. Thus if young carp hatched early in the year when water temperatures were cool, they were rapidly colonized by these species, and significant numbers began to build up on the gill surfaces. However, as water temperature increased, D. vastator became infective and colonization was much more extensive. This species caused gill damage and rapidly made conditions unsuitable for the early colonizers, and they were usually pushed to extinction on the majority of hosts (50 days after hatching the maximum number of D. extensus found on carp was one). They survived only 10 to 20 days in the presence of D. vastator. As the gills became seriously damaged by D. vastator, conditions on the host became unsuitable for it, and after a relatively brief tenure of 44 to 45 days per host, the populations became extinct. The gills gradually healed and again became available for colonization by the parasites. But a specific immune response to D. vastator eventually developed, the species was eliminated permanently from the community, and the gills were available to the other gill parasites. They very dynamic, non-equilibrium state of the populations and community where fish length is closely correlated with fish age.

This example contains the important elements of non-equilibrium coexistence within patches identified by Skellam (1951) and Hutchinson (1953) and is similar in some respects to the

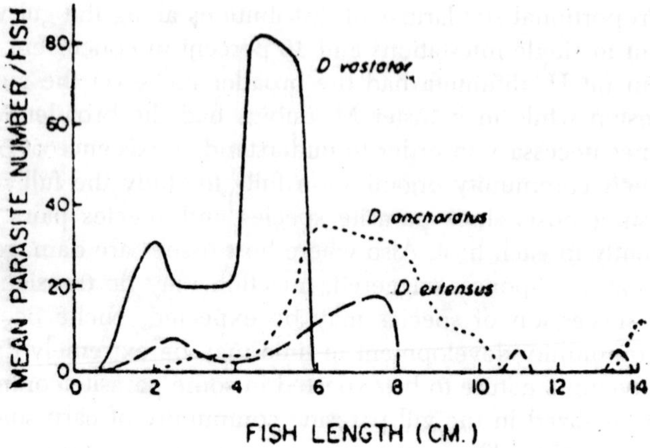

Fig. 5.7. Mean number of parasites per fish for the species Dactylogyrus vastator, D. anchoratus, and D. extensus in relation to fish length.

transient competitive displacement observed by Istock (1967). The competitively inferior species are the better colonizers and appear to have a lower reproductive capacity. New sites are continually becoming available for colonization. Sites rapidly become unsuitable for the competitive dominant, so its competitive edge is repeatedly lost. Many other examples in the parasitology literature many fit this scenario of transient competitive interactions, but in the absence of good experimental work interpretation of the mechanisms that produce complementary distributions is impossible. For example, Cucullanus heterochrous is evenly distributed in the flounder gut from November to May, but from June to October it is found predominantly in the rectum when C. minutus is present in the intestine (MacKenzie and Gibson, 1970).

As the authors point out, we cannot tell if this is a competitive effect or an age effect on C. heterochrous. Rohde has made a particularly intensive study on species packing in monogenean ectoparasites of fish. He has repeatedly concluded that many resources remain unexploited, that coexistence is non interactive, and that "segregation of sympatric species may be due to random selection of niches" (Rohde (1977a: 164, 1977b, c, 1976, 1978). His evidence is convincing. Monogenean flukes usually exist at low population densities (Rohde, 1977a), and although there is a latitudinal gradient of increasing diversity toward the tropics, there is no change in host specificity along this gradient or specificity in site selection on a host (Rohde, 1976). Monogenean species show highly restricted site selection on hosts even when only one species occurs or a host (Rohde, 1977a). When several species coexist, niche overlap is frequently minimal or nonexistent.

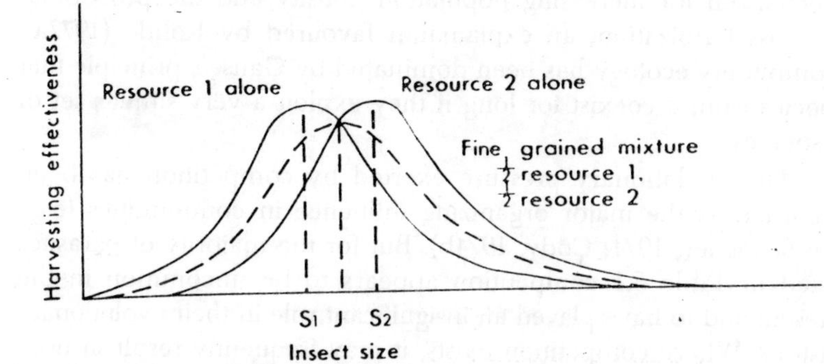

Fig. 5.8. Harvesting efficiency of two specialists and a single generalist on a simple resource continuum. The graph shows the efficiency of each specialist in its own resource niche (S_1 and S_2: solid curves) together with efficiency of the same specialists and a 'Jack-of-all-trades' (J) in exploiting a mixture of the two resource types (dotted curve).

A comparison of niche occupation on temperate versus tropical hosts yielded the conclusion that most microhabitats utilized by tropical parasites remain empty on fish in cold seas (Rohde, 1976). The mechanisms that result in such narrow niches on a host, even in the absence of competing species, are not fully understood, but

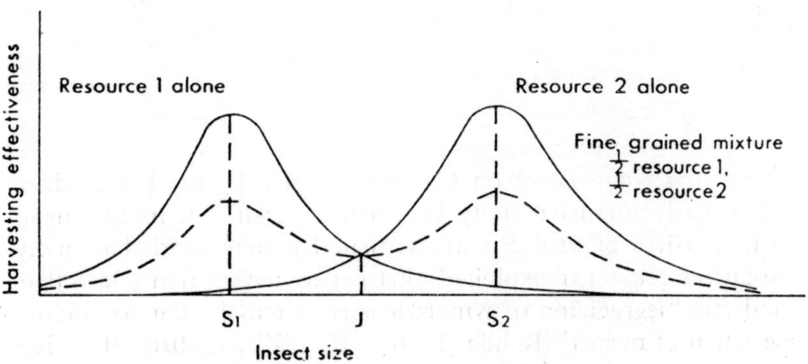

Fig. 5.9. As figure 5.8, Except that in this case the two specialists are now quite dissimilr in optimum resource position. In a mixed environment of this type, J is now inferior.

the great topographic diversity on fish, particularly on the gill where many monogenean species live, and the considerable complexities of water flow no doubt demand extreme site specificity. This seems to be at least as important as site specificity as a mechanism for increasing population density and the probability of cross fertilization, an explanation favoured by Rohde (1977a). Community ecology has been dominated by Gause's principle that species cannot coexist for long if they exploit a very similar set of resources.

The evolutionary pressure exerted by competition has been regarded as the major organizing influence in communities (e.g., see Schoener, 1974; Cody, 1974b). But for the majority of parasites listed in Table 5.1 competition appears to be uncommon in the present and to have played an insignificant role in their evolutionary history. Where competition exists, it may frequently result in non-equilibrium transient competitive displacement, a disruptive influence in community development. Based on parasites of plants and animals, fully support these generalizations. For such highly specialized organisms as parasites living in complex environments with steep gradients of resources, even small, randomly generated differences in physiology, behaviour, and morphology may result in optimal exploitation of very different sites in a host or very different hosts. Resources available are so diverse it is importable that a new colonist carries a niche exploitation pattern very similar to a resident species.

6

POPULATION ECOLOGY

A population is an interacting and interbreeding group of individuals of the same species, for example, all the perch in a, small lake. Although they may also occupy a particular area–in this example, the lake–certain populations, such as the elk, a migratory species, live in different locales at different times, of the year. Because the members of a population interbreed, they have a common gene pool; thus, populations are the units, upon which natural selection acts. Ecologists are interested primarily in *population dynamics*: the various factors contributing: to the increase, stability, decline, and extinction of populations including those of our own species. Understanding these factors is not only important in itself but is essential to an understanding of evolution. The term *population* is sometimes used to refer to larger or smaller groups than the fishes in a small lake. For instance, all the foxes in England or all the beech trees in North America can be regarded as populations.

Even such small groups as all the bacteria living in someone's mouth or all the ascarid worms parasitizing a dog can be considered populations. Although we often refer to the human population of a country, its members not only interact with but even interbreed with individuals of other countries. Moreover, within a single country, people may be divided into various subpopulations, some of which are quite isolated from each other. The term *population* thus has several connotations. The connotation intended is usually clear from the context in which the term is used.

Population Growth and Structure

The growth of a population in a given area is the result of two opposing forces:

1. the biotic potential or reproductive capacity, of the species and
2. the environment resistance. Biotic potential is the maximum rate of population increase under ideal conditions, which are seldom realized in nature for long periods of time.

Environmental-resistance includes all factors that limit population size, such as nutrient and habitat availability, disease, and predation.

Exponential Growth Versus Population size Stability

An important basis for the natural election theory of Darwin and Wallace was the fact that all organisms are capable of exponential, or logarithmic, growth in numbers. In a population growing exponentially, each successive total is a multiple of the one preceding it. Put another way, populations are capable of doubling at a constant rate, such as once every so many hours, weeks or years. (An example of logarithmic increase is provided by the series 2, 4, 8, 16, 32, 64, and so on. An arithmetic increase is an increase by a constant number, exemplified by the series 10, 20 30, 40, 50 and so on.) Many organisms have enormous reproductive potential. Some bacteria can double in Dumber every 15 minutes. A single mushroom produces hundreds of thousands of spores, each of which could, if provided the space, nutrients, and so on, give rise to a huge colony of mushrooms, each of

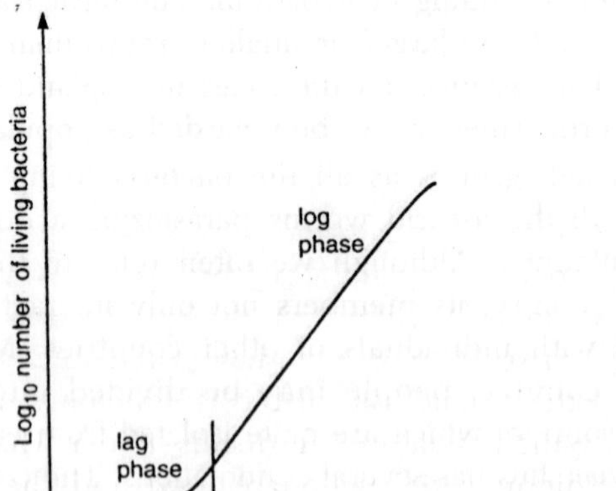

Fig. 6.1. Exponential growth of bacteria in laboratory culture.

which could produce many more mushrooms. In many insect and fish species, one individual can lay thousands of eggs. In some species of clams, one individual can produce 3 million eggs in a season. A single pair of mice, given the necessary food and space and an absence of disease and predators, could theoretically, have 3,000 descendants in one year–or 6.75 million in 3 years!

To emphasize the universality of the potential for logarithmic increase among organisms, Darwin analyzed the most slowly reproducing animal species, elephants, and found that even they were potentially capable, within a few years, of producing more offspring than the earth could accommodate. However, as we are well aware, this enormous potential is, except for brief periods, never realized. Darwin further observed that, with occasional fluctuations, populations of a given species tend to remain about the same size from one year to the next. He concluded that "this geometric tendency to increase must be checked by destruction at some period of life." In the more than a century since those words were written, we have learned much–but far from all–about the factors that regulate population growth.

Analysis of Exponential Growth

Some insight into factors that determine population size has been achieved by studying simplified ecosystems. For example, when a culture vessel containing a warm broth composed of a balanced mixture of all the nutrients needed for the growth of one particular species of bacterium is inoculated with a few hundred individuals derived from a thriving culture, the cells continue dividing at a steady rate. If the culture's growth is graphed as the logarithm of the number of cells against time, the result is a straight line. Such a culture is said to be in exponential growth, in a logarithmic growth phase, or, simply, in log growth. To predict the growth of a given population over a particular time period, we need to know the initial size of the population (symbolized by N) and the potential growth rate of that species, called its *intrinsic rate of increase* (symbolized r). Intrinsic rate is the difference between the population's *natality* the annual rate of births, hatchings, and the like–and its *mortality*–the annual rate of deaths–under what could be called ideal conditions, with adequate food and shelter, no predation, and so on. Both natality and mortality rates are largely genetically determined and can differ greatly from one species to another.

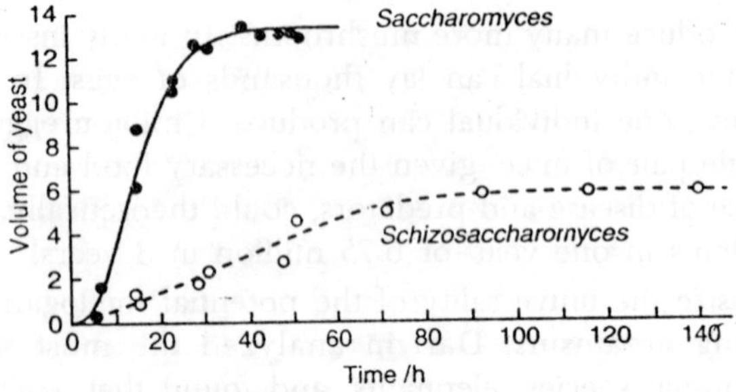

Fig. 6.2. Sigmoidal growth of pure cultures of the yeasts Saccharomyces and Schizosaccharomyces

Expressed mathematically the increase in size of a population undergoing exponential growth is stated as ΔN Δt = *r*N, in which ΔN represents the difference (Δ) in population size (N), Δt is the length of the time interval (difference in time) during which exponential growth occurred, and *r*N is the intrinsic rate of increase (r) for a starting population of size *N*. Stated in words, the equation reads as follows: the actual increase in numbers (ΔN) during a time interval of Δt (some fraction of a year) equals the potential maximum rate of increase per year (*r*) times the number of individuals (*N*) at the beginning of the time interval Δt. So it can be seen that, under ideal conditions, the actual amount of increase may equal the potential growth rate and, for brief periods, this may be case.

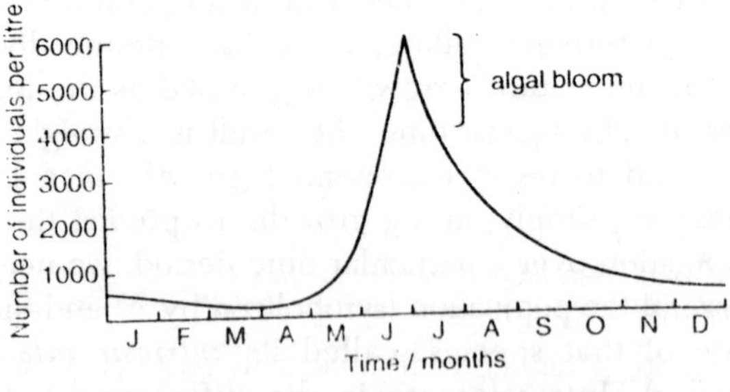

Fig. 6.3. J-shaped growth curve of planktonic algae.

Malthusiau Versus Logistic Growth

An early attempt to analyze population growth patterns and thus to derive from them some principles applicable to humans was made by the English economist Thomas R. Malthus in 1798.

He noted that populations tended to outgrow their food supply and predicted that human populations would eventually become drastically reduced by famine, disease, war, and similar catastrophes. Later, a differing concept, termed *logistic growth*, emerged. According to it populations eventually tend to level off at a relatively high level.

Malthusian growth

The Malthusian doctrine can be illustrated by the hypothetical growth curve which a period of exponential growth is climaxed by a series of catastrophic reductions in population size due to massive famine, disease, and other factors: This pattern is referred to as *irruptive growth* or *Malthusian growth*. Malthus idea about humanity's future was poorly received, not only at the time it was published but also in subsequent years. The concept was decried by religious leaders as an irreverent attempt to curb human procreation and by scientists as a lack of faith in the capabilities of science. Malthus was viewed as an alarmist, and his idea was generally ignored. Only in the latter half of this, century, as his predictions have begun to come true, has his theory been accorded much respect. The strongest reason for ignoring Malthus idea during the nineteenth century was cogent: the experience of so many observers to the contrary.

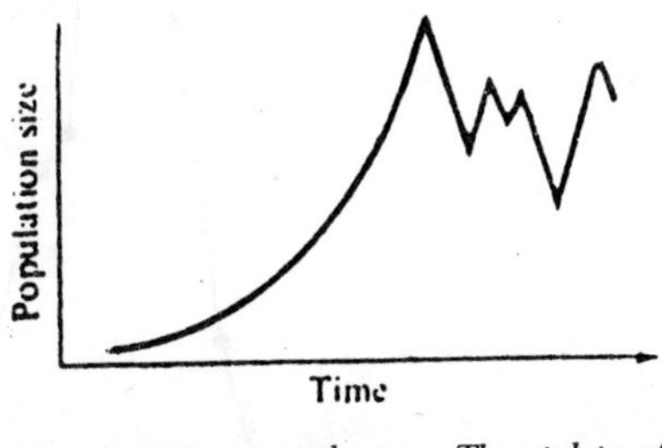

Fig. 6.4. Malthusian, or irruptive, growth curve. The upslope of the curve is steep and is followed by a series of partial population collapses.

By the mid-1800s, considerable data had accumulated on the growth of natural and laboratory populations of many species of organisms. In natural populations, in which predation and other limiting factors were operative, growth of the population often leveled off for an indefinite period, exhibiting a sigmoid (S-shaped) growth pattern. In laboratory cultures of microorganisms, a similar sigmoid growth curve was often obtained. Unfortunately, these bacterial experiments were terminated before what is called the death phase began; otherwise, the results would have been more typical of Malthusian predictions.

Logistic growth

The many examples of sigmoid growth curves in laboratory and natural populations led to the concept of logistic growth. This is a pattern typified by a slow beginning, a log growth phase, and a gradual slowing to a smooth, orderly, predictable pattern of population limitation at a level indefinitely supported by the ecosystem, said to be its *carrying capacity.* Although at first little attention was paid to this concept, it eventually received wide support, especially in the early 1900s. Some even hailed it as a universal law and applied it to the growth of all populations, including humans.

One enthusiast, invoking the principle of logistic growth, predicted that the world human population would naturally stabilize

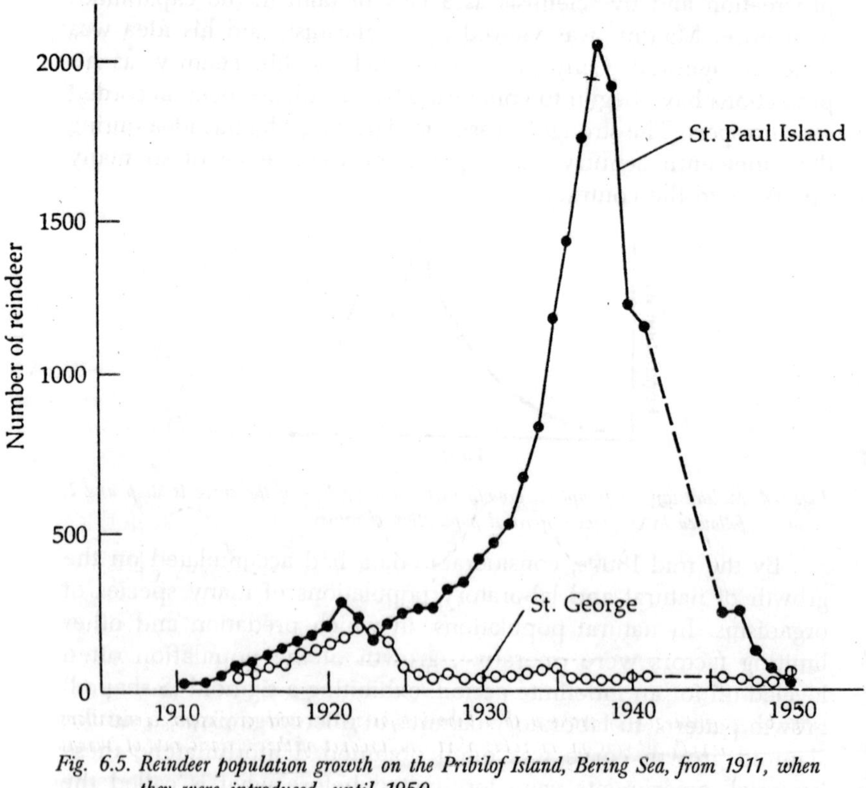

Fig. 6.5. Reindeer population growth on the Pribilof Island, Bering Sea, from 1911, when they were introduced, until 1950.

at 2.6 billion (2.6×10^9) people in AD 2100. As we know, due to changes in several factors since that time, this population size was exceeded before 1960, passed the 4 billion mark in 1975, and is still increasing rapidly. In a bacterial culture, after a period of log growth, the length of which depends on the size of the initial population and the total volume and nutrient concentration of the

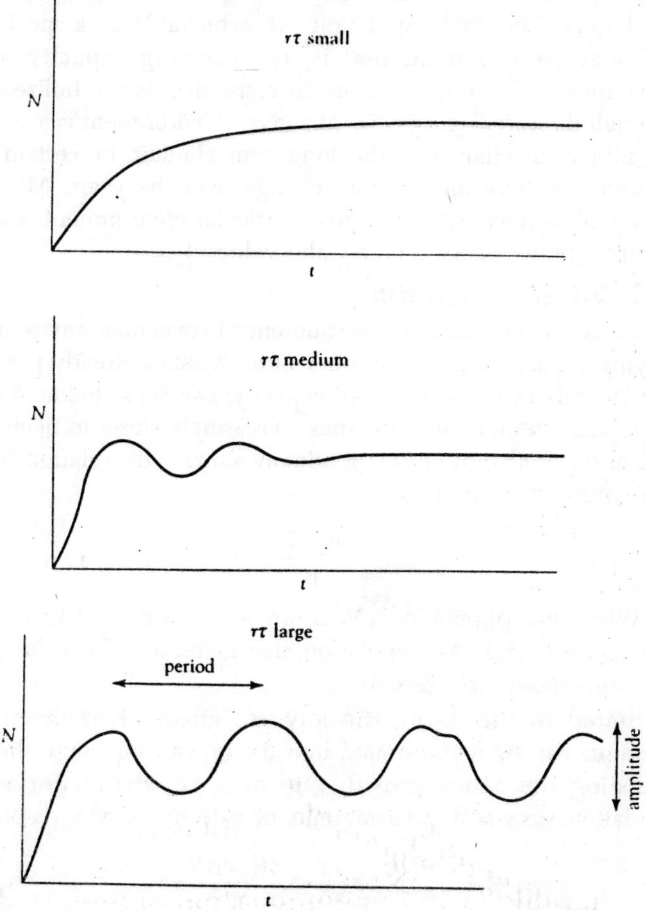

Fig. 6.6. Logistic growth curves with a time lag for species with overlapping generations. Growth depends on the value of rτ, the product of the intrinsic rate of increase and the time lag. In this figure, rτ increases from top to bottom. At a certain value of rτ, the population oscillation become stable (stable limit cycle) and no longer converge on the carrying capacity.

broth, growth of the culture slows. It then gradually enters a period of variable length in which the number of cells remains relatively constant.

In a bacterial culture, this is termed the maximum stationary phase. It represents a balance between the production of new cells and the death of older ones. A similar pattern is exhibited by populations of most species in most environments. The factors that control population size in such situations are examined later in this chapter. The final population size achievable by a species in a particular environment; that is, the carrying capacity of that environment, or the environmental resistance, is symbolized as *K*. Although the carrying capacity of a given environment is a constant, it is subject to change as the long-term climatic or certain other factors in an environment may change over the years. Moreover, by its evolutionary adaptation to a particular environment, a species can, over many years, increase the value of *K*.

The logistic growth equation

Stated mathematically, the difference between an environment's carrying capacity and the number of individuals already present in it (or the environment's unused capacity) can be stated as *K–N*. As the population increases, *N* comes increasingly closer to being equal to *K*, and population growth gradually slows. This relationship can be .expressed by the ratio.

$$\frac{K-N}{K}$$

When the population (*N*) is small, the value of this ratio is high, close to 1.0, As population size increases, the value of the ratio correspondingly lessens.

Stated in this form, the adverse effect of environmental resistance can be incorporated into the growth equation, there, by indicating the actual growth rate of a population for various population sizes and environments of various carrying capacities:

$$\Delta N / \Delta t = rN\frac{(K-N)}{K}$$

In this form, it is known as the logistic, or differential, growth equation. It expresses the negative feedback relationship between a population and the environmental resistance.

The Structure of Populations

Additional important information about a population's prospects for increase, decrease, or failure can be revealed by a survivorship curve. For natural populations, that is usually obtained by plotting data derived from repeated samplings of marked individuals. Figure depicts three types of curves, one approximating that for humans, another for the cnidarian *Hydra*, and a third for oysters. Because they are not cared for, the mortality rate for oyster eggs and larvae is extremely high, although any that survive early development have a good chance of living to maturity.

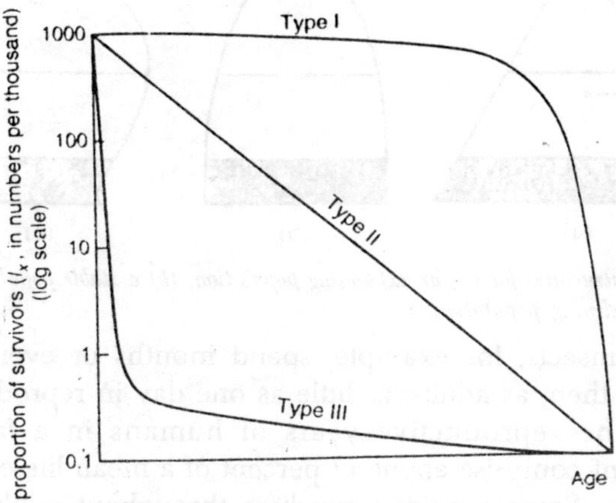

Fig. 6.7. Hypothetical survivorship curves.

Humans, on the other hand, with extended parental care, have a relatively high life expectancy at birth, with a sharp increase in mortality in the last 15 of their expected 70 years. Hydras exhibit the same death rate at every age. Although survivorship curves provide some insights into the prospects of a population of a given species for survival, much more useful in this regard are the age structures of particular populations. In one way of representing age structure, the population is divided into three age groups, each a percentage of the entire population:

1. pre-reproductives, or immature individuals;
2. reproductives, or fecund adults; and
3. post-reproductives, or nonfecund adults.

The proportions of a population in each class are key factors in determining whether that population can be expected to increase, decrease, or stay the same. The diagram for an expanding population, such as the present worldwide human population, has a pyramidal outline, its base being composed of pre-reproductives. The diagram for a stable population has a bell-shaped outline; that for a declining population has an urn-shape. An urn-shaped distribution, with its small proportion of pre-reproductives, usually indicates that a population is dying off. While greatly affected by environmental factors, age structures are also genetically influenced and can differ greatly from species to species.

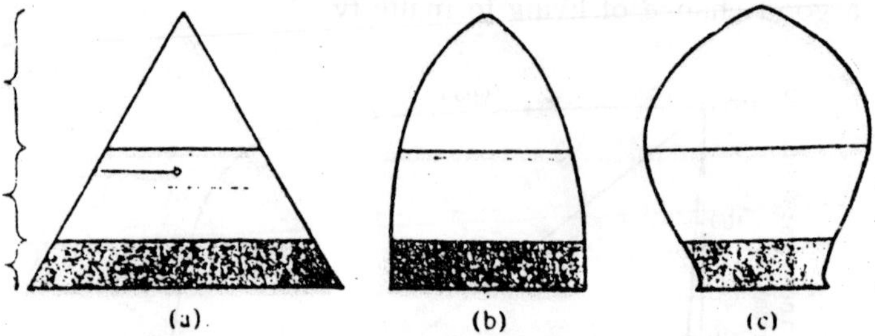

Fig. 6.8. Age structures for (a) an expanding population, (b) a stable population, and (c) a declining population.

Some insects, for example, spend months or even years as larvae and then, as adults, as little as one day in reproduction. In contrast, the reproductive years of humans in a favourable environment comprise about 42 percent of a mean life expectancy of 70 years. Some species reproduce throughout adult life, and many animals, including various worms, insects, and fishes, die almost immediately after reproducing. Populations of such species have no post-reproductive component in their age structure. The relative proportions of the component stages in the life history represented by each of the three classes can be greatly affected by both biotic and environmental factors and may therefore differ geographically for a given species. With many human populations, improvements in medicine, sanitation, and health care have increased the proportion of children living to reproductive age and the number of adults reaching old age. (Although life expectancy of North Americans in 1900 was less than 50 years, it is now about 70 years.

In the heyday of the Roman Empire, the mean life expectancy of its citizens was only 25 years.) In some economically depressed human populations today, only 50 percent of all children survive their fifth birth day, with very few living beyond their reproductive years. These examples emphasize that the growth rate of a particular population is the result of an interaction between its intrinsic growth rate and the resistance afforded by the local environment.

Life History Patterns

Organisms exhibit two basic approaches to reproduction. One involves an investment by the parents in large numbers of spores, seeds, eggs, or offspring that receive little or no care. In such species, sheer numbers tend to ensure that some individuals survive to reproductive age. The other approach involves the production of a few offspring that receive Substantial parental assistance. In the former case, much of the parental body is given over to reproductive structures; in the latter, most of the body is vegetative, that is, nonreproductive, biomass. This principle applies to all forms of life. Examples of plants that produce few individuals but invest much it each of them are those that send out either stolons (runners) or rhizomes (underground stems) from which new individuals bud off and reach maturity while still attached to the parent plant. An example of animals that produce few individuals are birds that lay one to three eggs and provide the hatchlings with food .and other parental care until they are almost fully grown.

Between the two extremes of one offspring that receives much care and many offspring that receive no care is a full spectrum of intermediate situations. For example, some organisms produce a moderately large number of offspring that receive some care and protection in the early stages of development. All of these behaviours, of causes, are genetically determined, although learned components are often incorporated into them. Like many other hereditary traits, they represent adaptations to particular environmental situations. What determines, for a particular species, which of the two opposite approaches is selected or whether the species will exhibit an intermediate pattern that incorporates elements of both? The answer is not always clear, and in some situations, it may not matter which approaches is followed; in many cases, probably either would serve. However, various environmental situations are most successfully taken advantage of by one or the other of the two approaches.

r Adaptation

Let us first consider the case of organisms most of whose developmental stages, such as spores, seeds, eggs, or larvae, are eaten by predators or infected by disease organisms. Many such species reproduce only once a year and have adapted to predation and disease by producing more spores, seeds, eggs, larvae, or hatchlings than the existing predators or disease organisms can eat or infect at one time. Such an approach is termed *r* adaptation or *r* selection because it depends on a rapid, exponential increase in numbers at a rate close to the organism's intrinsic rate of increase (*r*). Such high productivity at one time has the double effect of increasing the chances that some offspring will survive to maturity and of discouraging predators from specializing on a type of prey that is briefly available only once a year. However, because parasites and other disease organisms also can evolve, many have adapted their own life cycles to those of their hosts, multiplying and even becoming dormant in concert with the hosts. An elaboration on the once-a-year reproductive pattern is the *mast crop*, characteristic of many tree species in temperate zones.

In these species, 2 to several years may pass with little or no production of seeds or fruits, followed by a single season in which all such trees produce a great abundance of seeds. The populations of seed-eating animals, such as birds squirrels, and insects, that depend heavily on these particular seeds or fruits may decline in the area during the trees nonreproductive years. Then, suddenly, the reduced predator population is given far more food than it can eat; the next. Year's population of predators, now larger; faces the beginning of another cycle of several lean years. As might be expected, a selection pressure exists for various species of mast-crop-producing trees that inhabit the same environments to synchronize their mast crop production. Moreover, there would be less selection pressure for a tree to produce a mast crop if similar species in the area produced their mast crops in alternate years to its crop years.

For several tree species to synchronize their mast crops and thereby discourage predation of all of them permits a much higher population density of such trees and thus a greater overall efficiency in the capture of energy and the production of biomass. The trees do not need to communicate with each other in any way. Natural selection works against those that fail to synchronize their crops. Selection pressure for *r* adaptation is also exerted by environments that place a premium on the ability to produce large numbers offspring within a short time.

Two classic examples are the annual plants and the insects that live in deserts and those that live in the Arctic. These organisms develop from eggs, seeds, or other dormant stages whose development is triggered by the arrival of favourable conditions. They begin development grow to maturity, reproduce, and set seed or lay eggs all within the space of a few weeks each year. Thus, the *r* adaptation tends to be selected for by an environment with a brief growing season followed by conditions in which a high degree of mortality is independent of population density, such as occurs with a sudden, severe onset of desert aridity or an Arctic winter. Another situation that selects for *r* adaptation is the early stage of colonizing a new island or a burned-over area. Species that are the first to colonize such an area, or pioneer species, exhibit *r* adaptation. Organisms with a high intrinsic rate of population increase ale usually small, develop rapidly to reproductive maturity, and have short life spans.

K Adaptation

What kind of a situation might select for organisms that produce only one or a few offspring at a time? Most known examples occur in situations that exert strong negative-feedback controls on population growth. Under relatively constant year-round conditions in which an environment's carrying capacity for a particular species is often reached, competition for limited resources among species with overlapping needs tends to be high; in such circumstances *K* adaptation, or *K* selection, a term derived from the symbol for environmental resistance, tends to be favoured. Such are the conditions in tropical rain forests, where birds produce relatively small clutches of eggs. But other factors can also contribute to *K* adaptation. For example, larger, longer-lived organisms, such as the blue whale and the emperor penguin, are usually *K* adapted. A longer life span permits *iteroparity*, or repeated reproduction, which compensates for raising only one offspring at a time.

Population Regulation

Simple ecosystems represented by cultures of bacteria and :protozoa have proven useful in analyzing factors that regulate population size. Among the reasons that cells begin to die in thriving cultures of microorganisms is starvation. This, eventually results from depletion of essential nutrients and is an example of a *density-dependent factor* affecting population size. A factor, such as a disease epidemic, that exerts its effect on a population irrespective of its density is a *density-independent factor.* Density-dependent factors, such as nutrient supply depletion, constitute negative feedback controls over a population's absolute size and account for an increasing of mortality as density increases. The total amount of a key nutrient present in an environment may be quite unimportant when a population is small; it becomes increasingly important and eventually limiting as the population increases in density.

In a bacterial culture that has exhausted one or more of its essential nutrients, when the starved cells die, they disintegrate releasing their nutrient components into the medium. This new supply of nutrients permits some of the other starving cells to resume growth and division. Therefore, at this stage of the culture's existence, the number of new cells tends to equal the number of dying cells, resulting in the upper plateau of the sigmoid growth curve. If a new culture is begun by introducing into fresh broth medium several cells from an old culture that had reached maximum stationary phase, a lag phase at first occurs–that is, little or no growth takes place for a time–followed by a period of exponential growth and, as the nutrients become used up, culminating once more in a maximum stationary phase.

The maximum stationary phase of a bacterial culture may last for hours, days, or weeks. But eventually, if nothing is added or removed from the culture, this phase ends as the number of deaths begins to exceed the number of cell divisions. Thus begins the death, or negative growth, phase. What happens then depends on several factors that vary under different conditions and from one species to another. Some species exhibit a log death phase that can be as steep as or steeper than the long growth rate. In other cases, every cell dies in only a few days. In yet other instances, the number of cells drops to a new low level, but the culture continues to survive, its numbers oscillating irregularly at the lower level. Such a culture may persist in this condition for months or even years.

Density-Independent Factors

As a general rule, density-independent factors usually reduce a population by a certain constant percentage no matter what the size of the population is. One type of factor that, despite adequate nutrients, can slow, stop, or reverse the growth of a culture of microorganisms is the accumulation of toxic wastes. This factor can be much more serious for the ultimate survival of the culture can starvation. In old cultures, enough nutrients may be released from the disintegration of dead cells to support growth and reproduction of other cells indefinitely. But the concentration of inhibitory substances that can neither diffuse out of the culture nor be decomposed or detoxified by the organisms in the culture, although initially having no appreciable effect, may finally reach a level incompatible with the life of any of the cells. A toxic material that has reached a concentration that kills, most of the members of a culture is usually a density-independent factor for population growth.

The resulting irruptive growth curve is strikingly different from the sigmoid curve; it is said to be J shaped. In nature, examples of density-independent factors include droughts and cold spells which by arresting growth, may bring about a nearly total population collapse. Although a sudden drop in population size results in such cases, usually not all individuals die, and, when favourable conditions resume, a second period of log growth usually occurs. Massive death from brief disease epidemics, unfavourable climatic conditions, or temporary periods of starvation also produces J-shaped growth patterns. Not ail growth curves are readily classified as either sigmoid or J shaped. Nor are all factors affecting population growth .easily categorized. Both starvation and predation, which usually function as density-dependent factors, can sometimes cause population collapses resembling the downslope of a J-shaped growth curve.

Intraspecific Relationships and Carrying Capacity

From its first enunciation, some biologists criticized the concept of logistic growth as simplistic. Yet it is certainly true that natural populations maintain relatively stable levels year after year. A controversy also exists over whether density-dependent intraspecific competition or density-independent extrinsic factors are of primary importance in the regulation of population size. Careful analysis of stable populations, both in natural situations and under laboratory conditions, has revealed a wide variety of factors, both density-dependent and density-independent, that affect a population's size once it reaches a certain level.

Some of the density-dependent factors have also been revealed by laboratory experiments such as the, following on beetles, mice, and rats. In separate experiments, adults of several species of flour beetles were placed in sifted, humidified flour mixed with years. The beetles burrowed into the flour at once and laid eggs in the tunnels they made. The larvae, or meal worms, which hatched 5 to 10 days after egg deposition, also burrowed into the flour. Both the adults and the larvae ate any eggs and pupae they encountered. Male and female adult encounters, however, often resulted in copulation and the laying of more eggs. As the population increased, inhibitory substances released by the beetles accumulated. As the amount of these substances increased, females produced fewer eggs, larvae developed more slowly and adults grew to a smaller size.

A steady population level was finally reached, characteristic of the particular species of flour beetle being tested. The carrying capacity of that environment for that species had therefore been reached. A typical result was a steady increase of population to a stable level characterized by a fixed ratio of adults to larvae and a 90 per cent or higher mortality rate of eggs through cannibalism. But some species never equilibrated; instead, their population levels oscillated repeatedly in a pattern reminiscent of a Malthusian prediction. Such oscillations can result when large cohorts, groups of individuals of the same age, reduce the size of succeeding cohorts by cannibalism. The smaller cohort is, incapable of having the same effect on the large cohort that follows it. Such an effect can persist indefinitely.

Although other factors are also involved, the beetle experiment is primarily a case of population regulation by density-dependent intraspecific competition. In another series of experiments, separate mouse populations, with unlimited food and water were established in six 2-by-8- meter pens. In three of the pens, the food, water, and nesting boxes were scattered throughout the pens; in the other three pens, all the food and water was placed at one end, and all then nesting boxes were crowded together at the other. The maximum number to which each population rose differed from 25 to 138 among the six pens. Although all were adversely affected by the crowding, the particular level that was reached in each case seemed related more to individual temperamental differences in the animals than the immediate effects of the physical, conditions. As the population grew, so did the fighting among adults.

One population showed a decline in birth rate and a high mortality in the litters. In three of the six populations, the primary limiting factor was an increased mortality of newborn. Fighting often was followed by trampling of nests, which led to desertion or cannibalism of the litters by the females. Although food was abundant, food intake declined as the population increased. In some populations, food intake was so reduced as to lower normal female fecundity, the ability to produce offspring. But even in populations in which enough food was eaten to maintain fecundity, birth rates dropped off. This appeared to be due to less frequent copulation under the crowded conditions, increased spontaneous abortions (miscarriages), or both. Levels of female sex hormones were shown to be affected by increased odors resulting from crowding, with the result that ovulation ceased and sexual behaviour was inhibited. Similar studies were performed on rats in crowded conditions with unlimited food and water. In these experiments, population size proved to be regulated principally through deaths from fighting.

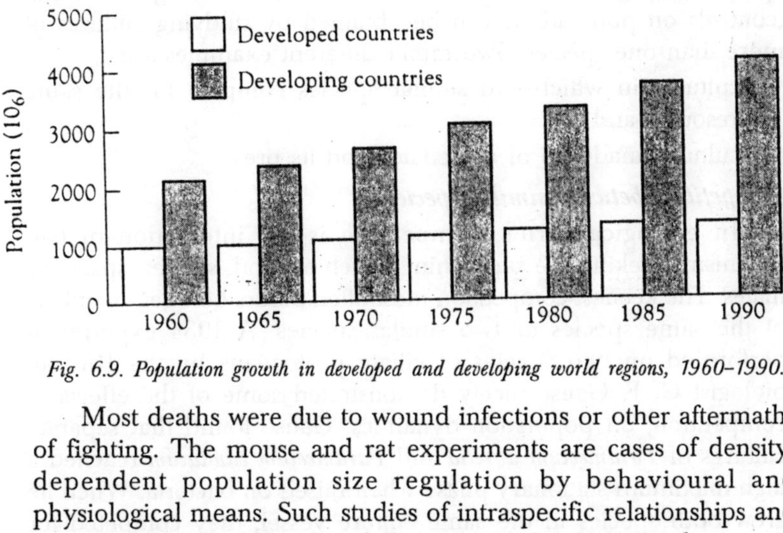

Fig. 6.9. Population growth in developed and developing world regions, 1960–1990.

Most deaths were due to wound infections or other aftermaths of fighting. The mouse and rat experiments are cases of density-dependent population size regulation by behavioural and physiological means. Such studies of intraspecific relationships and carrying capacity have revealed some important principles:

1. Even under controlled conditions, it is sometimes impossible to predict
   (a) whether a given population of any reach a stable level;
   (b) whether it will oscillate slightly about a certain level; or
   (c) whether it will exhibit a Malthusian pattern of' catastrophic declines alternating with periods of exponential growth.
2. The natural methods by which populations are limited vary greatly from species to species and sometimes from population

to population within a species. Besides such factors as limited food supplies and predation controls that limit population size in nature appear to include

(a) cannibalism of the young;
(b) deaths of adults from increased fighting; and
(c) reduced fecundity, birth rates, and rates of development as a result of the accumulation of toxic substances or other effects of crowding. Nature's way of regulating population sizes obviously includes methods that are not regarded as desirable or acceptable ways of limiting human populations.

Interspecific Relationships of Populations

We have seen how some insight into population dynamics can be gained by studying experimental ecosystems containing single species of organisms. An even better understanding of natural ,controls on populations can be obtained by studying cultures of more than one species. Two rather different examples are

1. cultures in which two similar species compete for the same resource and
2. cultures made up of a predator and its prey.

Competition between similar species

In ecological terms, competition is the interaction of two organisms seeking the same things, such as food, shelter, space, or mates. The keenest competition most often occurs between members of the same species or two similar species. A 1934 experiment performed on two species of ciliate protozoans by the Russian biologist G. F. Gause nicely demonstrated some of the effects of competition, on population dynamics. Gause found that separate cultures of *Paramecium aurelia* and *Paramecium caudatum* reached a high maximum stationary phase when raised on bacteria. When he grew both species in the same culture vessel, they competed for the available food.

As long as food was abundant, both populations grew exponentially; but as food became depleted, the growth of both populations dropped to a slower than normal rate. Later, the population of *P* caudatum severally declined, while that of *P aurelia* continued to increase. Eventually, the population of *P aurelia* reached about the same size it had achieved when grown alone. Gause obtained similar results when he grew other species of competing organisms. The experiments and various field studies led to the development of what came to be known as Gause's rule or the *competitive exclusion principle*, which states that in any one geographical area, an ecological niche cannot be fully and simultaneously occupied by stabilized populations of more than one species. This means that if two very similar species with very habits are ever found in the same locality–a rare occurrence–it can be presumed that the situation is unstable and that one of the species may soon disappear from the area.

Occasionally, seeming exceptions to this rule have been observed on closer examination, however, they have usually proved to be cases in which the niche of the two species did not overlap completely. For example, if one of two competing species of birds occasionally leaves its preferred food to its competitors and feeds on another food, the two do not occupy the same niche. In such a case, the two species may be able to coexist indefinitely in the same area. The longer two such species live in the same area, the less overlap they exhibit in appearance and behaviour, a phenomenon called *character displacement*. The less similar they become the less they will compete with each other. Such realizations led to the development of a corollary to the exclusion principle, namely, the *coexistence principle*, which states that different species that coexist indefinitely in the same ecosystem usually have different ecological niches. Exceptions to the competitive exclusion principle occur in at least two types of situations:

1. when a predator regularly feeds on two competing species, thereby preventing the potentially dominant one from excluding the other; and
2. when regular environmental disturbances, including those of human origin, favour the survival of the competing species that would have been excluded in the absence of the disturbance.

Predator-prey relationships

Other experiments performed by Gause shed light on predation as one of the checks on population size and density. Gause cultured *Paramecium caudatum* along with one of its most voracious predators, *Didinium nasutum*, another ciliate. First, Gause prepared several cultures and introduced five paramecia into each. (In the absence of predators, paramecia double in number as often as every 6 hours.) Two days later, he added three didinia to each culture. Within a few days, the didinia had eaten all the paramecia, and a few days after that, all the didinia had died of starvation Gause repeated this experiment but placed sediments at the bottom of the culture vessels to provide hiding places for the paramecia.

As before, the population of paramecia increased: when he added the didinia, they fed on the paramecia, drastically reducing their numbers. But in these cultures, some paramecia escaped by hiding. The didinia, which need a steady diet, eventually all starved and, in the total absence of didinia, the surviving paramecia increased exponentially and achieved the same dense concentration as they usually did in the predator's absence. Gause then repeated the first experiment, with no hiding places provided, but this time, he added one paramecium and one didinium to the culture every third day. This procedure was intended to simulate the occasional immigration of individuals from other area. The result was that both species, survived indefinitely in sizable numbers, although a series of' population oscillations of both occurred. One principle derived from these and similar experiments is: that it is not beneficial for a predator species to eat all of its prey individuals.

Field studies have shown repeatedly that, in nature, most predators feed on several prey species, but they almost always feed most heavily on whatever prey species is, most abundant at any given time. When the numbers of a particular type of prey decline, the predator usually switches to the prey species that is then most abundant. This behaviour pattern among some predators apparently enables both prey and predator populations to survive indefinitely. It probably accounts for some of the oscillations in population size so often observed in nature. Population oscillations are particularly evident when the numbers of all of a predator's prey species are so used as to be unable to support a large population of that predator; then the predator usually starves in massive numbers. But it seldom becomes extinct in the area because some individuals usually find enough food to survive.

The population oscillations that commonly occur in many prey and predator species are especially evident in Arctic and sub-Arctic regions, which are subject to extreme annual climatic changes and in which the animal communities are composed of relatively few species. Among the factors that help both predator and prey to survive are hiding places for the prey, a variety of prey species on which the predator can feed occasional declines in predator populations, and occasional immigration of both prey and predator from other areas.

Some oscillation of prey species also occurs in the absence of predation, indicating that predation is only one, and perhaps not even the most important, of the factors that keep natural populations in check. Oscillations in food availability for herbivores, for' example, can have dramatic effects, Nevertheless, an important key to the survival of both a predator and its prey in a particular area seems to be the safeguarding of the prey from predation when its populations are at their lowest level.

Predation of canadian lynxes on snowshoe hares

Although careful field studies of fluctuations in prey and predator population sizes have been made only in the last half century, data useful for such a study had been collected since the early nineteenth century across an extensive area of northern Canada. This fortunate circumstances came to light when it occurred to D. A. MacLulich to examine the old sales and purchase records of Canada's largest fur dealer, the Hudson Bay Company. He discovered that two of the animals long trapped for their fur were the Canadian lynx and its most important prey, the snowshoe hare. The number of pelts of each that were bought and sold in a given year represented trapped samples of these two populations in northern Canada. Pelt sales probably reflected the oscillations in the populations of the two species.

The fur company records suggested that the size of the lynx population was directly related to the size of the hare population. When there were many hares, the lynxes also thrived. Then, perhaps the lynxes ate the hares at a faster rate than the hares reproduced, many lynx began to starve, and their numbers felt this would have enabled the here population to increase. As the graph based on pelt sales indicates, the lynx population peaks always followed the here population peaks by about a year. Here population peaks occurred only after a decline in lynx numbers. The important principle illustrated by this example is that predation by a carnivore is probably an important immediate check on the population size of its prey species. However, the principle is not as applicable to predation on prey that is larger and more powerful than the predator. There is a growing amount of evidence to indicate that predation by carnivores on such animals as moose, mountain sheep, and deer is largely limited to the old, the sick, and the young. Any substantial limitation of the prey population by predation in such instances is apparently achieved primarily by reducing the number of young individuals.

The Effects of Predation

In the past half century, there has been a widespread eradication of many predatory animals in various parts of the world. This has been especially true where the predator is believed, rightly or wrongly, to be a threat to the lives or safety of humans or their livestock. In many cases, the belief that a particular "varmint" is a major threat to livestock has been altogether erroneous. As a consequence of, such fears, wolves, bears, coyotes, eagles, and many other-predators have been eliminated from large areas of the United States, Asia, and Africa.

Timber Wolves and White-Tailed Deer

After timber wolves, the natural predator of white-tailed deer deliberately eliminated from northern Wisconsin by unrestricted hunting, the deer population rose to unprecedented levels. Shortly after World War II, the available vegetation was no longer adequate to support such large numbers, and thousands of deer starved to death despite the fact that emergency feeding stations were provided by the state government. The deer population is now kept down by controlling the length of the hunting season and allowing both bucks and does to be shot in some years. Wolves are also being reintroduced in some areas.

The Dingo

Another example of a campaign to eliminate a predator is that waged against the dingo, the yellow dog of Australia. This wild dog was apparently introduced 8,000 to 10,000 years age by human immigrants from Asia. It is hated by ranchers, in part for preying on beef cattle, an important export commodity in Australia. One early attempt at dingo control following World War II involved dropping poisoned meat onto cattle ranges from airplanes. Wary adult dogs were seldom poisoned by this technique, although young ones were.

A more detailed study of the dingo's eating habits then showed that they subsisted mostly on small animals such as rabbits. It now appears that only during times of severe drought on the range do the dingoes, naturally solitary hunters, form packs to prey on cattle, and then they usually kill only calves. When interviewed, some ranchers noted that the killing of calves during a long drought improves the chances of a herd's survival until the drought is over. Most ranchers, however, continue to war against the dingo.

Starfishes Versus Barnacles and Mussels

Among the benefits of predation is that it sometimes prevents local elimination of one or more prey species, a fact already noted as an exception to the exclusion principle. This unanticipated effect on species diversity was revealed by a study of the populations of barnacles and mussels in seashore areas from which all sea stars (starfish) had been removed. The starfish normally feed on both barnacles and mussels. When the starfish were removed, the numbers of barnacles and mussels increased greatly. As they competed with each other for the available spaces on rocks in the intertidal zone, one species of mussel preempted the space, reducing the number of species of marine invertebrates in these areas from fifteen to eight. Because the predator had prevented competitive exclusion from running its course, it thereby enhanced species diversity.

Benefits of predation

Paradoxically, despite limiting its size predation usually tends to be beneficial to a prey species population. When the old, weak, diseased, and crippled are eliminated by predation, the surviving population tends to consist of the healthiest and most vigorous individuals. Eliminating large numbers of young during times of population, population expansion also reduces competition among the remaining individuals for existing food and other resources. The elimination of a predator from a region has sometimes been followed by a dramatic increase in the prey population, often beyond the carrying capacity of the area.

Different rules for small predators

Although large carnivores that prey on species that con injure them may usually limit their attacks to the old, sick, crippled, and young, this is not the case with small predators, especially when the prey is relatively defenseless. A snake is likely to eat any frog, an owl any mouse, and a spider a wide variety of insects. Moreover, Some small predators feed only on a single prey species, a situation that is quite unusual. Being limited to a single prey species means that the predator depends on the continued existence of that prey species for its own existence. If the predator is able to attack members of the prey species of all ages with impunity, both predators and prey could be liable to extinction in areas in which both are found.

The same principle also applies to species-specific parasites and to any animal that depends on a single plant species for food.

In most situations, a predator species that enters an area in which its only prey species exists may virtually eliminate that species from that area. Only chance migrants escape to establish colonies lese where. The predator then also tends to be eliminated from the area by starvation, with perhaps a few individuals surviving by encountering remaining prey or by migrating to other areas in which the prey is found. The result of such a relationship is a wide scattering of both the prey species and its predator, and both are difficult to find in anyone area. Occasionally, local population explosions of the prey species may occur, to be followed, sooner or later, by a corresponding increase in the predator species after some happen into the area. Both may then largely disappear again from the area for a time.

Symbiosis

Some organisms live much of the it lives in a very special ecological relationship an important part of their environment is a member of another species. In the traditional definition, a relationship in which members of two different species living in intimate relationship to each other, is termed *symbiosis* (from the Greek *sym*, "together," and *bios*, "life"). Symbiotic relation ships, which are always beneficial to at least one member of the ,association, may vary from associations of extreme intimacy and interdependency to relatively casual associations and from very temporary relationships to those involving the entire life span of the symbionts. Some relationships are so intimate that one lives on or in the other organism is termed the *symbiont*, while the one on or in which the symbiont lives is called the host. Such highly intimate symbioses can be classified as commensalism, mutualism, or parasitism.

Commensalism

Commensals (from the Latin *com*, "together," and mensa, *"table"*) may share the same food, and the symbiont may gain shelter by living on or in the host. But even if the symbiont shares the host's food, it causes no particular harm. For example, the bacterium *Escherichia coli* lives in the human colon it benefits from the nutrients, warmth, and shelter found there, but causes no disease or even any discomfort. Similarly, the protozoan *Opalina ranarum* lives harmlessly in the colon of the frog. In both eases the symbiont depends on the host for its supply of food.

Mutualism

In *mutualism*, some reciprocal benefit accrues to both partners. The flagellate protozoans that digest the cellulose of wood particles while living in the gut of termites are good examples. The termite needs the protozoans to digest the wood it eats the protozoans need the termite for the shelter and the nutrients it provides. When termites are experimentally rid of their protozoans, they soon begin to slarve, even when provided with all the wood they can eat. Many highly evolved mutualistic associations occur in nature, some of them very interesting. For instance, lichens are associations between certain kinds of algae and fungi. Other classic examples are provided by several species of tropical acacia trees that are hosts to particular species of ants.

The trees provide the ants not only with a safe home (their sharp thorns, which the ants hollow out) but with two types of food as well (special leaf products rich in proteins, fats, and vitamins,

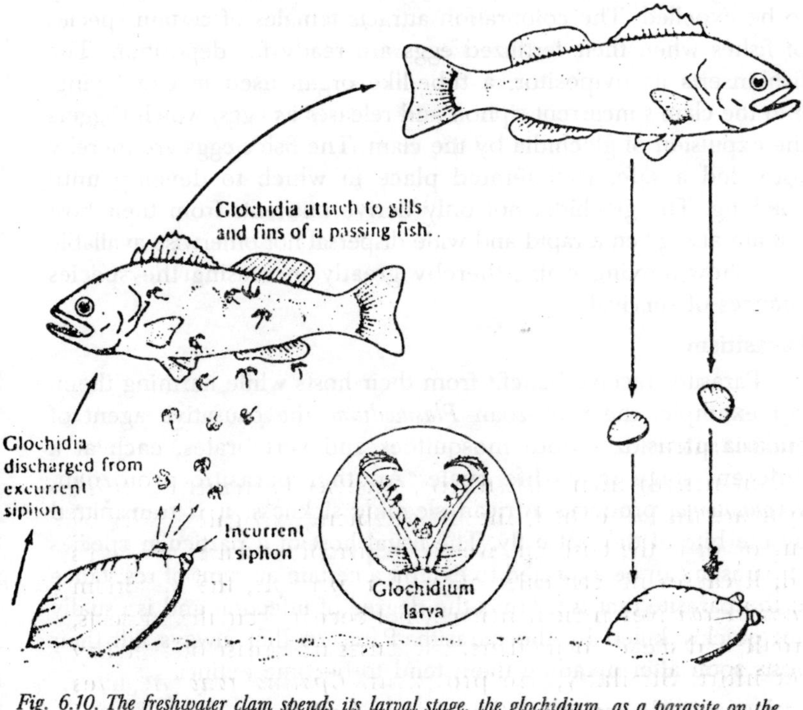

Fig. 6.10. The freshwater clam spends its larval stage, the glochidium, as a parasite on the gills and fins of a fish. At metamorphosis, it leaves the fish to take up its adult life.

and a special nectar rich in sugar). The ants, for their part, destroy story injurious insects that attack their tree host. Moreover, ant patrols destroy other plants, including other acacia trees, in the immediate vicinity, thereby ensuring their host tree access to adequate sunlight, soil nutrients, and moisture. Mutualism combined with parasitism exists between some fresh water calms and fish.

Most species of freshwater clams have a larval stage, the glochidium, which is a parasite of freshwater fish. The gloehidia are discharged from the female clam and usually settle to the bottom, where, upon encountering a fish of the right species, they attach to its gills or other parts of its body. After 10 to 30 days as a parasite, they metamorphose and leave the fish as juvenile clams. The parasitic infection does not greatly harm adult fish, and the lesions soon heal after the glochidia drop off.

With some species of freshwater clam and fishes, such a parasitic relationship also involves mutualism. The siphons of the female clams become conspicuously coloured when the glochidia are ready to be expelled. The colouration attracts females of certain species of fishes when their fertilized eggs are ready for deposition. The fish inserts its ovipositor, a tube-like organ used in egg laying, into the clam's incurrent siphon and releases its eggs, which triggers the expulsion of glochidia by the clam. The fish's eggs are thereby provided a safe, well-aerated place in which to develop until hatching. The glochidia not only derive nutrients from their host but are also given a rapid and wide dispersal not otherwise available to a show-moving clam, thereby greatly increasing the species chances of survival.

Parasitism

Parasites receive benefit from their hosts while harming them. For example, the protozoan *Plasmodium*, the causative agent of malaria parasitizes both mosquitoes and vertebrates, each at a different stage in its life cycle. Another parasitic protozoan, *Trypanosoma*, produces African sleeping sickness; it is transmitted by the bite of the tsetse fly. The usual host of a particular species of parasite can be expected to exhibit a certain amount of resistance to the parasite that is it limits the degree of infection and is usually not quickly killed by the parasite. Parasites that always kill their hosts soon after invading them tend to become extinct.

7

NUTRIENT CYCLES OF AN ECOSYSTEM

An ecosystem is a basic functional unit of nature comprising both organisms and their nonliving environment, intimately linked by a variety of biological, chemical and physical processes. The living and nonliving components interact among themselves and with each other; they influence each other's properties, and both are essential for the maintenance and development of the system. An ecosystem, then, can be visualized as a grouping of components–living organisms, organic debris, available nutrients, primary and secondary minerals and atmospheric gases-linked by food webs, flows of nutrients and flows of energy. A typical forest ecosystem might be visualized as a 1,000-hectare stand of mature deciduous forest (A hectare is 2.47 acres).

The lateral boundaries of the system can be either an edge of the stand or an arbitrarily determined line. The upper boundary is the treetop level, and the lower one is the deepest level of soil where significant biological activity takes place. Nutrients are found in four basic compartments of the ecosystem that are intimately linked by an array of natural Processes. The organic compartment consists of the living organisms and their debris. (There are probably more than 2,500 species of plants and animals in a 1,000-hectare system.)

The available-nutrient compartment is composed of nutrients held on the surface of particles of the clay-humus complex of the soil or in solution in the soil. Roots as they grow produce positively charged hydrogen ions that exchange with the nutrient ions (of calcium, magnesium and so on) held on the negatively charged particles, and the nutrients are then taken up by the roots. The third compartment consists of soil and rocks containing nutrients in forms temporarily unavailable to living organisms. The atmospheric compartment is made up of gases, which can be found on only in the air but also in the ground. Nutrients can flow between these compartments along variety of pathways.

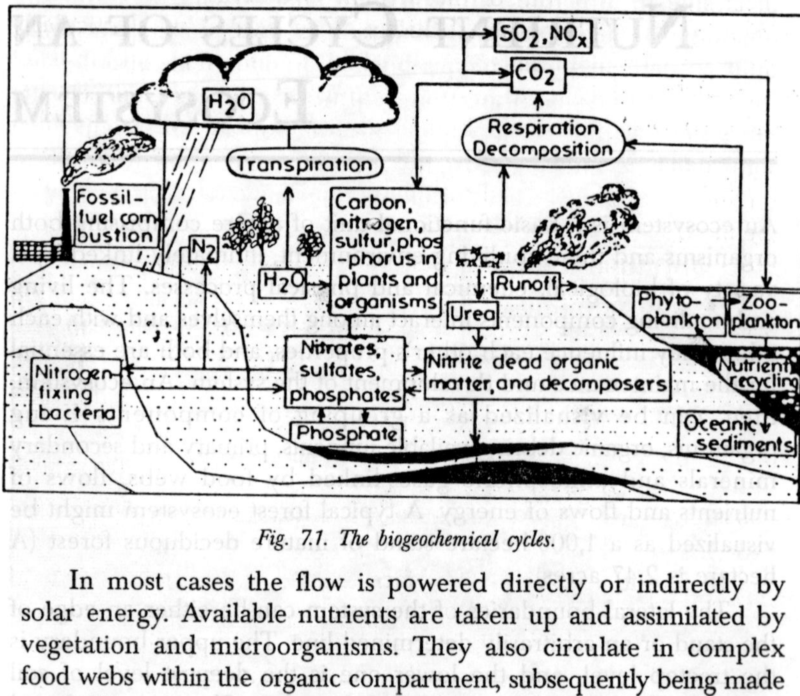

Fig. 7.1. The biogeochemical cycles.

In most cases the flow is powered directly or indirectly by solar energy. Available nutrients are taken up and assimilated by vegetation and microorganisms. They also circulate in complex food webs within the organic compartment, subsequently being made available again through decomposition or leaching. Minerals in soil and rock are decomposed by weathering, so that nutrients are made available to organisms. Sometimes available nutrients are returned to the soil-and-rock compartment through the formation of new minerals such as clay. Nutrients tend to cycle between the organic, available-nutrient and soil-and-rock compartments, forming an intrasystem cycle.

Nutrients in gaseous form are continually being transferred to and form other compartments by inorganic chemical reactions such as oxidation and reduction and by organic reactions related to such processes as photosynthesis, respiration and the fixation and volatilization of nitrogen. As ecosystem is connected to the surrounding biosphere by its system of inputs and outputs. They arrive or leave in such forms as radiant energy, gases, inorganic chemicals and organic substances. Inputs and outputs can be transported across ecosystem boundaries by meteorological forces such as precipitation and wind, geological forces such as running water and gravity and biological vectors involving the movement of animals in and out of the system. Ordinarily it is difficult to measure the input-output relations of an ecosystem, particularly those involving nutrients.

The nutrient cycle is closely connected to the water cycle: precipitation bring nutrients in, water leaches them from rocks and soil and stream flow carries them away. Hence one cannot measure the input and output of nutrients without simultaneously measuring the input and output of water. The problem usually is that subsurface flows of water, which can be a significant fraction of the hydrologic cycle, are almost impossible to measure. Several years ago it occurred to us that under certain circumstances the interaction of the nutrient cycle and the hydrologic cycle could be turned to good advantage in the study of an ecosystem. The requirements are that the ecosystem be a watershed underlain by tight bedrock or some other impermeable base. In that case the only inputs would be meteorological and biological, geological input need not be considered because there would be no transfer between adjacent watersheds.

In humid areas where surface wind is a minor factor losses from the system would be only geological and biological. Given an impermeable base, all the geological output would inevitably turn up in the streams draining the watershed. If the watershed is part of a larger and fairly homogeneous biotic unit, the biological output tends to balance the biological input because animals move randomly in and out of the watershed, randomly acquiring or discharging nutrients. Thus one need measure only the meteorological input and the geological output of nutrients in order to arrive at the net gain or loss of a given nutrient in the ecosystem. We measure the meteorological inputs to these watersheds by means of a network of gauging stations.

We measure the geological outputs by means of a weir built at the foot of each watershed, that is, at the point where the principal stream leaves the watershed. With the weir which also includes a ponding basin, one can both measure the water that is leaving the watershed and, by combining these data with frequent chemical measurements, ascertain the quantities of chemical substances that are leaving the watershed. In as much as the impermeable base prohibits deep seepage in these watersheds, the loss of water by evaporation and by transpiration through leaves is calculated by subtracting the hydrologic output from the hydrologic input.

Water budgets for the six watersheds from 1955 to 1968 indicate an average annual precipitation of 123 centimetres and a runoff of 72 centimetres. With evapotranspiration therefore averaging 51 centimetres. Precipitation is distributed rather evenly throughout the year, but runoff is uneven. Most of the runoff (57%), occurs during the snow-melt period of March, April and May; indeed, 35 percent of the total runoff occurs in April. In contrast, only .7 percent of the yearly runoff takes place in August. We accomplish chemical measurements by taking weekly samples of the water output (stream water at the weir) and collecting the total weekly water input (rain and snow) and analyzing them for calcium, magnesium, potassium, sodium, aluminum, ammonium, nitrate, sulphate, chloride, bicarbonate, hydrogen ion and silicate.

The concentrations of these elements in precipitation and in stream water are entered in a computing system, where weekly concentrations are multiplied by the weekly volume of water entering and leaving the ecosystem. In this way the input and output of chemicals is computed in terms of kilograms of an element per hectare of watershed. Knowing the input and output of chemicals, we have made nutrient budgets for nine elements. Considering four of the major ones, we find the following annual averages in kilograms per hectare entering the system and being flushed out of it; calcium, 2.6 and 11.8; sodium, 1.5 and 6.9; magnesium, .7 and 2.9 and potassium, 1.1 and 1.7. These inputs and outputs represent connections of the undistributed forest ecosystem with worldwide biogeochemical systems. The data also provide a comparative basis for judging the effects of managerial practices on biogeochemical cycle. Net losses of calcium, sodium and magnesium were recorded each year even though the period of measurement included wet and dry years as well as years of average precipitation.

Potassium, a major component of the bedrock, showed net gains in two years and a smaller average net loss than was recorded for

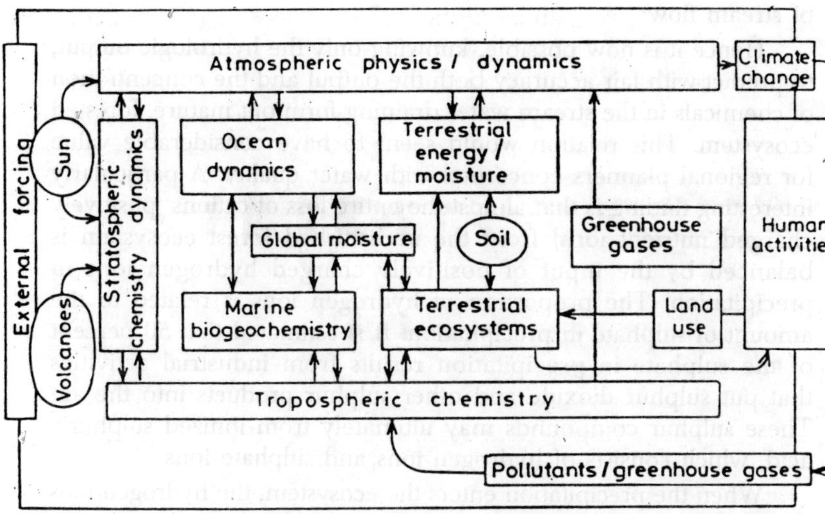

Fig. 7.2. Relations between biogeochemical cycles and physical climate system.

the other elements. Evidently potassium is accumulating in the ecosystem with respect to other elements. One reason may be that it is retained in the structure of illitic clays developing in the ecosystem. Perhaps also potassium is retained in proportionately larger amounts than other elements are in the slowly increasing biomass of the system. Highly predictable relations appear between the concentrations of the dissolved chemicals in stream water and the discharge rates of the stream. For example, the concentrations of sodium and silica are inversely related to discharge rates, whereas the concentrations of aluminum, hydrogen ion and nitrate increase as discharge rates increase. Magnesium, calcium, sulphate, chloride and potassium are relatively independent of discharge rate.

The magnitudes of change of concentration, however, are fairly small. The concentrations of potassium, calcium and magnesium hardly change at all, and the concentration of sodium decreases by only three times as the discharge rate increases by four orders of magnitude. These results were unexpected: we had thought that during the spring melt period there would be considerable dilution, making the concentrations of elements in stream water relatively low. All these relations show how strongly stream-water chemistry is under the control of processes inherent in the forest ecosystem. Because of the comparative constancy of chemical concentrations, the total output of elements is strongly dependent on the volume of stream flow.

Hence it is now possible, knowing only the hydrologic output, to predict with fair accuracy both the output and the concentration of chemicals in the stream water draining form out mature, forested ecosystem. This relation would seem to have considerable value for regional planners concerned with water quality. A particularly interesting finding is that almost the entire loss of cations (positively charged nutrient ions) from the undisturbed forest ecosystem is balanced by the input of positively charged hydrogen ions in precipitation. The proportion of hydrogen ions is related to the amount of sulphate in precipitation. It is estimated that 50 percent of the sulphate in precipitation results from industrial activities that put sulphur dioxide and other sulphur products into the air. These sulphur compounds may ultimately from ionized sulphuric acid, which consists of hydrogen ions and sulphate ions.

When the precipitation enters the ecosystem, the hydrogen ions replace the nutrient cations on the negatively charged exchange sites in the soil, and the cations are washed out of the system in stream water. Thus air pollution is directly related to a small but continuous loss of fertility from the land area of the ecosystem and a small but continuous chemical enrichment of streams and lakes. The relation appears to represent an important hidden cost of air pollution, made apparent through the analysis of ecosystems. So far we have mentioned only chemical losses appearing as dissolved substances.

Losses also arise when chemicals locked up in particulate matter such as rock or soil particles and in organic matter such as leaves and twigs are washed out of the ecosystem by the stream. We have measured these outputs and developed equations expressing the loss of particulate matter as a function of the discharge rate of the stream. The loss is highly dependent on the discharge rate. Losses of dissolved substances account for the great bulk of the chemical loss from our undisturbed ecosystem. Whereas they are largely independent of the discharge rate, losses of particulate matter are highly dependent on it. This point is of particular interest since forest management practices can either increase or decrease stream-discharge rates and thereby shift the balance between the loss of dissolved substances and the loss of particulate matter.

Weathering, or the release of elements bound in primary minerals, is another factor that must be considered in an ecosystem, since the elements thus released are made available as nutrients to the vegetation and animals. Based on net losses of elements from our ecosystem, a relatively uniform geology in the region and a knowledge of the bulk chemistry of the rock and soil, we estimate that the nutrients contained in some 800 kilograms per hectare of rock and soil are made available each year by watering. We now have for the undisturbed Northern hardwood ecosystem of Hubbard Brook estimates of chemical input in precipitation and output in stream water and the rates of generation of ions by the weathering of minerals within the system.

To complete the picture of nutrient cycling it is necessary to measures the nutrient content of the four compartments and the flow rates among them arising from uptake, decomposition and leaching and the formation of new minerals. The annual net loss of calcium from the ecosystem is 9.2 kilograms per hectare. This loss represents only about .3 percent of the calcium in the available nutrient and organic compartment of the system and only 1.3 percent of what is in the available-nutrient compartment alone: the data suggest that Northern hardwood forests have a remarkable ability to hold and circulate nutrients. It was against this background that we and the Forest Service embarked on the experiment of cutting down everything that was growing in one watershed.

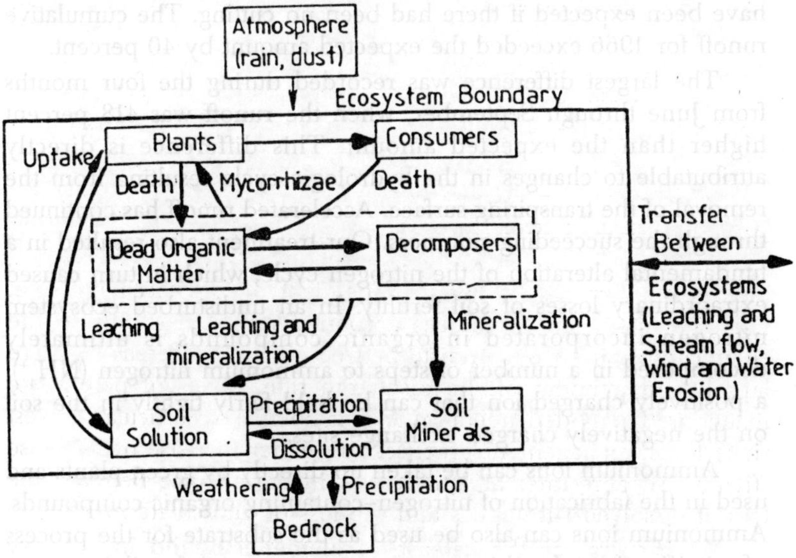

Fig. 7.3. A generalized diagram of mineral cycles in terrestrial ecosystems.

One of the objectives of this severe treatment was to block a major pathway of the ecosystem-the uptake of nutrients by higher plants-while the pathway of ultimate decomposition continued to function. We questioned whether or not the ecosystem had the capacity under these circumstances to hold the nutrients accumulating in the available-nutrient compartment. We also wanted to determine the effect of deforestation on stream flow, to examine some of the fundamental chemical relations of the forest ecosystem and to evaluate the effects of forest manipulation on nutrient relations and eutrophication of stream water. The experiment was begun in the winter of 1965-1966 when the forest of watershed NO_2, covering 15.6 hectares, was completely levelled by the forest service.

All trees, saplings and shrubs were cut and dropped in place; their limbs were removed so that no slash was more than 1.5 metres above the ground. No products were removed from the forest, and great care was taken to prevent disturbance of the surface of the soil that might promote erosion. The following summer regrowth of vegetation was inhibited by and aerial application of the herbicide Bromacil at a rate of 28 kilograms per hectare. Deforestation had a pronounce effect on runoff. Beginning in May, 1966 runoff from the cut watershed began to increase over the levels that would have been expected if there had been no cutting. The cumulative runoff for 1966 exceeded the expected amount by 40 percent.

The largest difference was recorded during the four months from June through September, when the runoff was 418 percent higher than the expected amount. This difference is directly attributable to changes in the hydrologic cycle resulting from the removal of the transpiring surface. Accelerated runoff has continued through the succeeding summers. Our treatment also resulted in a fundamental alteration of the nitrogen cycle, which in turn caused extraordinary losses of soil fertility. In an undisturbed ecosystem nitrogen incorporated in organic compounds is ultimately decomposed in a number of steps to ammonium nitrogen (NH_4^+), a positively charged ion that can be held fairly tightly in the soil on the negatively charged exchange sites.

Ammonium ions can be taken up directly by green plants and used in the fabrication of nitrogen–containing organic compounds. Ammonium ions can also be used as the substrate for the process of nitrification. In this process two genera of soil bacteria,

Nitrosomonas and Nitrobacter, oxidize ammonium to nitrate (NO_3^-). Two hydrogen ions are produced for every ion of ammonium oxidized to nitrate, as we have already mentioned, hydrogen ions can play a key role in the release of nutrient cations from the soil. Nitrate, being negatively charged, is highly leachable. If it is not taken up by higher plants, if can easily be removed from the ecosystem in drainage water. The available evidence indicates that nitrification is of minor importance in undisturbed forests such as ours, underlain by acid podzol soil.

The nitrate drained by our streams (invariably in low concentration) can be largely accounted for by its input in precipitation. In fact, our budgetary analyses show that undisturbed ecosystem are accumulating nitrogen at a rate of about two kilograms per hectare per year. The concentration of nitrate in stream water from undisturbed forests shows a seasonal cycle, being higher from November through April than it is from may through October. The decline in May and the low concentration in the summer are correlated with heavy demand for nutrients by the vegetation and generally increased biological activity associated with warming of the soil. Beginning in June, 1966, the concentration of nitrate in the deforested watershed rose sharply. At the same time the undisturbed ecosystem showed the normal spring decline.

The high concentration has continued in the deforested watershed during the succeeding years. Average net losses of nitrate nitrogen were 120 kilograms per hectare per year from 1966 through 1968. We estimate that the annual turn over of nitrogen in our undisturbed forests is about 60 kilograms per hectare. Therefore an amount of elemental nitrogen equivalent to double the amount normally taken up by the forest has been lost from the deforested watershed each year since cutting. The magnitude of nitrate loss is a clear indication of the acceleration of nitrification in the watershed. There is no doubt that the cutting drastically altered the conditions controlling the nitrification process. Another factor of interest about nitrification is the body of evidence from other regions that certain types of vegetations can inhibit nitrification chemically.

Presumably the effect is to inhibit production of the highly leachable nitrite ion. At the same time the positively charged ammonium ions may be held with in the system on the negatively charged exchange sites in the soil. If this inhibition process goes on at Hubbard Book, cutting the vegetation would promote nitrification. The effect may account for much of the nitrate loss from the deforested ecosystem. The export of nitrate to the small stream draining the cut watershed has resulted in nitrate concentrations exceeding the levels established by the U.S. Public Health Service for drinking water.

In general the deforestation has led to eutrophication of the stream and to the development of algal "blooms". This finding indicates that in some circumstances forest-management practices can contribute significantly to the eutrophication of streams Since nitrification produces hydrogen ions that replace metallic cations on the exchange surfaces in the soil, one would expect a loss of metallic nutrients from the deforested watershed. We have recorded substantial losses of this kind. The concentrations of calcium, magnesium, sodium and potassium in the stream water increased almost simultaneously with the increase in nitrate. About a month later the concentration of aluminium rose sharply.

Net losses of potassium were 21 times higher than those in an undisturbed watershed; of calcium, 10 times; of aluminium, nine; of magnesium, seven and of sodium, three. These figures represent a substantial loss of nutrients from the ecosystem. The finding suggests that commercial forestry should focus more on the effect of harvesting practices on the loss of nutrients, giving more consideration to such corrective procedures as selective cutting and the promotion of regrowth on cut areas. Our results indicate that the capacity of the ecosystem to retain nutrients is dependent on the maintenance of nutrient cycling within the system. When the cycle is broken, as by the destruction of vegetation, the loss of nutrients is greatly accelerated. This effect is related both to the cessation of nutrient uptake by plants and to the larger quantities of drainage water passing through the system.

The loss may also be related to increased rates of decomposition resulting from such changes in the physical environment as higher soil temperature and moisture soil. We also found a basic changes in the pattern of loss of particulate matter in the deforested watershed. The data for three years indicate an increase of some nine fold over a comparable undisturbed ecosystem. After an initial surge the loss of particulate organic matter has declined as a result of the virtual elimination of the production of primary organic matter in the ecosystem. In contrast, the loss of inorganic material from the stream bed has accelerated because of the greater erosive capacity of the augmented stream flow and also because several biological barriers to the erosion of surface soil and stream banks have been greatly diminished. The continuous layer of litter that once protected the soil surface is now discontinuous.

The extensive network of fine roots that tended to stabilize the stream bank is now dead, and the dead leaves that tended to plaster over exposed banks are gone. The erosive trend can be expected to rise exponentially as long as regrowth of vegetation is inhibited in the watershed. Our study clearly shows that the stability of an ecosystem is linked to the orderly flow of nutrients between the living and the nonliving components of the system and the production and decomposition of biomass. These processes, integrated with the seasonal changes in climate, result in relatively tight nutrient cycles within the system, a minimum output of nutrients and water and good resistance to erosion.

Destruction of the vegetation sets off a chain of interactions. Their net effect is an increase in the amount and flow rate of water and the breakdown of biological barriers to erosion and transportation, coupled with an increase in the export of nutrient capital and inorganic particulate matter. Three points, which are inherent in the ecosystem concept and are emphasized by the Hubbard Brook study, should be recognized as being basic to any wise scheme for managing the use of land. First, the ecosystem is highly complex natural unit composed of organisms (plants and animals, including man) and their inorganic environment (air, water soil and rock). Second all parts of an ecosystem are intimately linked by natural processes that are part of the ecosystem, such as the uptake of nutrients, the fixation of energy, the movement of nutrients and energy through food webs, the release of nutrients by the decomposition of organic matter, the weathering of rock and soil minerals to release nutrients and the formation of new minerals.

Third, individual ecosystems are linked to surrounding land and water ecosystems and to the biosphere in general by connections with food webs and the worldwide circulation of air and water. Failures in environmental management often result from such factors as failure to appreciate the complexity of nature, the assumption that it is possible to manage one part of nature alone and the belief that somehow nature will absorb all types of manipulation. Good management of the use of land-good from the viewpoint of society at large–requires that managerial practices be imposed only after a careful analysis and evaluation of all the ramifications. A focus for this type of analysis and evaluation is the ecosystem concept.

Nutrient Cycles

Nutrient Pools and Exchanges

Nutrients can be used as an organizing focus in ecosystem studies. We can view the biological community as a complex processor in which individuals more nutrients from one site to another within the ecosystem. These biological exchanges of nutrients interact with physical and chemical exchanges, and for this reason, nutrient cycles are also called biogeochemical cycles. All nutrients reside in compartments, which represents defined space in nature. Compartments can be defined very broadly or very specifically. It includes all of the plants in the ecosystem as one compartment, but we could recognize each species of plant as a separate compartment or even the leaves and the stem of a single plant as separate compartments.

A compartment contains a certain quantity, or pool, of nutrients. The phosphorus dissolved in the water is one pool, and the phosphorus contained in the herbivores is another pool. Compartments exchange nutrients, and thus we must measure the uptake and outflow of nutrients for each compartment. The rate of movement of nutrients between two compartments is called the flux rate and is measured as the quantity of nutrient passing from one pool to another per unit of time. The flux rate and pool sizes together define the nutrient cycle within any particular ecosystem.

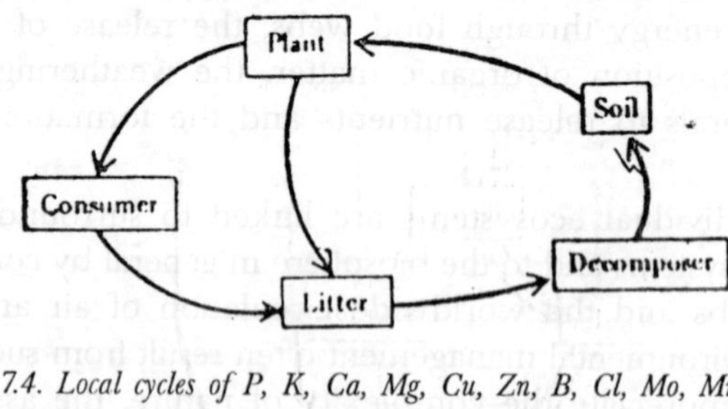

Fig. 7.4. Local cycles of P, K, Ca, Mg, Cu, Zn, B, Cl, Mo, Mn, and Fe.

Ecosystems are not isolated from one another, and nutrients come into an ecosystem through meteorological, geological or biological, transport mechanisms and leave an ecosystem via the same routes. Meteorological inputs include dissolved matter in rain and snow, atmospheric gases, and dust blown by the wind; geological inputs include elements transported by surface and subsurface drainage; and biological inputs include movements of animals between ecosystems. Nutrient cycles can be studied by the introduction of radioactive tracers into laboratory or natural ecosystems. Studies on the movement of radioactive phosphorus in small aquariums illustrate this approach. There are changes that followed the introduction of 100 microcuries (μc) of ^{32}P-labelled phosphoric acid in a 200-litre aquarium.

There is a very rapid initial uptake of ^{32}P by the phytoplankton. One-half of the ^{32}P had been taken up by the phytoplankton with in 2 hours, and within 12 hours there was an equilibrium of uptake and excretion of ^{32}P between the phytoplankton and the water. Filamentous algae attached to the sides and bottom of the aquarium slowly picked up ^{23}P, and crustaceans grazing on phytoplankton began to accumulate ^{32}P even more slowly. As the experiment progressed, an increasing fraction of the radio active tracer began to accumulate in the bottom mud and was tried up in less active or bound form the sediments. Some ^{32}P does more from the sediment back into the water column, but more moves down into the sediment and thus accumulates.

The nutrient cycle of phosphorus in aquariums is broadly similar to that in natural lakes. Phosphorus and other nutrients

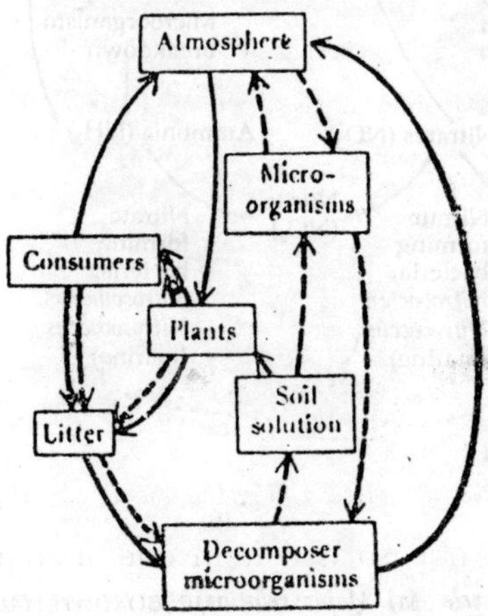

Fig. 7.5. Global cycles of C, N, O and H.

tend to accumulate in the sediment of lakes so that continual nutrient inputs are required to maintain high productivity. The pattern of movement of phosphorus helps to explain the design of

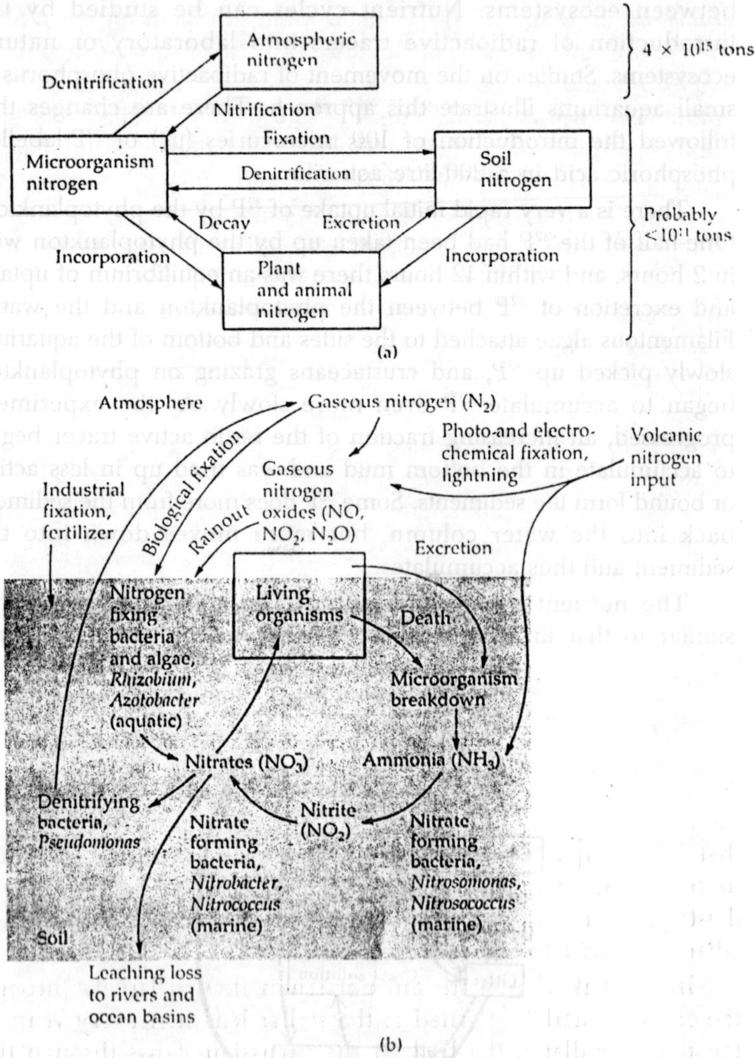

Fig. 7.6. The nitrogen cycle. (a) Major relationships between the very large atmosphere pool of gaseous nitrogen and the biosphere. (b) The complex interrelationships of the soil-based portion of the cycle.

the lake-fertilization experiments described. A continued input of phosphate in needed to sustain high availability for phytoplankton. These results are also critical for understanding how lakes can recover from the effects of nutrient additions from pollution. Thus the sediments of lakes become nutrient rich deposits. Nutrient cycles may be subdivided into two broad types. The phosphorus cycle we have just described is an example of a sedimentary or local cycle, which operates within an ecosystem.

Local cycles involve the less mobile elements that have no mechanism for long-distance transfer. By contrast, the gaseous cycles of nitrogen, carbon, oxygen, and water are called global cycles because they involve exchanges between the atmosphere and ecosystem. Global nutrient cycles link together all of the world's living organisms in one giant ecosystem called the biosphere, the whole earth ecosystem. The nitrogen cycle is a good example of a gaseous global cycle. Gaseous nitrogen is the most abundant element in the atmosphere, and the atmospheric N_2 provides a large reservoir for nitrogen-fixing organisms. The quantity of nitrogen tried up in living organisms is very small compared with the total capacity of the atmosphere. Almost all nitrogen available for plants comes form nitrogen-fixing bacteria or algae. In many tertestrial ecosystems, the fixation of nitrogen may limit plant growth, and there can be intense competition for soil nitrogen.

Nutrient Cycles in Forests

The harvesting of forest trees removes nutrients from a forest site, and this continued nutrient removal could result in a long-term decline in forest productivity unless nutrients are somehow returned to the system. Because of the economic question of forest productivity, an increasing amount of work is being directed toward the analysis of nutrient cycles in forests. Figure shows the factors that must be quantified in order to describe the nutrient cycle in a forest. Some examples of nutrient cycles in forest stands will illustrate these concepts. Figure shows the potassium cycle in two adjacent woodlands in Great Britain.

In the oak woodland, the potassium flow is largely through the oak trees and is returned to the soil as leaf litter every year. In the pine woodland, the bulk of the potassium flows through the ground flora (particularly bracken fern), and the pines take up relatively little potassium. Thus different species of trees have different nutrient demands, in the same way that different agricultural

Table 7.1. Biomass of Trees

		Fertilizer Treatment							
		Control	*PK*	*N_1*	*N_2PK*	*N_2^a*	*N2PK*	*N_4^b*	*N4PK*
Mean height (cm)	1956	186	163	158	181	167	172	165	172
Stem volume (o.b.m.^{3})^c.	1960	27	20	26	35	31	33	31	33
	1970	143	125	130	171	145	160	137	152
Dry weight (kg/ha)									
Stem wood	1960	8,520	6,920	8,740	10,530	9,680	10,760	10,340	10,700
	1970	40,060	35,380	38,490	49,090	43,520	41,570	41,860	45,740
Stem bark	1960	1,720	1,400	1,760	2,120	1,960	2,170	2,100	2,160
	1970	6,400	5,720	6,260	7,760	6,970	7,560	6,770	7,260
Branches	1960	7,510	6,570	9,200	10,480	9,920	10,880	10,610	10,680
	1970	12,860	11,190	11,020	15,190	12,470	14,490	12,110	13,980
Needles	1960	8,680	7,560	10,840	12,310	11,670	12,790	12,480	12,540
	1970	13,400	11,770	11,580	15,530	12,940	14,900	12,680	14,360
Total above stumps	1960	26,430	22,450	30,540	35,540	33,230	36,600	35,530	36,080
	1970	72,720	64,060	67,350	87,570	75,900	84,520	73,420	81,340
Stumps	1970	2,390	2,160	2,380	2,880	2,610	2,820	2,570	2,700
Roots>5mm diam.	1970	11,070	9,180	9,850	13,150	10,910	12,130	10,780	12,170
Roots<5mm diam.	1970	4,740	4,740	4,740	4,740	4,740	4,740	4,740	4,740
Total biomass	1970	90,920	81,140	84,320	108,340	94,160	104,210	91,510	100,950

Note: Plots were fertilized annually with nitrogen alone or with mixtures of nitrogen, phosphorus, and potassium starting in 1957. $^a N_2$= twice the amount of nitrogen as N_1. $^b N_4$=four times the amount of nitrogen as N_1. o.b.m^3=Volume "outside the bark," i.e., including the bark and the stemwood.

crops have different nutrient requirements. Fertilization experiments in forests have provided additional data on nutrient cycling within forest sites.

Some forest sites respond to artificial fertilization by increasing production, but other sites do not. Tamm (1975) describes a 14-year fertilization experiment on a Norway spruce stand in Sweden. The addition of nitrogen caused the trees to build a crown of green needless more rapidly, so that during the first few years of the experiment, tree growth was improved, but the effects of nitrogen addition were nullified as time went on. Stem growth was reduced at high nitrogen levels, and total primary production differed little among treatments (Table 7.1). In this forest stand, the nitrogen cycle was operating efficiently, and thus nitrogen was not limiting primary production of the trees. One of the most extensive studies of nutrient cycling in forests has been carried out at the Hubbard Brook Experimental Forest in New Hampshire.

The Hubbard Brook forest is a nearly mature, second-growth hardwood ecosystem. The area is underlain by rocks that are relatively impermeable to water, so all runoff occurs in small streams. The area is subdivided into several small watersheds that are distinct yet support similar forest communities, and these watersheds are good experimental units for study and manipulation. Nutrients enter the Hubbard Brook forest ecosystem in precipitation, and the precipitation input was measured in rain gauges scattered over the study area.

Nutrients leave the ecosystem primarily in stream runoff, and this loss was estimated by measuring stream flows. Table 7.2 gives the concentration of dissolved substances in precipitation and streamwater for the Hubbard Brook watersheds. For most dissolved nutrients, the streamwater leaving the system contains more nutrients than the rainwater entering the system. About 60 percent of the water that enters as precipitation turns up as stream flow; most of the remaining 40 percent is transpired by plants or evaporated. The chemical composition of the precipitation and the stream discharges changed very little from year to year.

Table 7.3 gives the annual nutrient budgets for watersheds in the Hubbard Brook system, based on the difference between precipitation input and stream outflow. Eight elements show net losses from the ecosystem: calcium, magnesium potassium, sodium, aluminium, sulphate, silica, and bicarbonate. Three elements showed and average net gain: nitrate, ammonium, and chloride. If we assume that these nutrient budgets should be in equilibrium in this undisturbed ecosystem, the net losses must be made up by chemical decomposition of the bedrock and soil. With this background, Bormann et al., studied the effect of logging on the nutrient budget of a small watershed at Hubbard Brook. One 15.6-hectare watershed was logged in 1966, and the logs and branches were left on the ground so that nothing was removed from the area.

Table 7.2. Average Concentrations of Various Dissolved Substances in Bulk Precipitation and Streamwater for Undisturbed Watersheds

	Precipitation (mg/L)	*Streamwater (mg/L)*
Calcium	0.21	1.58
Magnesium	0.06	0.39
Potassium	0.09	0.23
Sodium	0.12	0.92
Aluminium	([a])	0.24
Ammonium	0.22	0.05
Sulphate	3.10	6.40
Nitrate	1.31	1.14
Chloride	0.42	0.64
Bicarbonate	([a])	1.90[b]
Dissolved silica	([b])	4.61

[a] Not determined, but very low.
[b] Watershed 4 only.
Source : After Likens and Bormann (1972).

Great care was taken to prevent disturbance of the soil surface to minimize erosion. For the first three years after logging the area was treated with a herbicide to prevent any regrowth of vegetation. This deforested water shed was then compared with an adjacent intact watershed. Runoff in the small streams increased immediately after the logging, and annual runoff in the deforested watershed was, respectively, 41, 28, and 26 percent above the control in the 3 years after treatment. Detritus and debris in the stream outflow increased greatly after deforestation (Table 7.4), particularly 2 to 3 years after logging. Correlated with this was a large increase

Table 7.3. Annual Nutrient Budgets (kg/ha) for Watersheds

	1963-1964	1964-1965	1965-1966	1966-1967	1967-1968	1968-1969	Mean 1963-1969
1	2	3	4	5	6	7	8
Calcium							
Input	3.0	2.8	2.7±0.07	2.7±0.03	2.8±0.05	1.6±0.02	2.6
Output	12.8±0.8	6.3±0.4	11.5±0.6	12.3±0.7	14.2±0.7	13.8±1	11.8
Net	–9.8	–3.5	–8.8	–9.6	–11.4	–12.2	–9.2
Magnesium							
Input	0.7	1.1	0.7±0.02	0.5±0.005	0.7±0.013	0.3±0.004	0.7
Output	2.5±0.06	1.8±0.09	2.9±0.07	3.1±0.08	3.7±0.08	3.3±0.12	2.9
Net	–1.8	–0.7	–2.2	–2.6	–3.0	–3.0	–2.2
Potassium							
Input	2.5	1.8	0.6±0.02	0.6±0.008	0.7±0.01	0.6±0.009	1.1
Output	1.8±0.1	1.1±0.08	1.4±0.09	1.7±0.1	2.2±0.15	2.2±0.16	1.7
Net	+0.7	+0.7	–0.8	–1.1	–1.5	–1.6	–0.6
Sodium							
Input	1.0	2.1	2.0±0.04	1.3±001	1.7±0.03	1.1±0.02	1.5
Output	5.9±0.3	4.5±0.3	6.9±0.5	7.3±0.6	9.1±0.3	7.6±0.6	6.9
Net	–4.9	–2.4	–4.9	–6.0	–7.4	–6.5	–5.4
Aluminium							
Input	–	(*)	(*)	(*)	(*)	(*)	(*)
Output	–	1.2±0.18	1.7±0.75	1.9±0.87	2.1±1.00	2.2±1.04	1.8
Net	–	–	–1.7	–1.9	–2.1	–2.2	–1.8

Ammonium							
Input	–	2.1	2.6±0.06	2.4±0.04	3.2±0.06	3.1±0.07	2.7
Output	–	0.27±0.03	0.92±0.03	0.45±0.07	0.24±0.02	0.16±0.06	0.4
Net	–	+1.83	+1.7	+2.0	+3.0	+2.9	+2.3
Nitrate							
Input	–	6.7	17.4±0.3	19.9±0.2	22.3±0.3	15.3±0.3	16.3
Output	–	5.6±0.4	6.5±0.02	6.6±0.4	12.7±0.3	12.2±04	8.7
Net	–	+1.1	+10.9	+13.3	+9.6	+3.1	+7.6
Sulphate							
Input	–	30.0	41.6±0.3	42.0±0.3	46.7±0.3	31.2±0.3	38.3
Output	–	30.8±0.4	47.8±0.4	52.5±0.4	58.5±0.4	53.3±0.3	48.6
Net	–	–0.8	–6.2	–10.5	–11.8	–22.1	–10.3
Dissolved Silica							
Input	–	(*)	(*)	(*)	(*)	(*)	(*)
Output	–	20.8±0.1	36.1±4.8	41.6±4.8	42.1±5.7	55.0±6.0	35.1
Net	–	–	–36	–42	–42	–35	–35
Bicarbonate							
Input	–	–	(*)	(*)	(*)	(*)	(*)
Output	–	–	12.0	16.8	16.5*	13.1*	14.6
Net	–	–	–12	–17	–17	–13	–14.6
Chloride							
Input	–	–	2.6±0.04	6.7±0.12	5.0±0.09	6.4±0.05	5.2
Output	–	–	4.3±0.05	4.8±0.1	5.3±0.01	5.2±0.08	4.9
Net	–	–	–1.7	–1.9	–0.3	+1.2	+0.3

Note: Error limits are 1 standard deviation of the mean.
*Not measured, but very small.

Table 7.4. Annual Losses of Particulate Matter in Streams on Watershed 6 (Control and Watershed 2 (Deforested) at Hubbard Brook, New Hampshire

Source of Output	Watershed 6 (Undisturbed Forest)			Watershed 2 (Deforested)		
	Organic	*Inorganic*	*Total*	*Organic*	*Inorganic*	*Total*
1	2	3	4	5	6	7
1965-1966						
Ponding basin	2.12	1.77	3.89	5.97	7.16	13.13
Net	0.34	0.01	0.35	0.19	0.00	0.19
Filter	1.37	1.28	2.65	1.44	1.44	2.88
Total	3.83	3.06	6.89	7.60	8.60	16.20
1966-1967						
Ponding basin	13.41	17.07	30.48	24.96	41.83	66.79
Net	0.39	0.01	0.40	0.27	0.01	0.28
Filter	2.72	2.95	5.67	4.81	5.49	10.30
Total	16.52	20.03	36.55	30.04	47.33	77.37
1967-1968						
Ponding basin	3.83	5.93	9.76	28.56	63.38	91.94
Net	0.43	0.01	0.44	0.28	0.01	0.29
Filter	2.61	2.82	5.43	4.59	5.12	9.71
Total	6.87	8.76	15.63	33.43	68.51	101.94

1968-1969						
Ponding basin	4.61	8.31	12.92	36.31	158.34	194.65
Net	0.42	0.01	0.43	0.26	0.01	0.27
Filter	2.57	2.81	5.38	4.21	4.74	8.95
Total	7.60	11.13	18.73	40.78	164.09	203.87
1969-1970						
Ponding basin	11.28	30.67	41.90	45.16	320.15	365.31
Net	0.40	0.01	0.41	0.25	0.01	0.26
Filter	3.30	3.69	6.99	6.16	7.07	13.23
Total	14.98	34.37	49.30	51.57	327.23	378.80
Average per Year Based on 5 Years						
Ponding basin	7.05	12.75	19.79	28.19	118.17	146.36
Net	0.40	0.01	0.41	0.25	0.01	0.26
Filter	2.51	2.71	5.22	4.24	4.77	9.01
Total	9.96	15.47	25.42	32.68	122.95	155.63

Notes :Data are kilograms of oven-dry weight of materials per hectare of watershed. Losses are separated by size as ponding basin (coarse materials), net (finer materials), and filter (very fine materials).

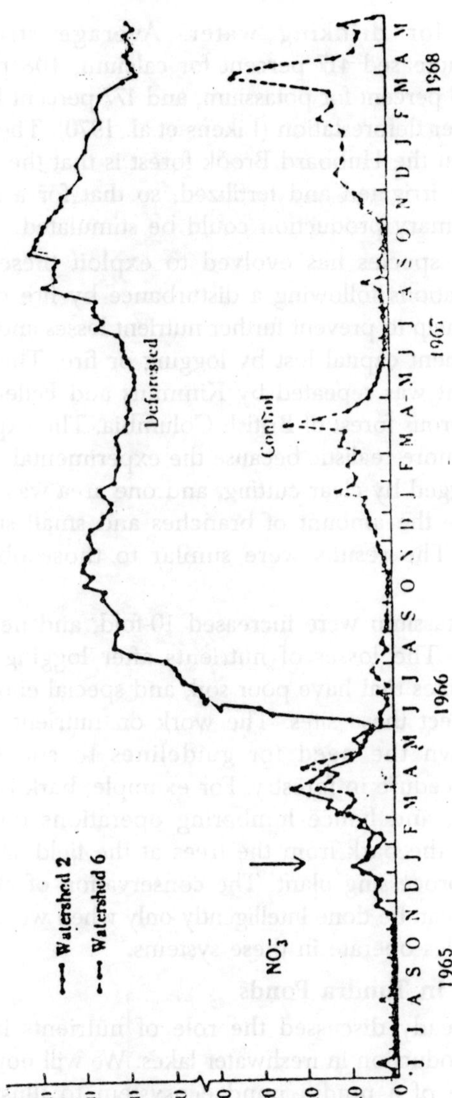

Fig. 7.7. Streamwater concentrations of nitrate in two watersheds at the Hubbard Brook Experimental Forest, New Hampshire. The arrow marks the completion of the cutting of the trees on watershed.

in streamwater concentrations of all major ions in the deforested watershed.

Nitrate concentrations in particular increased 40-to 60-fold over the control values. For 2 years the nitrate concentration in the streamwater of the deforested site exceeded the health levels recommended for drinking water. Average streamwater concentrations increased 417 percent for calcium, 408 percent for magnesium, 1558 percent for potassium, and 177 percent for sodium in the 2 years after deforestation (Likens et.al. 1970). The net result of deforestation in the Hubbard Brook forest is that the ecosystem is simultaneously irrigated and fertilized, so that for a short time after logging, primary production could be stimulated.

An array of species has evolved to exploit these transient nutrient-rich situations following a disturbance by fire or logging. These transients help to prevent further nutrient losses and to restore some of the nutrient capital lost by logging or fire. The Hubbard Brook experiment was repeated by Kimmins and Feller (1976) in the coastal coniferous forest of British Columbia. The experimental conditions were more realistic because the experimental areas were commercially logged by clear cutting, and one area was later slash burned to reduce the amount of branches and small stems lying on the ground. The results were similar to those observed at Hubbard Brook.

Losses of potassium were increased 10-fold, and net losses of nitrate occurred. The losses of nutrients after logging could be severe on forest sites that have poor soil, and special efforts should be taken to protect these sites. The work on nutrient cycling in forests has shown the need for guidelines to specify sound management procedure in forestry. For example, bark is relatively rich in nutrients, and hence lumbering operations ought to be designed to strip the bark from the trees at the field site and not at some distant processing plant. The conservation of nutrients in forest ecosystems can be done intelligently only when we understand how nutrient cycles operate in these systems.

Nutrient Cycles in Tundra Ponds

We have already discussed the role of nutrients in limiting phytoplankton production in freshwater lakes. We will now describe the carbon cycle of a tundra pond ecosystem to illustrate how nutrient cycling can be described in detail for an aquatic system. We will also use this example to illustrate the steps involved in the construction of a systems model for part of the tundra pond ecosystem. The arctic coastal plain of northern Alaska contains thousands of small ponds, typically about 50 metres in diameter and 20 cm deep. They are frozen for 9 months of the year and

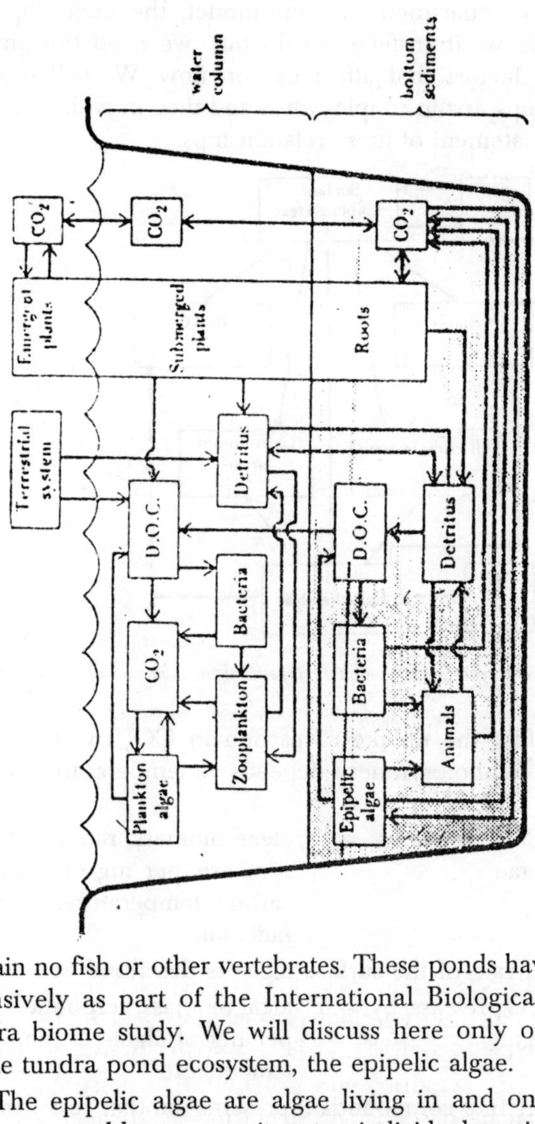

Fig. 7.8. Major biological compartments and carbon-flow pathways of a tundra pond ecosystem. D.O.C. is dissolved organic carbon.

contain no fish or other vertebrates. These ponds have been studied intensively as part of the International Biological Programme's tundra biome study. We will discuss here only one part of this whole tundra pond ecosystem, the epipelic algae.

The epipelic algae are algae living in and on the sediments and are treated here as a unit, not as individual species. Live epipelic algal cells are distributed throughout the upper 5 cm of sediments in the ponds, but photosynthesis is limited to the top few millimetres because of light extinction in the sediment. For this reason, the epipelic algae have been divided into two compartments in the systems model–surface algae and buried algae. Once we have identified the compartments in our model, the next step is to quantify the arrows (transfers). To do this, we need the physical and biological factors that affect carbon flow. We will describe these relationships first in simple words and then provide a detailed mathematical statement of these relationships.

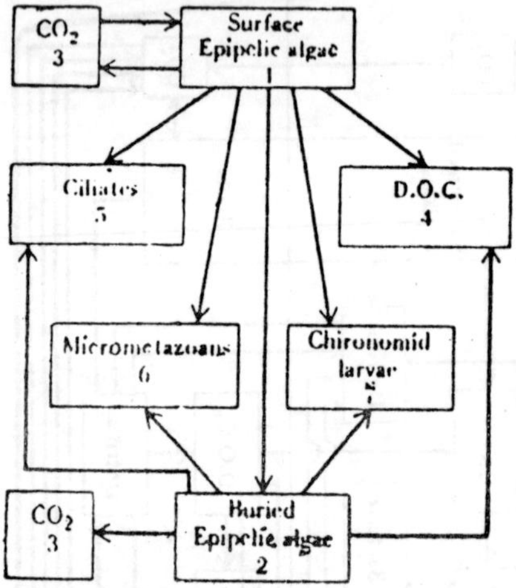

Fig. 7.9. Compartment representation of the epipelic algae submodel of the tundra pond ecosystem.

Consider first the uptake of carbon in CO_2 by the surface algae. The rate of photosynthesis depends on temperature and solar radiation:

1. Rate of CO_2 uptake by surface algae = f algae biomass, rate of photosynthesis per mg of algal carbon, temperature, solar radiation

The respiration rate of the surface algae is described as

2. Rate of CO_2 release by surface algae = f algal biomass, respiration rate per mg of algal carbon, temperature

Excretion of organic compounds represents another loss from the algae and a gain for this solved organic carbon (DOC) compartment. This process can be described by the equation

3. Rate of release of dissolved organic carbon by surface algae = f rate of phtosynthesis, algal biomass

Algae at the sediment surface are continually transported downward into the sediment by a variety of physical disturbances, primarily from aquatic invertebrates burrowing in the mud:

4. Rate of burial of surface algae	= f	algal biomass, density of aquatic invertebrates, activity of benthic invertebrates

Grazing by benthic invertebrates removes some of the epipelic algae and can be described by the general equation

5. Rate of grazing of epipelic algae	= f	invertebrate biomass, algal biomass, temperature

This rate of grazing has to be applied for each of the three main classes of invertebrate grazers Finally, we can calculate the rate of change in biomass of the surface and the buried algae as follows:

6. Rate of change in surface algal carbon	=	(gross photosynthesis)–(respiration)– re-excretion)–(burial)–(grazing losses)
Rate of change in buried algal carbon	=	(burial)–(respiration)–(excretion)–(grazing losses)

These questions are given in more precise mathematical form in Table 7.5. The main point to note is that in this systems model we have achieved a mathematical description of each of the arrows shown in Figure. We can now put all these equations together and see if they provide an accurate description of the system.

The field observation on three tundra ponds, the model's predictions based on the equations in Table 7.5, and the temperature and solar radiation data measured in the field. It is clear that the model does provide an accurate description of carbon flow into the epipelic algae compartment of these tundra ponds. Two steps need to be taken next. First, this model is only part of a larger model, which needs to be integrated together. Second, the model is presently not capable of dealing with the effects of increased nutrient-loading on the tundra ponds.

Because this model has been constructed for carbon flow, we do not know how to translate increased loadings of phosphorus or nitrogen into the carbon cycle. Note that in its present form, this model (Table 7.5) does not contain any explicit reference to phosphorus, nitrogen, or any other nutrient. The interrelationships of nutrient cycles are poorly understood, and more experimental studies on controlling nutrients must be done.

Table 7.5. Detailed Equation of the System Model of the Epipelic Algal Compartment of a Tundra Pond

Definition: X_i = biomass of carbon in the ith compartment

Gij = transfer rate of carbon (mg C/hr) from compartment i to compartment j.

T = temperature (°C) on surface of sediment, measured as degrees above the threshold temperature of 5°C

R = solar radiation (langleys /hr)

S = solar radiation level at which photosynthesis rate of algae is half of maximum.

(1) $G_{3,1} = (X_1)(0.05)(2.2^{t/10})\{R/(R+S)\}$

(2) $G_{1,3} = (X_1)(0.01)(2.0^{t/10})$

$G_{2,3} = (X_2)(0.01)\{2.0^{(t-2/10)}\}$

(3) $G_{1,4} = (0.02)(G_{3,1})+(0.001)(X_1)$

$G_{2,4} = (0.001)(X_2)$

(4) $G_{1,2} = (X_1)(0.0001)(X_7)\{3.0^{(t-8)/10}\}$

(5) $G_{1,5} = (X_5)(0.0025)\{2.0^{(t-8)/10}\}\{X_1/(X_1+X_2)\}$

$G_{2,5} = (X_5)(0.0025)\{2.0^{(t-8)/10}\}\{X_8/(X_1+X_2)\}$

$G_{1,6} = (X_6)(0.003)(X_1)(1.0)\{2.0^{(t-8)/10}\}$

$G_{1,7} = (X_7)(0.003)(X_1)(1.0)\{2.0^{(t-8)/10}\}$

(6) $dX_1/dt = G_{3,1}-G_{1,3}-G_{1,4}-G_{1,8}-G_{1,5}-G_{1,6}-G_{1,7}$

$dx_2/dt = G_{1,2}-G_{2,3}-G_{3,4}-G_{2,5}.$

Notes: This system model for the epipelic algae is provided with hourly data on temperature and radiation, and the starting value of the surface algae (40 mg carbon/m^2) and buried algae (80 mg carbon/ m^2) are given for the beginning of the summer season. The equation are then applied sequentially for each hour of each day during the growing season to provide the results. The equations here are numbered as they are in the text.

Nutrient Recovery Hypotheses

One of the most striking biological events on tundra areas of North America and Eurasia is the lemming cycle. Every 3 to 4 years these small rodents build up to high densities, only to decline and become rare again in a never-ending cycle. This biological rhythm has been studied in detail on the arctic coastal-plain tundra near point barrow, Alaska. Lemmings exert a dominant effect on the tundra ecosystem at point Barrow and cause striking yearly changes in primary production, nutrient concentrations in plants, decomposition rates, and abundances of vertebrate predators. From early observation on the effects of lemmings on the tundra ecosystem of arctic Alaska, Pitelka (1964) and Schultz (1964) proposed the nutrient-recovery hypothesis to explain the lemming cycle. This hypothesis linked together in an ingenious fashion the entire tundra ecosystem and has formed the basis of an extensive analysis of the Alaskan tundra as part of the international Biological Programme (Brown et al. 1980). The arctic tundra ecosystem can be described as a set of interlocking gears, or feedback loops, in which each compartment either counteracts of amplifies the change of state of the next compartment.

Thus, for example, high nutritional quality of forage may help the lemming population to increase, but by grazing, a high-density lemming population will reduce the plant biomass. The feedback loop model of the arctic tundra is oversimplified but makes an important point-ecosystem compartments are usually part of several interlocking feedback loops. Many different casual pathways can be traced, and no one factor governs the system. The ecosystem of the arctic coast of Alaska is simple compared with more temperate or tropical ecosystem, There are about 100 species of vascular plants, but 10 species comprise 90 percent of the plant biomass. There is only one major herbivore, the brown lemming, whose diet is largely comprised of three species of vascular plants.

There are two major bird predators and six minor mammal and bird predators that eat lemmings. The tundra ecosystem is underlain by perennially frozen soil, or permafrost, and only the top 25 to 40 cm of soil thaws during the short summer. The permafrost stops water movement down into the soil, and the surface layers of the tundra are often saturated with water during the summer. Nutrient cycles in the tundra ecosystem contrast sharply with those in temperate and tropical habitats. The total amount of carbon in the tundra ecosystem of northern Alaska is about the same as that in tropical rain forest. But only about 2 percent of the carbon is held in living materials in the tundra, compared with 64 percent in living organisms in a tropical rain forest.

The same conclusion seems to hold for nitrogen and phosphorus. In the northern Alaska tundra, 96 percent of the carbon is bound up in peat, and the activity of decomposers seems to limit the rate of nutrient cycling in tundra ecosystem. Of the nutrients held in living tissue in the tundra ecosystem, the majority is held underground. About 75 percent of the carbon in living plants is tied up in roots, rhizomes, and stem bases. Nutrient cycles in the arctic coastal tundra are not in a steady state because organic matter is slowly accumulating as peat. Little nutrient input occurs though precipitation at point Barrow since precipitation averages only 110 mm per year. Because of the flat terrain, there is little runoff, and most of the precipitation evaporates. Thus the tundra ecosystem shows extreme conservation of nutrients, in contrast to temperate forest ecosystems such as Hubbard Brook.

Primary production in the arctic coastal tundra near Barrow is restricted to the short summer season by temperature, but within the growing season, production is limited by nutrients (Schultz 1969). Fertilization of 6 acres of tundra with nitrogen, phosphorus, potassium, and calcium increased primary production to three to four times that of control plots. The calcium and phosphorus levels of green plants increased dramatically and remained high on the fertilized plot. The fertilization experiment clearly demonstrated that in spite of the high nutrient capital in the tundra ecosystem (Table 7.6), little is directly available to plants. Lemmings have an important effect on nutrient cycles in the Barrow tundra ecosystem, and this has been the basis of the nutrient-recovery hypothesis. The interactions proposed by Schultz (1969) between the lemming, vegetation, soil compartments of the ecosystem are as follows:

1. Intensive grazing occurs during the winter buildup of lemming numbers, and this releases a pulse of nutrients on the tundra surface by decomposition of litter, urine, and feces.
2. These nutrients are rapidly leached and absorbed by plants after snow melts and cause high nutrient concentration early in the growing season.
3. Grazing removes plant materials, and thus reduces insulation of soil, allowing more radiation to warm the soil so that the depth of soil thaw increases.
4. Plant roots are allowed to penetrate lower in the soil, where they are exposed to lower concentrations of nutrients, and this results in lower nutrient uptake by plants in late summer.
5. Low nutrient uptake by plants results in poor-quality forage in the summer after a lemming density peak.
6. Lemming density declines for reasons unrelated to nutrient cycling perhaps because of predation, social strife, or some other factors.

Table 7.6. Total Quantity of Carbon, Nitrogen, and Phosphorus to a Depth of 20 cm Present in Various Ecosystem and the Percentage of Each Element Contained in Vegetation

Ecosystem	*Latitude*	*Site*	*Total Quantity in System (g/m2)*			*Presence in Vegetation (%)*		
			Carbon	*Nitrogen*	*Phosphorus*	*Carbon*	*Nitrogen*	*Phosphorus*
Wet meadow tundra	71°	Barrow, Alaska	19,760	972	63.4	1.7	1.0	1.5
Tussock tundra	65°	Eagle Cr., Alaska	18,767	964	48.7	0.6	0.4	0.3
Subalpine Heath	47°	Washington						
Residual soil			13,379	979	–	2.2	0.5	–
Organic soil			32,096	1,276	–	2.7	0.9	–
Grassland								
Pawnee	40°	Colorado	–	140	–	–	5.5	–
Cotton-wood	43°	South Dakota	–	275	–	–	4.0	–
Osage	36°	Oklahoma	–	261	–	–	2.1	–
Red alder (20 yr)	65°	Fairbanks, Alaska	4,276	154	146	76.1	36.0	3.3
Red alter (35 yr)	46°	Cascades, Wash,	19,527	701	–	49.3	9.9	–
Douglas fir forest (35 yr)	46°	Cascades, Wash	12,276	160	166	73.7	20.4	4.0
Oak forest	35°	Oklahoma	12,884	250	37.7	75.0	42.3	25.2
Tropical rain forest	18°	El Verde, Puerto Rico	16,966	–	–	64.1	–	16.7

7. Breeding by lemmings is inhibited in the year following peak density because of low nutrient supplies, particularly phosphorus, in the forage.
8. During the next 2 to 3 years the process is reversed: vegetation slowly recovers, and the standing crop of plant matter increases. Depth of thaw decreases due to increasing insulation, and active roots are confined to the surface soil. Nutrient quality of the forage increases, and lemming breeding improves.

A considerable amount of evidence supports some of the causal links in the nutrient-recovery hypothesis, but others now seem doubtful. Phosphorus changes in the forage at Barrow do, indeed, rise and fall with lemming densities. Heavy grazing by lemmings in the peak year reduces the standing crop of grasses and sedges about 50 percent, and this decrease in the plant canopy does increase the depth of thaw in the soil by approximately 20 to 30 percent. But the increased depth of thaw by itself dose not seem to reduce nutrient content of the forage as suggested in points 4 and 5 in the list, and this part of the nutrient-recovery hypothesis seems to be in error.

Nevertheless, nutrients in vegetation in the coastal tundra are often reduced in the years after a lemming peak, and the possibility exists that these nutrient levels affect lemming reproduction during the winter. In summer, primary production exceeds the lemmings requirements, and lack of food cannot explain the continued population decline. The value of the nutrient-recovery hypothesis is that it has focused on the entire ecosystem and has attempted to integrate the complex interchanges that occur through the compartments of this ecosystem. The arctic coastal tundra is a relatively simple ecosystem, but its nutrient cycles are still difficult to disassociate and measure. The dominant dynamic feature of the tundra system is its low energy input, and this affects the biological community mostly through the physical effects of permafrost. Permafrost affects the availability of nutrients, which in turn affect primary and secondary production and thus ultimately the rates of nutrient cycling.

Efficiency of Nutrient Use

Large areas of the Northern Hemisphere have been glaciated, and the soil, derived from till in which the bedrock has been pulverized are very fertile, with a high availability of nutrients.

Areas of volcanic activity can also have rich soils. But in much of the world, soils are very old, highly weathered, and basically infertile. For example, the continents derived from Gondwana land–Australia, South America, and India–have large areas covered with very old, poor soils.

The vegetation supported on these soils has adapted remarkably well to efficient nutrient use by recycling within the plant and by leaf fall and reabsorption. Australian soils are typical of very old, highly weathered soils and contain almost no phosphorus. Eucalyptus are adapted to grow on soils of low phosphorus content (Attiwill 1981) Table 7.7 gives the biomass and nutrient content of forest ecosystems from temperate areas growing on good soils and from Australian sites on poor soils. There is no suggestion from Table 7.7 that eucalyptus growing on poor soils have lower amounts of nutrients than forests in other countries, with the single exception of phosphorus.

Eucalyptus have only one-half to one-fifth the amount of phosphorus in their tissues as do Northern Hemisphere Forests. Plants growing in nutrient-poor soils, one would think, ought to use nutrients more efficiently and contain less nutrients than plants in fertile soils. In fact, the opposite is true. Plants from infertile habitats consistently have higher nutrient concentrations than plants from fertile habitats when grown under the same controlled conditions. Plants from nutrient-poor habitats may achieve this by being more efficient than plants from nutrient-rich habitats (Vitousek 1982). An abundance of data is available from forests to investigate nutrient use efficiency. For a forest we can define:

$$\text{Nutrient use efficiency} = \frac{\text{grams of organic matter lost from plants}}{\text{grams of nutrient lost}}$$

These are two measures for a variety of forest from tropical and temperate areas. Two extreme hypothesis are the line of equal of nutrient-use efficiency and constantly decaying nutrient-use efficiency. The data fit neither extreme hypothesis but fall closer to line Nutrient-use efficiency is clearly above expectations in habitats with low nitrogen levels and below expectations in habitats rich in nitrogen. One consequence of this is that forest productivity may be high on soils with low nutrient levels. A classic example is the tropical rain forest of the Amazon basin, which represents one type of nutrient cycling strategy.

Table 7.7. Above ground Living Biomass and its Nutrient Content of Temperate Forest and Australian Eucalyptus Forest Ecosystems

Forest	*Location*	*Age (yr)*	*Biomass (t/ha)*	*Nutrient Content (kg/ha)* N	P	K	Na	Mg	Ca
1	*2*	*3*	*4*	*5*	*6*	*7*	*8*	*9*	*10*
Coniferous									
1. Abies amabilis-Tsuga meretensiana	Washington	175	469	372	67	980	–	160	1046
2. Abies balsamea–Picea rubens	Quebec	various	132	387	52	159	–	36	413
3. Abies mayriana	Japan	?	129	527	63	273	–	415	515
4. Cryptomeria japonica	Japan	?	114	386	36	244	–	118	314
5. Larix leptolepis	Japan	?	100	230	28	212	–	40	107
6. Picea abies	England	47	140	331	37	161	5	39	212
7. Pieca abies	Sweden	55	308	770	87	437	38	69	459
8. Picea glauca	Minnesota	40	151	383	58	230	–	41	720
9. Picea mariana	Quebec	63	107	167	42	84	–	27	276
10. Pinus banksiana	Ontario	30	81	171	15	85	–	19	114
11. Pinus banksiana	Minnesota	40	150	276	27	106	–	40	226
12. Pinus muricata	California	88	409 (mean)	510	418	321	83	140	589
13. Pinus radiata	New Zealand	35	305	319	40	324	–	–	187
14. Pinus resinosa	Minnesota	40	204	373	46	205	–	62	335

15.	Pinus sylvestris	Southern Finland	28	21	82	9	40	–	–	50
			45	79	194	19	98	–	–	115
			47	45	136	14	58	–	–	63
16.	Pinus sylvestris	England	47	164	364	34	314	8	43	204
17.	Pseudotsuga menziesi	Washington	36	173	326	67	227	–	–	342
18.	Pseudotsuga menziesi	Oregon	450	540	371	49	265	–	–	664
19.	Sequoia sempervirens	California (slopes)	?	1155	1474	616	941	102	232	1068
20.	Sequoia sempervirens	California (flats)	?	3190	3846	1695	2412	271	569	2574
Hardwood										
1.	Ainus rubra	Washington	34	185	516	40	224	–	114	334
2.	Betula verrucosa	England	22	63	264	33	76	3	36	327
3.	Betula verrucosa - Betula pubescens	Southern Finland	40	91	232	24	136	–	–	185
4.	Betula platyphylla	Japan	?	114	265	20	84	–	95	401
5.	Evergreen broadleaf forest	Japan	?	114	329	52	249	–	94	259
6.	Faqus sylvatica	England	39	134	285	38	187	4	42	151
7.	Faqus sylvatia	Sweden	a. 90	314	800	53	458	24	118	980
			b. 90	324	1050	84	452	32	105	620
			c. 100	226	640	65	318	17	85	478
8.	Populus tremuloides	Minnesota	40	170	383	48	297	–	62	881
9.	Quercus alba	Missouri	35-92	100	204	20	115	–	35	601

10. Quercus alba-Quercus rubra-Acer saccharum-Carya glabra	Illinois	150	190	478	29	310	–	105	1603
11. Quercus robur	England	47	130	393	35	246	5	45	257
12. Quercus robur-Carpinus betulus-Faqus sylvatica	Belgium	30-75	121	406	32	245	–	81	868
13. Quercus robur-Fraxinus excelsior	Belgium	115-160	328	947	63	493	–	126	1338
14. Quercus stellata-Quercus marilandica	Oklahoma	various	195	902	75	1093	–	230	3895
Eucalyptus									
1. E. regnans	Australia	38	654	399	38	1389	138	192	849
2. E. regnans	Australia	27	831	–	17	–	–	–	–
3. E. obliqua- E. dives	Australia	38	373	426	17	111	103	71	264
4. E. obliqua	Australia	51	316	–	31	256	–	204	336
5. E. sieberi	Australia	27	929	–	14	–	–	–	–
6. Mixed dry sclerophyll	Australia	?	176	395	–	–	–	–	–
7. E. signata-E. umbra	Australia	?	104	456	18	192	169	77	344
8. E. diversicolor	Australia	37	263	473	27	296	82	211	1133
9. E. diversicolor-E. calophylla	Australia	?	305	449	31	424	125	344	1266

The oligotrophic strategy occurs on nutrient-poor soils, like the Amazon Basin, and the eutrophic strategy occurs on nutrient-rich soils. In the temperate zone, where most forest research has been done, forests are usually of the eutrophic type. Oligotrophic ecosystems constitute a much greater proportion of the tropics. But exceptions occur, and not all tropical forests are oligotrophic nor all temperate forests eutrophic. Table 7.8 compares two eutrophic and two oligotrophic forest ecosystems. There are some striking differences between these forest types. Oligotrophic systems have a larger biomass in the humus layer of the soil, and this layer of fine roots and humus is critical for nutrient conservation in these systems. Productivity and nutrient cycling do not differ significantly in oligotrophic and eutrophic forests, as long as these ecosystems are not disturbed. But when the forest is cleared for agriculture, the nutrient-poor systems quickly lose their productive potential, while the nutrient-rich ones do not. Once the humus and root layer on top of the mechanism of efficient nutrient recycling is lost, and nutrients are leached out of the system. Oligotrophic ecosystems cannot be used for crop production unless critical nutrients are supplied in fertilizers.

Acid Rain : The Sulphur Cycle

Human activity through the combustion of fossil fuels has altered the sulphur cycle more than any of the other nutrient cycles. While human-produced emissions of carbon dioxide and nitrogen are only about 5 to 10 percent of the level of natural emissions, for sulphur we produce about 160 percent of the level of natural emissions. One clear manifestation of this alteration of the sulphur cycle is the widespread problem of acid rain in Europe and North America. Acid precipitation is defined as rain or snow that has a pH of less than 5.6. Low pH values are caused by strong acids (sulphuric acid, nitric acid) that originate as combustion products from fossil fuels.

Over large areas of Western Europe and eastern North America, annual pH values of precipitation average between 4.0 and 4.5, and individual storms may produce acid rain of pH 2 to 3. Ore smelters and electrical generating plants have increased emissions during the past 50 years. To offset local pollution problems smelters and generating plants have built taller stacks, which reduce pollution at ground level. Tall stacks (over 300 m) now are the standard, and they have exported the pollution problem downwind. The effects

Table 7.8. Comparison of Biomass, Production, and Calcium Distribution in Eutrophic and Oligotrophic Forest Ecosystems from Temperate and Tropical Areas

	Eutrophic		*Oilgotrophic*	
	Temperate Mixed Mesophytic Oak Ridge, Tenn.	*Trophical Montane Tropical Rain Forest, Puerto Rico*	*Temperate Oak-Pine, Brock haven, Long Island*	*Tropical Amazonian Rain Forest, Venezuela*
Soil				
1. Exchangeable calcium (meq/100g)	3.3	1-5	<0.22	0.38
2. Exchangeable calcium (kg/ha: soil depth, cm)	3784:75	1900:40	<990:150	306:40
Biomass				
3. Leaves (t/ha)	4.0	8.0	4.0	8.0
4. Stems and branches (t/ha)	158.0	190.0	61.0	158.0
5. Roots (t/ha)	17.0	65.0	36.0	132.0
6. Living total (t/ha)	179.0	263.0	101.0	298.0
7. Litter (t/ha)	5.0	1-6	16.0	6.0
8. Humus (t/ha)	–	–	47.0	137.0
9. Total organic (t/ha)	184.0	266.0	164.0	441.0
Production				
10. Wood ($g/m^2/yr$)	432.0	486.0	451.0	440.0

11. Litter (g/m^{2}yr)	358.0	547.0	406.0	572.0
Calcium Concentration				
12. Leaves (mg/g dry wt)	14.5	6.46	5.9	8.7
13. Stems and branches (mg/g dry wt)	8.0	2.06	0.9	1.9
14. Roots (mg/g dry wt)	7.5	5.0	2.1	1.2
Calcium Standing Crop				
15. Leaves (kg/ha)	72.0	55.0	26.0	70.0
16. Stems and branches (kg/ha)	589.0	380.0	55.0	206.0
17. Roots (kg/ha)	128.0	325.0	76.0	153.0
18. Living total (kg/ha)	796.0	760.0	157.0	529.0
19. Litter (kg/ha)	72.0	11.0	148.0	3.0
20. Humus (kg/ha)	–	–	36.0	33.0
21. Total organic (kg/ha)	861.0	771.0	341.0	565.0
Calcium Flux				
22. Precipitation (P) (kg/ha/yr)	10.5	21.8	3.3	16.0
23. Subsurface runoff (R) (kg/ha/yr)	27.4	43.1	9.7	2.8
24. P–R (kg/ha/yr)	–16.9	–21.3	–6.4	13.2

of acid rain on the environment are the subject of much current research.

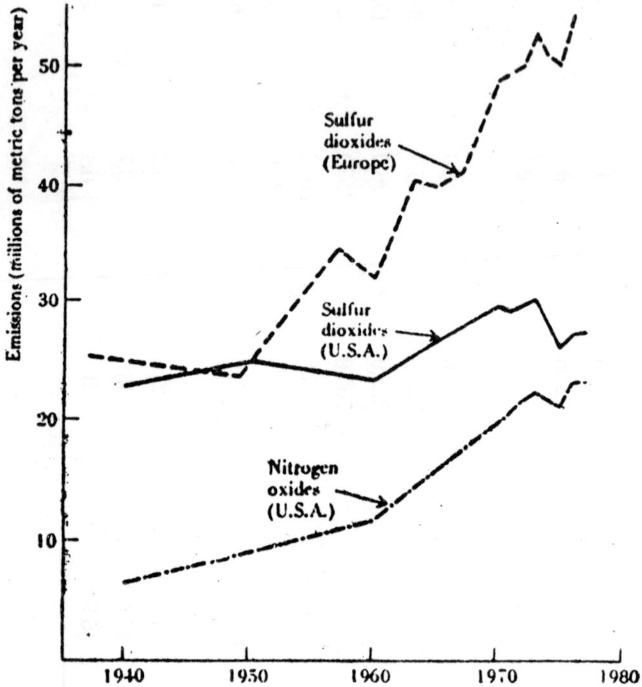

Fig. 7.10. Rise in the emissions of sulphur and nitrogen oxides in the United State and Europe, 1937-1977. The curve reflect increases in the burning of fossil fuels and the smelting of ores. Sulphur and nitrogen may be converted into strong acids in precipitation.

Some effects are very clear already. Fresh water ecosystems seem to be particularly sensitive. In areas underlain by granite and granitoid rocks, which are highly resistant to weathering, the acid rain is not neutralized in the soil, so lakes and streams become acidified. Lakes in these bedrock can be an initial guide to sensitive areas. The Precambrian Fennoscandian Shield in Scandinavia, the Canadian Shield, all of New England, the Rocky Mountains, and other areas are thus potential trouble spots. The clearest effects of acid precipitation have been on fish populations in Scandinavia and eastern Canada.

Fish populations have been reduced or eliminated in many thousands of lakes in Southern Norway and Sweden once the pH in these water has fallen below pH 5. The effect on terrestrial ecosystems is more complex and difficult to unravel. Highly acidic precipitation can reduce primary production in terrestrial plants by direct toxicity or leaching of nutrients from the foliage. The rate of decomposition in the soil may be slowed by lower pH, and this could interfere with nutrient cycling of other chemical elements. In some cases, productivity of terrestrial systems might increase if a limiting nutrient like nitrogen or phosphorus is added in the precipitation. The main point is that acidic precipitation has the potential to change nutrient cycling in natural ecosystems in a great variety of ways we cannot yet understand, much less predict. We cannot continue this aerial bombardment of ecosystems in the native belief that nutrient cycles have infinite resilience to human inputs.

8

OXYGEN CYCLE

The history of our planet, as recorded in its rocks and fossils, is reflected in the composition and the biochemical peculiarities of its present biosphere. With a little imagination one can reconstruct from that evidence the appearance and subsequent evolution of gaseous oxygen in the earth's air and water, and the changing pathways of oxygen in the metabolism of living things. Differentiated multicellular life (consisting of tissues and organs) evolved only after free oxygen appeared in the atmosphere. The cells of animals that are truly multicellular in this sense, the Metazoa, obtain their energy by breaking down fuel (produced originally by photosynthesis) in the presence of oxygen in the process called respiration.

The evolution of advanced forms of animal life would probably not have been possible without the high levels of energy release that are characteristic of oxidative metabolism. At the same time free oxygen is potentially destructive to all forms of carbon-based life (and we know no other kind of life). Most organisms have therefore had to "learn" to conduct their oxidations anaerobically, primarily by removing hydrogen from foodstuff rather than by adding oxygen. Indeed, the anaerobic process called fermentation is still the fundamental way of life underlying other forms of metabolism. Oxygen in the free state thus plays a role in the evolution and present functioning of the biosphere that is both pervasive and ambivalent.

The origin of life and its subsequent evolution was contingent on the development of systems that shielded it from, or provided chemical defenses against, ordinary molecular oxygen (O_2), ozone (O_3) and atomic oxygen (O). Yet the energy requirements of higher life forms can be met only by oxidative metabolism. The oxidation of the simpler sugar glucose, for example, yields 686 kilocalories per mole; the fermentation of glucose yields only 50 kilocalories per mole. Free oxygen not only supports life; it arises from life. The oxygen now in the atmosphere is probably mainly, if not wholly, of biological origin. Some of it is converted to ozone, causing certain high-energy wavelengths to be filtered out of the radiation that-reaches the surface of the earth.

Oxygen also combines with a wide range of other elements in the earth's crust. The result of these and other processes is an intimate evolutionary interaction among the biosphere, the atmosphere, the hydrosphere and the lithosphere. Consider where the oxygen comes from to support the high rates of energy release observed in multicellular organisms and what happens to it and to the carbon dioxide that is respired. The oxygen, of course, comes from the air, of which it constitutes roughly 21 percent. Ultimately, however, it originates with the decomposition of water molecules by light energy in photosynthesis. The 1.5 billion cubic kilometres of water on the earth are split by photosynthesis and reconstituted by respiration once every two million years or so. Photosynthetically generated oxygen temporarily enters the atmospheric bank, whence it is itself recycled once every 2,000 years or so (at current rates).

The carbon dioxide that is respired joins the small amount (0.3 percent) already in the atmosphere, which is in balance with the carbon dioxide in the oceans and other parts of the hydrosphere. Through other interactions it may be removed from circulation as a part of the carbonate ion (CO_3) in calcium carbonate precipitated from solution. Carbon dioxide thus sequestered may eventually be returned to the atmosphere when limestone, formed by the consolidation of calcium carbonate sediments, emerges from under the sea and is dissolved by some future rainfall. Thus do sea, air, rock and life interact and exchange components. Before taking up these interactions in somewhat greater detail let us examine the function oxygen serves within individual organisms.

Oxygen plays a fundamental role as a building block of practically all vital molecules, accounting for about a fourth of the atoms in living matter. Practically all organic matter in the present biosphere originates in the process of photosynthesis, whereby plants utilize light energy to react carbon dioxide with water and synthesize organic substances. Since carbohydrates (such as sugar), with the general formula $(CH_2O)_0$, are the common fuels that are stored by plants, the essential reaction of photosynthesis can be written as $CO_2 + H_2O + \text{light energy} \rightarrow CH_2O + O_2$. It is not immediately obvious from this formulation which of the reactants serves as the source of oxygen atoms in the carbohydrates and which is the source of free molecular oxygen.

In 1941 Samuel Ruben and Mar demonstrate that the molecular oxygen is derived from the splitting of the water molecule. This observation also suggested that carbon dioxide is the source of the oxygen atoms of the synthesized organic molecules. The primary products of photosynthesis undergo a vast number of chemical transformations in plant cells and subsequently in the cells of the animals that feed on plants. During these processes changes of course take place in the atomic composition and energy content of the organic molecules. Such transformations can result in carbon compounds that are either more "reduced" or more "oxidized" than carbohydrates.

The oxidation-reduction reactions between these compounds are essence of biological energy supply and demand. A more reduced compound has more hydrogen atoms and fewer oxygen atoms per carbon atom; a more oxidized compound has fewer hydrogen atoms and more oxygen atoms per carbon atom. The combustion of a reduced compound liberates more energy than the combustion of a more oxidized one. An example of a molecule more reduced than a carbohydrate is the familiar alcohol ethanol (C_2H_6O); a more oxidized molecule is pyruvic acid ($C_3H_4O_3$). Differences in the relative abundance of hydrogen and oxygen atoms in organic molecules result primarily from one of the following reactions:

1. the removal (dehydrogenation) or addition (hydrogenation) of hydrogen atoms,
2. the addition of water (hydration), followed by dehydrogenation;
3. the direct addition of oxygen (oxygenation).

The second and third of these processes introduce into organic matter additional oxygen atoms either from water or from molecular oxygen. On decomposition the oxygen atoms of organic molecules are released as carbon dioxide and water. The biological oxidation of molecules such as carbohydrates can be written as the reverse of photosynthesis:

$$CH_2O + O_2 \rightarrow CO_2 + H_2O + \text{energy}.$$

The oxygen atom of the organic molecule appears in the carbon dioxide and the molecular oxygen acts as the acceptor for the hydrogen atoms. The three major nonliving sources of oxygen atoms are therefore carbon dioxide, water and molecular oxygen, and since these molecules exchange oxygen atoms, they can be considered as a common pool. Common mineral oxides such as nitrate ions and sulphate ions are also oxygen sources for living organisms, oxides circulate through the biosphere their oxygen atoms are exchanged with water. The dynamic role of molecular oxygen is as an electron sink, or hydrogen acceptor, in biological oxidations. The biological oxidation of organic molecules proceeds primarily by dehydrogenation: enzymes remove hydrogen atoms from the substrate molecule and transfer them to specialized molecules that function as hydrogen carriers. If these carriers become saturated with hydrogen, no further oxidation can take place until some other acceptor becomes available.

In the anaerobic process of fermentation organic molecules serve as the hydrogen acceptor. Fermentation therefore results in the oxidation of some organic compounds and the simultaneous reduction of others, as in the fermentation of glucose by yeast: part of the sugar molecule is oxidized to carbon dioxide and other parts are reduced to ethanol. In aerobic respiration oxygen serves as the hydrogen acceptor and water is produced. The transfer of hydrogen atoms (which is to say of electrons and protons) to oxygen is channelled through an array of catalysts and cofactors. Prominent among the cofactors are the iron-containing pigmented molecules called cytochromes, of which there are several kinds that differ in their affinity for electrons. This affinity is expressed as the oxidation-reduction, or "redox," potential of the molecule; the more positive the potential, the greater the affinity of the oxidized molecule for electrons. For example, the redox potential of cytochrome b is .12 volt, the potential of cytochrome c is .22 volt and the potential of cytochrome a is .29 volt. The redox potential for the reduction of oxygen to water is .8 volt.

The passage of electrons from one cytochrome to another down a potential gradient, from cytochrome b to cytochrome ć to the cytochrome a complex and on to oxygen, results in the alternate reduction and oxidation of these cofactors. Energy liberated in such oxidation-reduction reactions is coupled to the synthesis of high-energy phosphate compound such as adenosine triphosphate (ATP). The special copper-containing enzyme cytochrome oxidase mediates the ultimate transfer of electrons from the cytochrome a complex to oxygen. This activation and binding of oxygen is seen as the fundamental step, and possibly dative system of enzymes and electron carriers is located in the special organelles called mitochondria. These organelles can be regarded as efficient low-temperature furnaces where organic molecules are burned with oxygen.

Most of the released energy is converted into the high-energy bonds of ATP. Molecular oxygen reacts spontaneously with organic compounds and other reduced substances. This reactivity explains the toxic effects of oxygen above tolerable concentrations. Louis Pasteur discovered that very sensitive organisms such as obligate anaerobes cannot tolerate oxygen concentrations above about 1 percent of the present atmospheric level. Recently the cells of higher organisms have been found to contain organelles called peroxisomes, whose major function is thought to be the protection of cells from oxygen.

The peroxisomes contain enzymes that catalyze the direct reduction of oxygen molecules through the oxidation of metabolites such as amino acids and other organic acids. Hydrogen peroxide (H_2O_2) is one of the products of such oxidation. Another of the peroxisome enzymes, catalase, utilizes the hydrogen peroxide as a hydrogen acceptor in the oxidation of substrates such as ethanol or lactic acid. The rate of reduction of oxygen by the peroxisomes increases proportionately with an increase in oxygen concentration, so that an excessive amount of oxygen in the cell increases the rate of its reduction by peroxisomes. Christian de Duve of Rockefeller University has suggested that the peroxisomes represent a primitive enzyme system that evolved to cope with oxygen when it first appeared in the atmosphere.

The peroxisome enzymes enabled the first oxidatively metabolizing cells to use oxygen as a hydrogen acceptor and so reoxidize the reduced products of fermentation. In some respects this process is similar to the oxidative reactions of the mitochondria. Both make further dehydrogenation possible by liberating oxidized hydrogen carriers. The basic difference between the mitochondrial oxidation reactions and those of peroxisomes is that in peroxisomes the steps of oxidation are not coupled to the synthesis of ATP.

The energy released in the peroxisomes is thus lost to the cell; the function of the organelle is primarily to protect against the destructive effects of the molecular oxygen. Oxygen dissolved in water can diffuse across both the inner and the outer membranes of the cell, and the supply of oxygen by diffusion is adequate for single cells and for organisms consisting of small colonies of cells. Differentiated multicellular organisms, however, require more efficient modes of supplying oxygen to tissues and organs. Since all higher organisms depend primarily on mitochondrial aerobic oxidation to generate the energy that maintains their active mode of life, they have evolved elaborate systems to ensure their tissues an adequate supply of oxygen, the gas that once was lethal (and still is, in excess). Two basic devices serve this purpose: special chemical carriers that increase the oxygen capacity of body fluids, and anatomical structures that provide relatively large surfaces for the rapid exchange of gases. The typical properties of an oxygen carrier are exemplified by those of hemoglobin and of myoglobin, or muscle hemoglobin.

Hemoglobin, in blood readily absorbs oxygen to near-saturation at oxygen pressures such as those found in the lung. When the blood is exposed to lower oxygen pressures as it moves from the lungs to other tissues, the hemoglobin discharges most of its bound oxygen. Myoglobin, which acts as a reservoir to meet the sharp demand for oxygen in muscle contraction, gives up its oxygen more rapidly. Such reversible bonding of oxygen in response to changes in oxygen pressure is an essential property of biochemical oxygen carriers. Lungs and gills are examples of anatomical structures in which large wet areas of thin membranous tissue come in contact with oxygen. Body fluids are pumped over one side of these membranes and air, or water containing oxygen, over the other side. This ensures a rapid gas exchange between large volumes of body fluid and environment. How did the relations between organisms and gaseous oxygen happen to evolve in such a curiously complicated manner? The atmosphere under which life arose on the earth was almost certainly devoid of free oxygen.

The low concentration of noble gases such as neon and krypton in the terrestrial atmosphere compared with their cosmic abundance, together with other geochemical evidence, indicates that the terrestrial atmosphere had a secondary origin in volcanic outgassing from the earth's interior. Oxygen is not known among the gases so released, nor is it found as inclusions in igneous rocks. The chemistry of rocks older than about two billion years is also inconsistent with the presence of more than trival quantities of free atmospheric oxygen before that time.

Moreover, it would not have been possible for the essential chemical precursors of life-or life itself-to have originated and persisted in the presence of free oxygen before the evolution of suitable oxygen-mediating enzymes. On such grounds we conclude that the first living organism must have depended on fermentation for its livelihood. Organic substances that originated in non-vital reactions served as substrates for these primordial fermentations. The first organism, therefore, was not only an anaerobe; it was also a heterotroph, dependent on a preexisting organic food supply and incapable of manufacturing its own food by photosynthesis or other autotrophic processes.

The emergence of an autotroph was an essential step in the onward march of biological evolution. This evolutionary step left its mark in the rocks as well as on all living forms. Some fated eobiont, as we may call these early life forms whose properties we can as yet only imagine, evolved and became an autotroph, an organism capable of manufacturing it owns food. Biogeological evidence suggests that this critical event may have occurred more than three billion years ago. If, as seems inescapable, the first autotrophic eobiont was also anaerobic, it would have encountered difficulty when it first learned to split water and release free oxygen.

John M. Olson of the Brookhaven National Laboratory recently suggested biochemical arguments to support the idea that primitive photosynthesis may have obtained electrons from substances other than water. He argues that large-scale splitting of water and release of oxygen may have been delayed until the evolution of appropriate enzymes to detoxify this reactive substance. We nevertheless find a long record of oxidized marine sediments of a peculiar type that precedes the first evidence of atmospheric oxygen in rocks about 1.8 billion years old; we do not find them in significant amounts in more recent strata.

These oxidized marine sediments, known as banded iron formations, are alternately iron rich and iron-poor chemical sediments that were laid down in open bodies of water. Much of the iron in them is ferric (the oxidized form, Fe^{+++}) rather than ferrous (the reduced form, Fe^{++}), implying that there was a source of oxygen in the column of water above them. Considering the problems that would face a water-splitting photosynthesizer before the evolution of advanced oxygen-mediating enzymes such as oxidases and catalases, one can visualize how the biological oxygen cycle may have interacted with ions in solution in bodies of water during that time. The first oxygen-releasing photoautotrophs may have used ferrous compounds in solution as oxygen acceptors–oxygen for them being merely a toxic waste product. This would have precipitated iron in the ferric form ($4FeO + O_2 \rightarrow 2Fe_2O_3$) or in the ferro-ferric form (Fe_3O_4). A recurrent imbalance of supply and demand might then account for the cyclic nature and differing types of the banded iron formations. Once advanced oxygen-mediating enzymes arose, oxygen generated by increasing populations of photoautotrophs containing these enzymes would build up in the oceans and begin to escape into the atmosphere.

There are ultraviolet component of the sun's radiation would dissociate some of the molecular oxygen into highly reactive atomic oxygen and also give rise to equally reactive ozone. Atmospheric oxygen and its reactive derivatives (even in small quantities) would lead to the oxidation of iron in sediments produced by the weathering of rocks, to the greatly reduced solubility of iron in surface waters (now oxygenated), to the termination of the banded iron formations as an important sedimentary type and to the extensive formation of continental red beds rich in ferric iron.

The record of the rocks supports this succession of events: red beds are essentially restricted to rocks younger than about 1.8 billion years, whereas banded iron formation is found only in order rocks. So far we have assumed that oxygen accumulated in the atmosphere as a consequence of photosynthesis by green plants. How could this happen if the entire process of photosynthesis and respiration is cycle, representable by the reversible equation $CO_2 + H_2O +$ energy $= CH_2O + O_2$? Except to the extent that carbon or its compounds are somehow sequestered, carbohydrates produced by photosynthesis will be reoxidized back to carbon dioxide and water, and no significant quantity of free oxygen will accumulate.

The carbon that is sequestered in the earth as graphite in the oldest rocks and as coal, oil, gas and other carbonaceous compounds in the younger ones, and in the living and dead bodies of plants and animals, is the equivalent of the oxygen in oxidized sediments and in the earth's atmosphere! In attempting to strike a carbon-oxygen balance we must find enough carbon to account not only for the oxygen in the present atmosphere but also for the "fossil" oxygen that went into the conversion of ferrous oxide to ferric oxides, sulphides to sulphates, carbon monoxide to carbon dioxide and so on. Interestingly, rough estimates made some years ago by William W. Rubey, now of the University of California at Los Angeles, do imply an approximate balance between the chemical combining equivalents of carbon and oxygen in sediments, the atmosphere, the hydrosphere and the biosphere. The relatively small excess of carbon in Rubey's estimates could be accounted for by the oxygen used in converting carbon monoxide to carbon dioxide. Or it might be due to an underestimate of the quantities of sulphate ion or ferric oxide in sediments. (Rubey's estimates could not include large iron formations recently discovered in western Australia and elsewhere.)

The carbon dioxide in carbonate rocks does not need to be accounted for, but the oxygen involved in converting it to carbonate ion does. The recycling of sediments through metamorphism, mountain-building and the movement of ocean-floor plates under the continents is a variable of unknown dimensions, but it probably does not affect the approximate balance observed in view of the fact that the overwhelmingly large pools to be balanced are all in the lithosphere and that carbon and oxygen losses would be roughly equivalent. The small amounts of oxygen dissolved in water are not included in this balance. Nonetheless, water does enter the picture. Another possible source of oxygen in our atmosphere is photolysis, the ultraviolet dissociation of water vapour in the outer atmosphere followed by the escape of the hydrogen from the earth's gravitational field. This has usually been regarded as a trivial source, however.

Although R.T. Brinkmann of the California Institute of Technology has recently argued that nonbiological photolysis may be a major source of atmospheric oxygen, the carbon-oxygen balance sheet does not support that belief, which also runs into other difficulties. When free oxygen began to accumulate in the atmosphere some 1.8 billion years ago, life was still restricted to sites shielded from destructive ultraviolet radiation by sufficient depths of water or by screens of sediment. In time enough oxygen built up in the atmosphere for ozone, a strong absorber in the ultraviolet, to form a shield against incoming ultraviolet radiation.

The late Lloyd V. Berkner and Lauriston C. Marshall of the Graduate Research Centre of the Southwest in Dallas calculated that only 1 percent of the present atmospheric level of oxygen would give rise to a sufficient level of ozone to screen out the most deleterious wavelengths of the ultraviolet radiation. This also happens to be the level of oxygen at which Pasteur found that certain microorganisms switch over from a fermentative type of metabolism to an oxidative one. Berkner and Marshall therefore jumped to the conclusion (reasonably enough on the evidence they considered) that this was the stage at which oxidative metabolism arose. They related this stage to the first appearance of metazoan life somewhat more than 600 million years ago.

The geological record has long made in plain, however, that free molecular oxygen existed in the atmosphere well before that relatively late date in geologic time. Moreover, recent evidence is consistent with the origin of oxidative metabolism at least twice as long ago. Eucaryotic cells–cells with organized nuclei and other organelles-have been identified in rocks in eastern California that are believed to be about 1.3 billion years old. Since all living eucaryotes depend on oxidative metabolism, it seems likely that these ancestral forms did too. The oxygen level may nonetheless have still been quite low at this stage. Simple diffusion would suffice to move enough oxygen across cell boundaries and within the cell, even at very low concentrations, to supply the early oxidative metabolizers.

A higher order of organization and of atmospheric oxygen was required, however, for advanced oxidative metabolism. Perhaps that is why, although the eucaryotic cell existed at least 1.2 billion years ago, we have no unequivocal fossils of metazoan organisms from rocks older than perhaps 640 million years. In other words, perhaps Berkner and Marshall were mistaken only in trying to make the appearance of the Metazoa coincide with the onset of oxidative metabolism: Once the level of atmospheric oxygen was high enough to generate an effective ozone screen, photosynthetic organisms would have been able to spread throughout the surface waters of the sea, greatly accelerating the rate of oxygen production.

The possible episodes in geological history to correlate with this development are the secondary oxidation of the banded iron formations and the appearance of sedimentary calcium sulphate (gypsum and anhydrite) on a large scale. These events occurred

Years before Presents	*Lithosphere*	*Biosphere*	*Hydrosphere*	*Atmosphere*
20 million	Glaciation	Mammals diversify grasses appear		Oxygen approaches present level
50 million	Coal formation volcanism			
100 million		Social insects flowering plants Mammals		Atmospheric oxygen increases at fluctuating rate
200 million	Great volcanism Coal formation		Oceans continue to increase in volume	
		Insects appear Land plants appear		
500 million		Metazoa appear		Oxygen at 3-10 percent of present atmospheric level
	Glaciation Sedimentary Calcium Sulphate	Rapid increase in phytoplankton	Surface waters opened to phytoplankton	Oxygen at 1 percent of present atmospheric levels ozone screen effective
1 billion	Volcanism	Eucaryotes		Oxygen increasing carbon dioxide decreasing
2 billion	Red beds........	Advanced oxygen-mediating enzymes	Oxygen diffuses into atmosphere	Oxygen in atmosphere
	Glaciation Banded iron formations Oldest sediments Oldest earth rocks	First oxygen-generating photosynthetic cells Prokaryotes Abiogenic evolution	Start of oxygen generation with ferrous iron as oxygen sink	
5 billion	(Origin of solar system)			No free oxygen

just as or just before the Metazoa first appeared in early Paleozoic time. The attainment of a suitable level of atmospheric oxygen may thus be correlated with the emergence of metazoan root stocks from pre-metazoan ancestors beginning about 640 million years ago.

The fact that oxygen could accumulate no faster than carbon (or hydrogen) was removed argues against the likelihood of a rapid early buildup of oxygen. That subsequent biospheric and atmospheric evolution were closely interlinked can now be taken for granted. What is not known are the details. Did oxygen levels in the atmosphere increase steadily throughout geologic time, marking regular stages of biological evolution such as the emergence of land plants, of insects, of the various vertebrate groups and of flowering plants, as Berkner and Marshall suggested? Or were there wide wings in the oxygen level? Did oxygen decrease during great volcanic episodes, as a result of the oxidation of newly emitted carbon monoxide to carbon dioxide, or during times of sedimentary sulphate precipitation? Did oxygen increase when carbon was being sequestered during times of coal and petroleum formation? May there have been fluctuations in both directions as a result of plant and animal evolution, of phytoplankton eruptions and extinctions and of the extent and type of terrestrial plant cover? Such processes and events are now being seriously studied, but the answers are as yet far from clear.

9

Mineral Cycles

The periodic table lists more than 100 chemical elements. Yet ecologists have defined the biosphere as the locus of interaction of only four of them: hydrogen, carbon, nitrogen and oxygen. In the periodic table these four are numbered 1, 6, 7 and 8. This definition, although it deals handsomely with much of the chemistry of life, turns out to be a little too restrictive. But when we enlarge it to include phosphorus and sulphur, as we do here, we have gone no farther up the table than element No. 16. From this it should be apparent that no element lighter than sulphur can be ignored, either by ecologists or by anyone else. The fact is that most human problems–all, environmental ones, anyway–arise from the exceptional reactivity of six of the 16 lightest elements. Because our definition of the biosphere is based more on reactivity than on atomic number, it is a minimum definition.

It is not intended to exclude heavier elements that react with the primary six. As a matter of empirical fact it is known that no element lighter than iron and cobalt, elements No. 26 and No. 27, is unimportant to the biosphere. Beyond copper, No. 29, there are a few conspicuously reactive elements such as the heavy halogens bromine and iodine. Most of the heavies are metals, such as gold, mercury and lead (Nos. 79, 80 and 82), however, and their main effect on the lightweight biosphere is to depress it. Toward the end of the periodic table are some famously overweight metals whose tendency to lighten themselves has disastrous effects on any light substances that get in the way. In order to understand how it is that many elements interacts with the essential six, one must briefly reflect on the biosphere as a whole. Because the biosphere is so reactive, its influence on the hydrosphere, the lithosphere and the atmosphere is inversely proportional to its mass. This mass is very small.

An average square centimetre of the earth's surface supports a tiny amount of biosphere: 580 milligrams, less than the weight of two aspirin tablets. A roughly equivalent mass is found in the same area of hydrosphere a single centimetre deep, or in a paper-thin slice of lithosphere. Still, from a worm's-eye view the biosphere has real substance, particularly on land, where it amounts to 200 oven-dry tons on an average hectare. A glance at a partial list of the elements that compose the biosphere shows why hydrogen, oxygen, carbon and nitrogen dominate conceptions of biosphere chemistry. Together these elements constitute all but a tiny fraction of the average terrestrial vegetation, which in turn constitutes more than 99 percent of the world's standing crop.

The quantities are shown in the chart on the next page, based on a splendid compilation by L. E. Rodin and N. I. Basilevich. What I have done is to weight their chemical analyses in proportion to the kinds of land area they represent. The weighting factors, for desert, forest, tundra and so on, are the same ones I used to calculate the earth's production of carbon in an earlier article. Incidentally, on the basis of this new calculation terrestrial carbon production comes out at 65×10^9 tons of carbon per year, about 15 percent more than the computed before. What chemical compounds do these elements form? The standard way to determine the chemical composition of an organic substance is to burn it and collect the products. The list of components that results from this destructive procedure expresses some obvious facts, such as the familiar one that the biosphere is mainly carbon dioxide and water.

Nitrogen, a major constituent of protein, seems surprisingly scarce (about five parts per 1,000 by weight) until we remember that the biosphere is chiefly wood, that is, not protein but the carbohydrate cellulose. The destructive procedure would also leave a smudge, about 12 parts per 1,000 of the total, loosely called ash. Its dominant elements calcium, potassium silicon and magnesium have important biochemical functions. One atom of magnesium, for instance, lies at the centres of every molecule of chlorophyll, and silicon, the stuff of sand, is obviously useful for building hard structures. Iron and manganese also play central roles in the biosphere, a fact that could not be guessed from their position in our chart.

In biochemistry as in geochemistry the importance of these elements is in governing oxidation-reduction reactions, but the masses involved are small. As for the major cations–ions of such elements as calcium, potassium, magnesium and sodium–new insights have just begun to flood in with their discovery in rainwater. There are many other metallic elements that appear in trace amounts. Not all of them are listed in the chart because some could be accidental contaminants. There remain two, sulphur and phosphorus, each amounting to more than 10 percent of the nitrogen, that do not look like contaminants. To ignore these elements as "traces" or even to think of them as "ash" or "inorganic" elements is to misconstrue the chemical architecture of the biosphere. A listing of elements and compounds does not reveal that architecture. There is a big difference between a finished house and a pile of building materials.

Nevertheless, a list is a useful point of departure. If it is made with care, it can protect ecologists from the kind of mistake that architects sometimes make, such as forgetting the plumbing. When a list contains as much information as a shopping list–when it shows amounts as well as kinds of materials–some conclusions can be drawn from the relative proportions. (As a former bureaucrat I have learned that a "laundry list" contains even more ambiguous information than a "shopping list"; good bureaucrats keep both.) If a housewife's shopping list showed a pound of coffee, four pork chops and 100 pounds of sugar, for example, we would know that madame is either hoarding or running a private business. If she also wants a ton of flour, she is evidently baking, not distilling.

The inclusion of two dozen light bulbs would suggest that she works mainly at night, but the listing of 10 dozen light bulbs would point to a faulty generator. As it happens, this kind of semi-quantitative ratiocination was applied to ash, and to biogeochemistry, by the master of non-obvious deduction, Sherlock Holmes. Unfortunately no copy of his analytical results has yet come to light. If the charts on the opposite page can serve as a substitute, we can cast a Holmesian eye over the list. Our thinking will be more productive if we compare the composition of the biosphere with the composition of the lithosphere, the hydrosphere and the atmosphere. For this comparison all four of the "spheres" in the chart are converted from parts by weight to atoms per 100 atoms. (The masses of the four spheres being very different, these percentages will give no idea of the earth's mean or total composition.) At first glance the four spheres do not seem to belong in the same universe.

Not surprisingly, the lithosphere turns out to be a slightly metallic aluminium silicate. The biosphere, in sharp contrast, is both wet and carbonaceous. A single class of compounds, formaldehyde (CH_2O) and its polymers, including cellulose, could make up more than 98 percent of the total (by weight). Still, even when it is dried in an oven at 110 degrees Celsius, life is mainly hydrogen and oxygen, in close approximation to the proportions known as water. In other words, the biosphere is notably carboxylated: it is both more hydrated and chemically more reduced (hydrogenated than is the lithosphere from which, in some sense, it came. Among the 10 most abundant elements of the lithosphere there is no obvious source for life's carbon.

Hydrogen is also fairly far down the list for rock (and would be farther down if I had not copied some old figures from Frank W. Clarke's The Data of Geochemistry, which overweight the acidic rocks of continents). Even the elementary Dr. Watson might conclude that life's hydrogen comes from some inorganic hydrate–water, for instance–and indeed the hydrosphere provides an ample and ready supply. This will not work for carbon, though, and in trying to account for carboxylation we can make a deduction that is truly elementary in the Holmesian, or non-obvious sense. We begin again by nothing that life is mainly aqueous, and also that it concentrates carbon in proportions for greater than those in any accessible source. Is it possible that these facts are related? If they are, what do we know about water that throws any light on this relation and on the behaviour of carbon? (At this point a lesser detective might reach for the carbonated water and pause for a reply).

Al	Aluminium	Cl	Chlorine
Ar	Argon	Fe	Iron
B	Boron	H	Hydrogen
C	Carbon	K	Potassium
Ca	Calcium	Mg	Magnesium
Mn	Mangnese	P	Phosphorus
N	Nitrogen	S	Sulphur
Na	Sodium	Si	Silicon
Ne	Neon	Ti	Titanium
O	Oxygen		

Relative amounts of elements in the biosphere, the lithosphere, the hydrosphere and the atmosphere are presented in the charts on the opposite page. Here, however, amounts are given not as kilograms per hectare but as atoms per 100 atoms. Here again scale is logarithmic to show less abundant elements, which otherwise could not be compared.

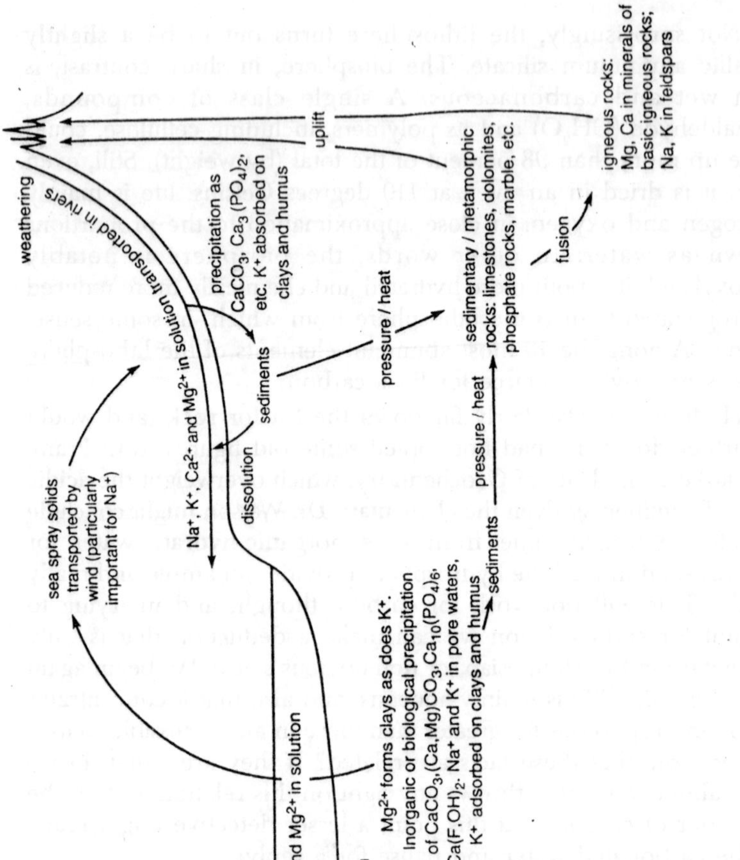

Fig. 9.1. The geochemical cycles of sodium (Na), potassium (K), calcium (Ca) and magnesium (Mg).

Instead of guessing, Holmes would proceed with his review of the evidence. Water, of course, is continuously recycled near the earth's surface, by runoff, evaporation and condensation. That is, it flows in rivers from the lithosphere to the hydrosphere, and it returns to rewash the land by way of the atmosphere. Any water-soluble elements are certain to track this cycle at least partway, from land to sea, although they may find the sea to be a sink, as boron does. If they are to get out, they can reach the land as part of an uplifted sea bottom, but that is a chancy mechanism. Recycling is both faster and surer if the element is volatile as well as soluble, so that one of its compounds can move landward through the atmosphere as water does. In the biosphere there are at least three elements besides those of water-carbon, nitrogen and sulphur–that fall in this doubly mobile class. Among their airborne compounds

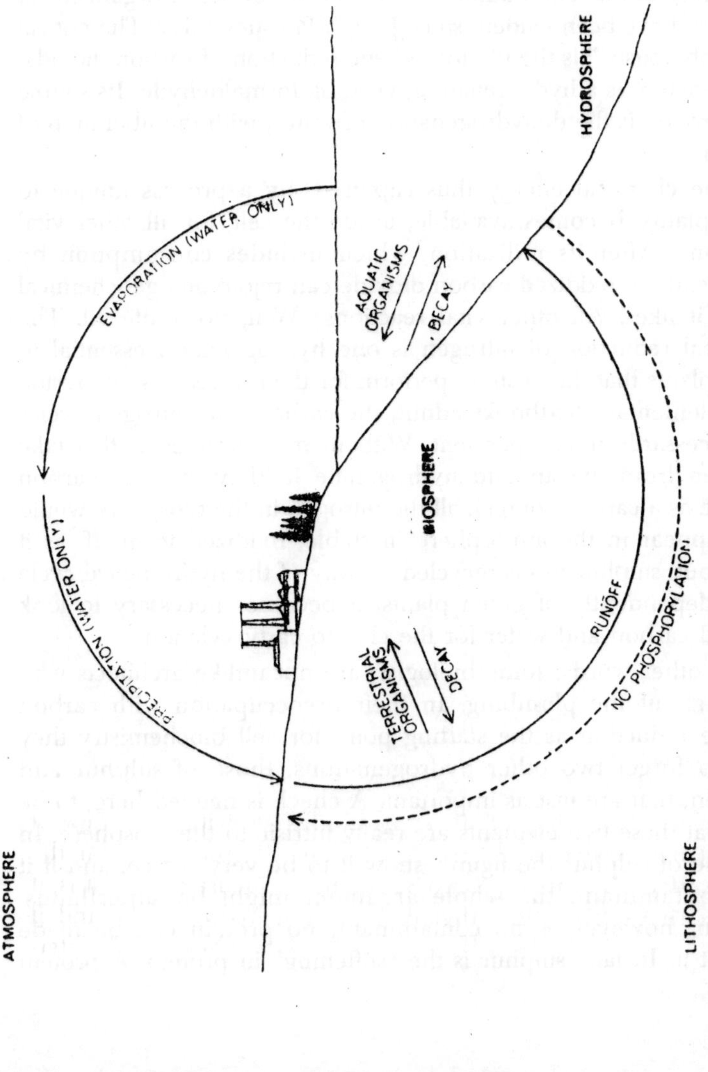

Fig. 9.2: Soluble-element cycle is followed by mineral such as phosphorus that dissolve in water but are not volatile, that is, they are not carried into the air by evaporation.

are carbon dioxide (CO_2), methane (CH_4), free nitrogen (N_2), ammonia (NH_3), hydrogen sulphide (H_2S) and sulphur dioxide (SO_2).

It is interesting that when carbon, nitrogen and sulphur are recycled, their valence changes. It may not be an accident that all three are more reduced in the biosphere than they are in the external world. Be that as it may, they all seem to belong to the biosphere, which is otherwise mainly water. Hence all three must be recycled together, along with the water, if the earth is to sustain its most unusual hydrate. The outlines of the carbon cycle, in organisms at any rate, have been evident since Joseph Priestley's day. The critical step, "obviously," is the photosynthetic reduction of carbon dioxide. That reaction is a hydrogenation, yielding formaldehyde. Its source of hydrogen is the dehydrogenation of water, with the liberation of oxygen.

The chemical energy thus captured, by a process unique to green plants, becomes available, inside the cell, for all other vital reactions. After its utilization, which includes consumption by animals, the reoxidized carbon dioxide can rejoin any geochemical cycles it likes. All other vital reactions? Well, not quite all. The chemical reduction of nitrogen is one hydrogenation essential to green plants that they cannot perform for themselves. As one result, even elementary textbooks admit, the carbon and nitrogen cycles are necessarily interdependent. Without microorganisms that take nitrogen from the air and hydrogenate it (they can use carbon dioxide as a carbon source), all the nitrogen in the biosphere would soon appear in the atmosphere in stable, oxidized form. If, as it turns, out, sulphur too is recycled by way of the hydrological cycle but independently of green plants, it becomes necessary to look beyond carbon and water for the clue to carboxylation.

In other words, some biologists are not unlike architects who forget about the plumbing. In their preoccupation with carbon dioxide reduction as the starting point for cell biochemistry they tend to forget two other hydrogenations, those of sulphur and nitrogen, that are just as important. A check is needed here, to be sure that these two elements are really intrisic to the biosphere. In the case of sulphur the figures show it to be very scarce, and if it is a contaminant, the whole argument might be superfluous. Sulphur, however, is no contaminant; no protein can be made without it. In fact, sulphur is the "stiffening" in protein. A protein

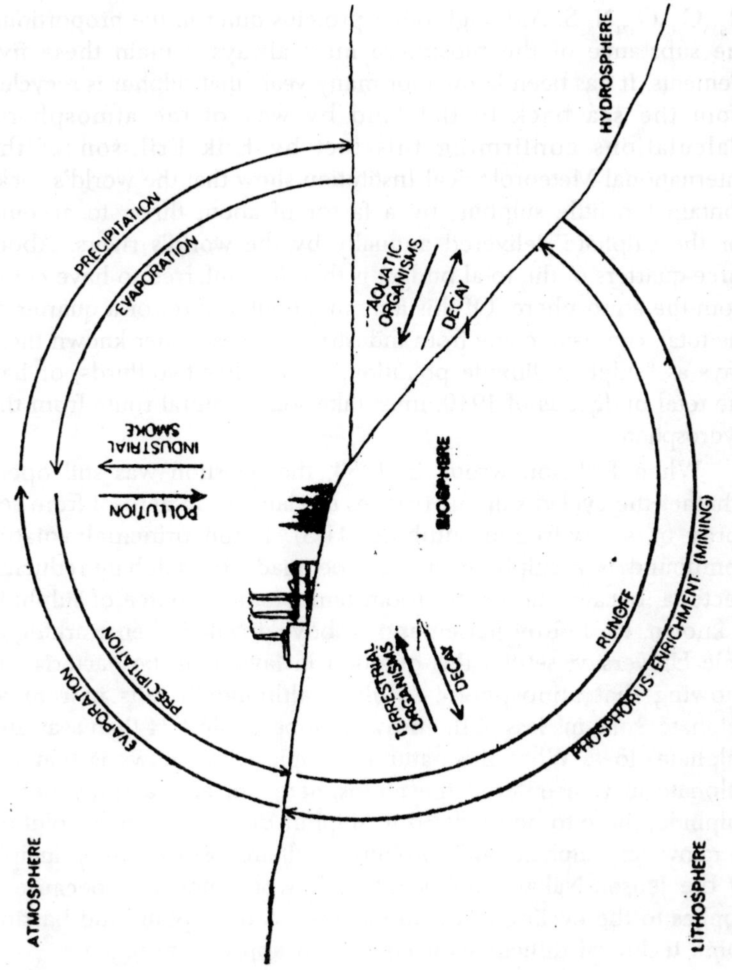

Fig. 9.3. Eutrophication of the Biosphere is the intensive cycling of phosphorus, nitrogen and sulphur.

cannot perform its function unless it is folded and shaped in a particular way. This three-dimensional structure is maintained by bonds between sulphur atoms that link one segment of a protein molecule to another. Without these sulphur bonds a protein would coil randomly, like a carelessly dropped rope.

The reason for the apparent scarcity of sulphur is the low protein content of woody tissue; any animal body contains much more. Cod-meal protein, for example, with 2.26 percent of the sulphurous amino acid methionine, has the empirical formula

$H_{555}C_{265}O_{174}N_{83}S$. Although other proteins differ in the proportions, the substance of the biosphere must always contain these five elements. It has been known for many years that sulphur is recycled from the sea back to the land by way of the atmosphere. Calculations confirming this fact by Erik Eriksson of the International Meteorological Institution show that the world's rocks contain too little sulphur, by a factor of about three, to account for the sulphate delivered annually by the world's rivers. About three-quarters of the total budget is therefore inferred to have come from the atmosphere. Of this amount about a third, or a quarter of the total, can have come from industrial sources–better known these days as "sulphur dioxide pollution." The other two-thirds, or half the total budget as of 1940, must take some natural route from the hydrosphere.

When Eriksson wrote, in 1959, the question was still open, whether the cycled sulphur reaches the land as an aerosol from sea spray or as hydrogen sulphide (H_2S). If the principal volatile compound is a sulphide, it must be made by sulphate-reducing bacteria, because no other "room temperature" source of sulphide is known. M. LeRoy Jensen and Noboyuki Nakai, then working at Yale University, settled this question in favour of the bacteria, by showing that atmospheric sulphur, although it falls in rain as sulphate, contains less of the heavy isotope sulphur 34 than seawater sulphate does. What the natural isotopic label shows is that the sulphate in rain entered the atmosphere not as sea spray but as sulphide, there to be oxidized to sulphur dioxide. After dissolution in rainwater, sulphate (and sulphuric acid) are formed. The principle of the Jensen-Nakai demonstration is worth noticing, because it applies to the cycling of carbon as well as of sulphur, and barring some technical difficulties it could also apply to nitrogen.

The route followed through oxidation-reduction reaction by ordinary sulphur (sulphur 32) is analogous to the route followed by ordinary carbon (carbon 12) in photosynthesis. These lighter, more mobile isotopes appear preferentially in reduced compounds such as hydrogen sulphide, methane and formaldehyde. At equilibrium in a closed system the oxidation products (carbon dioxide or sulphate), have correspondingly more of the heavier isotopes carbon 13 and sulphur 34 without change in the total mass. If, however, a reduced and isotopically light product escapes, as hydrogen sulphide does from the hydrosphere, equilibrium is not attained, and if the gaseous product is trapped and reoxidized in a separate system, the oxide (sulphur dioxide in this case) remains light. Exactly where within the hydrosphere most sulphate-reducers reside is still unclear. The known ones are obligate anaerobes, and their habitat is mud.

Swamps, marshes and the floor of eutropic lakes must all be important, and they may be quantitatively more important than the blue mud of estuaries and continental shelves. The sulphur metabolism of such large systems is not easy to study, even with isotopic tools. Minze Stuiver, now at the University of Washington, injected radioactively labelled sulphate ions into one eutrophic lake, Linsley Pond in Connecticut. Sulphate reduction proved to be intense, as had been expected. This lake, however, has quite a bit of ferrous iron in its deeper waters, and more in the mud itself.

In the presence of the ferrous iron all the labelled sulphide was firmly held in the mud as ferrous sulphide, and no hydrogen sulphide escaped. At least for the duration of the radioactive label, with its half-life of 89 days, this mass of mud was not a source of atmospheric sulphur but a sink. It follows from all of this that the cycling of sulphur in nature is no less relevant to carboxylation than the cycling of carbon and nitrogen.

Without downgrading photosynthesis, we can say that carbon fixation is only one of at least three critical steps in the global synthesis of protein. All three are hydrogenations, achieved with the aid of enzymes, which are themselves proteins, and therefore occur only in the biosphere. Of the three reductions, however, only the reduction of carbon calls for green plants and sunlight. The other two, the reduction of nitrogen and of sulphur, are accomplished anaerobically, by microbes. Thus the locus of the nitrogen and sulphur reductions is, broadly speaking, oxygen-deficient soil and mud. Both loci are separated spatially from that airy, sunlit world where green plants (addicted, like human societies, to the external disposal of wastes) are thoughtlessly liberating oxygen.

With three critical steps for five elements, moving through four "spheres" of abstract space, one feels the need for a picture–a "systems model" –just to keep track of the relations. The two-dimensional analogue is simple but adequate. Although it fails to specify fluxes, or any chemical quantities, it provides a mental framework for the movement of five elements: hydrogen, oxygen, carbon, nitrogen and sulphur, either alone or in combinations such as water, nitrate, the dioxides of carbon and sulphur, and carbohydrate. The synthetic output is the biosphere, with the empirical composition of protein. The central or regulatory position of the biosphere in this model follows from the fact that for all five elements it is both a source and a temporary sink. For any element that might be tempted to cycle around the edges of the model, the biosphere provides several high-energy alternatives. The most interesting of these are the reductions of carbon, nitrogen and sulphur, each concentrated at a different interface, two being out of a immediate contact with air.

Input	*Output (Kilograms Per Hectare)*	*Net Export (Kilograms Per Hectare)*	*(Kilograms Per Hectare)*
Calcium (Ca^{++})	2.8→	3.0→	+0.2
Magnesium (Mg^{++})	1.1→	1.8→	+0.7
Sodium (Na^{+})	2.1→	4.2→	+2.1
Potassium (K^{+})	1.8→	1.1→	–0.7
Chloride (Cl^{+})	2.8→	4.1→	+1.3
Sulphate (SO_4^{++})	30.0→	29.4→	–0.6
Ammonium (NH_4^{+})	2.1→	0.3→	–1.8
Nitrate (NO_3^{-})	6.7→	4.8→	–1.9
Silicon Dioxide (S_1O_2)	1.9→	20.9→	+19.0
Alluminium (Al^{++} -)	–	1.4	–
Bicarbonate $(HCO_3)^{-}$	0	0.7→	+0.7
Total	51.4→	71.7→	+20.3

Eutrophication of Dry Land is indicated by the imbalance between the quantity of certain ions falling from the atmosphere on the forest at Watershed No. 6 at Hubbard Brook in New Hampshire and the output of these ions in the brook itself. Input (smaller arrows at left) of some elements such as calcium, magnesium and sodium is smaller than the output (larger arrows at right). The input of potassium, ammonium, sulphate and nitrate, however, is larger (larger arrows at left) than output of these substances (smaller arrows). The excess of input indicates that the forests is utilizing these four substances as it grows.

Water, although it is able in principle to cycle independently, is the source of the hydrogen that energizes the biosphere, and cannot long avoid the biospheric loop as long as the biosphere functions. It is easy to be bemused by so fascinating a model. Its function, however, is to clarify thought. If further thought disrupts the model, nothing is lost but a few lines on paper. More or less instantly, by reference to the table of biospheric composition, we can see that the model is incomplete. Phosphorus has been left out, along with calcium, potassium, silicon and magnesium, four elements that are commoner in the biosphere than sulphur is. Will any or all of these cycle tamely through the model, or will they disrupt it beyond repair? For phosphorus, but not yet for the others, the answer is clear:

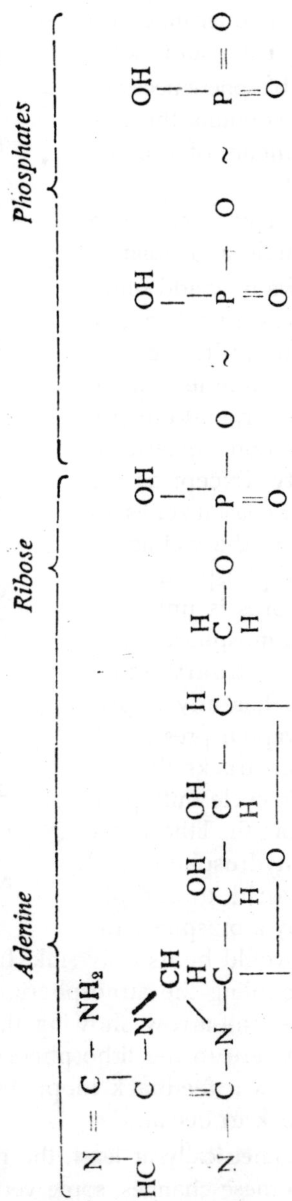

Universal fuel of living matter is adenosine triphosphate (ATP). High-energy phosphate bonds of ATP (~) each store 12,000 calories and release 7,500 calories when broken.

With one significant modification, the model can accommodate phosphorus. First, let us be sure, as be made sure for sulphur, that phosphorus is necessary to the biosphere. It is not a constituent of protein, but no protein can be made without it. The "high-energy phosphate bond," reversibly moving between adenosine diphosphate (ADP) and adenosine triphosphate (ATP), is the universal fuel for all biochemical work within the cell. The photosynthetic fixation of carbon would be a fruitless tour de force if it were not followed by the phosphorylation of the sugar produced. Thus although neither ADP nor ATP contains much phosphorus, one phosphorus atom per molecule of adenosine is absolutely essential. No life (including microbial life) is possible without it. Out provisional model of the biosphere has been constructed on two explicit assumptions:

1. the biosphere necessarily contains the five elements of protein, and
2. all five are both soluble and volatile.

If we now add phosphorus as a sixth necessary element, we can safely assume its solubility in water, and the crucial question concerns its volatility. Except as sea spray in coastal zones, or as dust in the vicinity of exposed phosphate rock, phosphorus is unknown in the atmosphere; none of its ordinary compounds has any appreciable vapour pressure. It therefore tracks the hydrologic cycle only partway, from the lithosphere to the hydrosphere, and in a world uncomplicated by a biosphere the ocean would be its only sink. In terms of my model this amounts to uncoupling the atmospheric reservoir (except for water), omitting the half-arrow showing the return of phosphorus from the hydrosphere to the lithosphere and leaving the biosphere's phosphorus as a feedback loop, diverting some of the oneway flow from rock to ocean.

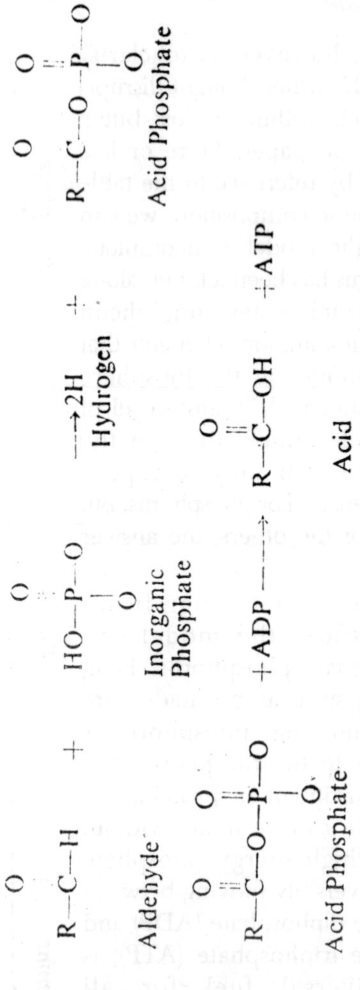

Fig. 9.4. Production of ATP, shown in generalized form, consists of two stages. The first stage begins as aldehyde reacts with an inorganic phosphate to produce hydrogen and acid phosphate. In second stage (bottom) acid phosphate reacts with ADP (adenosine diphosphate) to make an organic acid and ATP. R stands for radicals, or side groups.

Geometrically at least, the model is general enough to accommodate these changes, some version of which will be needed if any permanent sinks are discovered in the system. For any soluble but nonvolatile element a closed natural cycle is possible only through the biosphere. The model hints at the reason why many elements–

vanadium, cobalt, nickel and molybdenum among them–are best known in aquatic organisms and cycle mainly within the hydrosphere. Now, however, the model, nonquantitative though it is, suggests something else. If the biosphere demands such an element as phosphorus (and the cases of iron and manganese should be similar), two alternative inferences are permissible, depending on the magnitudes of reservoirs and fluxes. If the lithosphere contains an ample supply of phosphorus, or if the flux to the hydrospheric sink is large, the biosphere can take off what it needs and waste the rest.

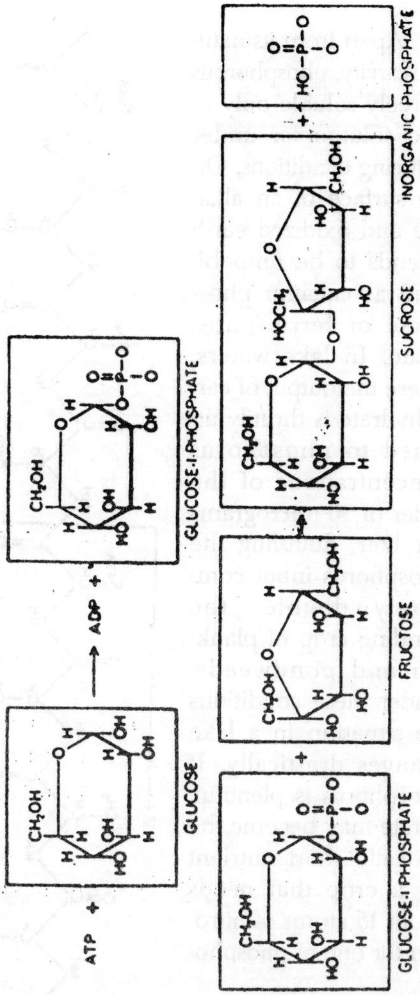

Fig. 9.5. Synthesis of Sucrose is an example of a reaction for which ATP supplies energy.

It is commonly believed such elements as sodium and calcium are thus wasted by terrestrial vegetation, although ecologists are beginning to doubt it. On the other hand, if the quantity is scanty or the flux small, the element will be in critically short supply. And if short supply is chronic, the output of the entire system could be expected to be adjusted to the rate of exploitation of one critical element, much as the performance of bureaucracy is closely geared to the supply of paper clips. In undisturbed nature of chronic shortage of phosphorus is notorious; that it what most people mean by "soil infertility." In the lithosphere phosphorus is scarcer than carbon, and in the hydrosphere, because phosphorus falls in the parts-per-billion range, it fails to show up at all in the chart of constituents.

Apart from its natural scarcity, phosphorous is freely soluble only in acid solution or under reducing conditions. On the surface of an alkaline and oxidized earth it tends to be immobilized as calcium phosphate or ferric phosphate. In lake waters, where the output of carbohydrate is thriftily attuned to phosphorus concentrations of the order of 50 micrograms per liter, doubling the phosphorus input commonly doubles the standing crop of plankton and pondweeds. Under these conditions the situation in a lake changes drastically. If phosphorus is plentiful, nitrate may become the critically short nutrient for a crop that needs about 15 atoms of nitrogen for one of phosphorus.

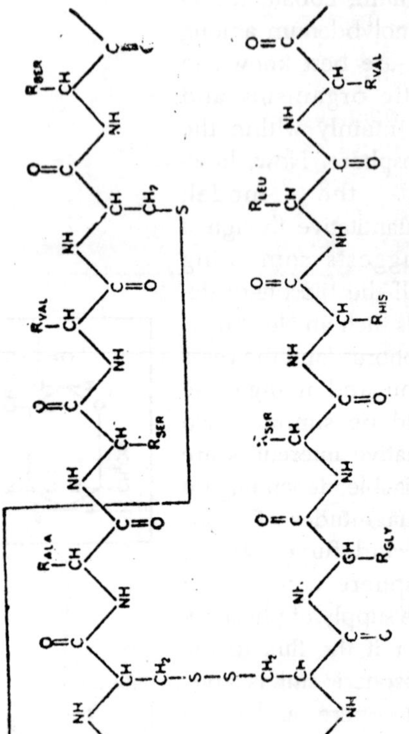

Fig. 9.6. Basic function of sulphur in living matter appears to be provide a linkage between the polypeptide chains in a protein molecule.

Blue-green algae may then take over the plankton because by reducing atmospheric nitrogen they escape the dependence that other algae have on nitrate. Meanwhile, judging from much recent experience, the phosphorus input will probably have doubled again–but the subject under discussion is no longer undisturbed nature. What started as "cottage entrophication," by seepage from a few septic tanks, has been escalated into a noisome mess by "treated" sewage and polyphosphate detergents. To conserve biogeochemical parity the atmosphere has begun to deliver into lakes nitrate and sulphate from the combustion of fossil fuels. It would be wrong to read too much into a systems model. "Conserving biogeochemical parity" it just a figure of speech, technically hyperbole, and ironical at that. After all, the pollution of the air by nitrate and sulphate is quite independent–technologically, spatially and politically–of the pollution of water by phosphorus.

If the accelerated phosphorus cycle in lakes takes advantage of these added inputs, we dare not say that nitrate and sulphate have been drawn into the biosphere from the atmosphere, as U.S. power was drawn to a Canadian circuit breaker in the Northeast blackout of 1965. What the model tells us is that matters can look that way from the standpoint of the biosphere. If the lake segment of the phosphorus cycle is accelerate to the point where nitrogen and sulphur are as critical as phosphorus used to be, and if there is a new source of nitrate and sulphate in the atmosphere, the atmosphere is adequately coupled to all other subsystems to ensure the success of the newly accelerated loop. The loop, known as eutrophication, is thus amplified from a lacustrine nuisance to a system problem, and around such lakes as Lake Erie it threatens to become a cancer in the global ecosystem. The trouble started, of course, when the world's one-way phosphorus cycle was first reversed and then accelerated by human activity.

Since bird guano was discovered on desert islands, later to be supplemented in fertilizers by phosphate rock, marine phosphate has been restored to the lithosphere in ever increasing amounts. As a device for growing people in ever increasing numbers the practice cannot be faulted, but if people are to continue to flourish in the biosphere, they will have to pay more attention to scarce resources. Phosphorus is much too valuable to be thoughtlessly shared with blue-green algae. The term eutrophication, which means enrichment, usually inadvertent, is not ordinarily applied to forests and deserts. I dare to extent it to the terrestrial biosphere because two new lines of evidence have suddenly appeared to suggest that the known pollution of air by nitrate and sulphate also encourages the bloom on dry land. The first line of evidence comes from Hubbard Brook, N.H., where F. Herbert Bormann of the Yale School of Forestry and Gene E. Likens of Cornell University have had six forested watersheds under close study since 1963. What interests us here is the difference, per hectare of ecosystem, between the input of ions in rainfall (plus dry fallout, if any) and the output as measured at a dam at the foot of each drainage basin.

Among the common ions entering and leaving Watershed No. 6 at Hubbard Brook, chloride and three positive ions–calcium, magnesium and sodium–show an excess of output over input, presumably derived from the local rocks and soil. These four ions conform, if any barely, to the idea that the biosphere wastes excess salts on their way to the sea. In contrast, potassium and ammonium (NH_4) and the two major negative ions, sulphate and nitrate, are avidly held by this segment of the biosphere, as it indicated by the fact that their input exceeds their output. In the case of potassium all but 700 grams per hectare is captured.

Collectively the "nonvolatile" minerals, including silica, that fall from the clear New Hempshire sky amounted to some 13 kilograms per hectare in a typical year. With sulphate, nitrate and ammonium added, the total reached 51.4 kilograms per hectare. The second line of evidence indicates that the biosphere as a whole is becoming larger. Ecologists expect to find growth in secondary forests, but climax vegetation should be in a steady state, with annual gains balancing losses. According to figures I have recompiled from Robin and Basilevich, the mean world vegetation is not yet at climax. After the known quantity of dead leaves, branches and other litter is subtracted from the net production of new tissue, the difference is always positive, at an average 2,148 kilograms of new biomass per hectare of land per year.

With ash making up 1.2 percent of this biomass, about 26 kilograms of ash is annually withdrawn from an average hectare to sustain the increment of carbohydrate. The input of airborne elements at Hubbard Brook could provide this ash twice over, with no contribution from the local lithosphere. This comparison is impressionistic, and it may be misleading. Apart from industrial sulphate, which (as sulphuric acid) is perhaps as likely to corrode the biosphere as to nourish it, the world's vegetation may be in no danger of instant eutrophication. (If the biosphere is really becoming larger, the input of industrial carbon dioxide may constitute another major nutrient). The modes of recycling discovered at Hubbard Brook are nonetheless astonishing.

Added to what we know or can safely infer about other volatile elements, such studies underscore the necessity of a global view of biochemistry. What can be said with assurance is that there is a unique and nearly ubiquitous compound, with the empirical formula $H_{2960}O_{1480}C_{1480}N_{16}P_{1.8}S$, called living matter. Its synthesis, on an oxidized and uncarboxylated earth, is the most intricate feat of chemical engineering ever performed-and the most delicate operation that people have ever tempered with.

10

Water Cycle

The most abundant single substance in the biosphere is the familiar but unusual inorganic compound called water. The earth's oceans, ice caps, glaciers, lakes, rivers, soils and atmosphere contain 1.5 billion cubic kilometres of water in one form or another. In nearly all its physical properties water is either unique or at the extreme end of the range of a property. Its extraordinary physical properties, in turn, endow it with a unique chemistry. From these physical and chemical characteristics flows the biological importance of water. It is the purpose of this article to describe some of water's principal qualities and their significance in the biosphere. Water remains a liquid within the temperature range most suited to life processes, yet in due season there are occasions when liquid water exists in equilibrium with its solid and gaseous form, for example as ice on the top of a lake with water vapour in the air above it. Freezing starts at the surface of the water and proceeds downwards; this follows from one of water's many peculiar attributes.

Like everything else, ice included, liquid water contracts when it is cooled, but the shrinkage ceases before solidification, at about four degrees Celsius. From the temperature down to the freezing point the water expands, and because of its decreased density the cooler water floats on top of the warmer. Ice has a density of .92 with respect to the maximum density of water and hence an unconstrained block of ice will float in water with about an eleventh of its volume projecting above the surface. The biological significance of freezing from the surface downward, rather than from the bottom upward, is too well known to need repetition here. Among its other thermal properties water has the greatest specific heat known among liquids (the ability to store heat energy for a given increase in temperature). The same is true of water's latent heat of vaporization: at 20 degrees C. (68 degrees Fahrenheit), 585 calories are required to evaporate one gram of water.

Finally, with the exception of mercury, water has the greatest thermal conductivity of all liquids. Some consequences of water's large latent heat of evaporation, which is a major energizer of the atmosphere, will be considered below. Its great specific heat means that, for a given rate of energy input, the temperature of a given mass of water will rise more slowly than the temperature of any other material. Conversely, as energy is released its temperature will drop more slowly. This slow warming and cooling, together with other important factors, affects yearly, daily and even hourly changes in the temperature of oceans and lakes, which are quite different from the corresponding changes in the temperature of land.

Among other things, this can lead to differences in the thermal regimes of soils that are of major importance in ecology. The type of soil, interacting with water, determines the earliness or lateness of plant growth at a given site; the interaction may also affect the local risk of frost. In basic structure the water molecule has a small dipole moment and is freebly ionized. Water will dissolve almost anything to some extent (fortunately the extent is extremely small for many substances). The dissolved material tends to remain in solution because of another of water's exceptional attributes. The values given by the inverse-square law for the force that attracts separated positive and negative ions are determined by multiplying the square of the distance separating the ions by a constant that varies according to the nature of the separating medium. Known as the dielectric constant, this constant is greater for water than for any other substance. To get the same attractive force in water as in air, for example, the water separation has to be cut down to a ninth of the separation in air. Because of its extreme dielectric constant liquid water in the biosphere is not chemically pure (unlike water vapour, which is always pure, or ice, which can be and often is pure).

Instead liquid water is an ionic solution and one that always contains some hydrogen ions because the water itself can supply them. The concentration of hydrogen ions, expressed as a degree of dilution, gives the physical chemist a numerical index that describes the state of various water samples. The number is the logarithm (to the base 10) of the degree of dilution; the chemist labels pH. For his tests he is armed with a pH meter, calibrated from zero to 14. Fourteen orders of magnitude is an enormous range for any terrestrial property, yet the water content of the soil may give a reading anywhere from pH 3 (very acid) to pH 10 (very alkaline), which is equivalent to a range of from one to 10 million. These are extremes, however, and most terrestrial plant growth-including much of the world's agriculture proceeds in soil with a water content that ranges only a few units on each sides of pH 6.

The range for marine organisms is even more restrictive: coastal waters are about pH 9 and the general oceanic average is just over pH 8. Below pH 7.5 many marine animals die; eggs are particularly vulnerable. Below pH 7 the carbonate in seawater would remain in solution, rendering production of any kind of skeleton impossible. Another method of describing the state of a given water sample is independent of hydrogen-ion content. Material in solution, whether it is ionized or not, disturbs the liquid structure of the water; in thermodynamic terms the presence of solutes decreases the free energy of the water. Many soil and plant workers find it convenient to use the symbol pF for such changes in free energy, with the steps between units also representing one order of magnitude. As with the pH range, the range of pF values is very great. The quantity being measured in pF units is basically a potential, with the same dimensions as pressure. If all the water problems in soils, plants and animals were problems of solutions, it would be sufficient to describe the consequent variations in free energy as variations in osmotic potential, expressed in any of the conventional units of pressure.

The free energy of water, however, can be decreased in other ways, notably in capillary systems. The energy to lift the water into a notably in capillary tube (or in nature into the porous and cellular systems of soils and plants) comes out of the free energy of the water. Today this is known as "matric" potential, a term that has replaced the earlier "capillary" potential. In soils and plants the matric potential may be more than the osmotic potential. A comparison of numbers will give an idea of the pF scale and its ranges. The pressure is expressed as the height in centimetres of an equivalent column of water; thus one bar equals one atmosphere, which equals a 1,000-centimetre water column. This is equivalent to pF 3. In a waterlogged soil, beginning to drain, the matric potential may be between pF 0 and pF 1; in a fully drained soil the potential may be near pF 1.7.

In a soil that is as dry as plant uptake and the transpiration of water from leaves can make it, the matric potential will be about pF 4.2, which is close to 16 atmospheres of suction. The osmotic potential of seawater is near pF 4.5, which makes seawater too "dry" for plant roots: the salt content of plant cells might be anywhere in a range from less than pF 4 to pF 4.5. Here once again water is extreme. Associated with the matric potential in a capillary system there is a curved liquid-air interface; the value of the potential is found by doubling the known value of the liquid's surface tension and dividing the product by the radius of curvature.

Water has the greatest surface tension of any liquid known, so that at any given matric potential the radius of curvature of a water meniscus will be greater than it could be for another liquid. The greater the radius of curvature, the greater the total water content. In a soil this means that more liquid can be retained as water just because it is water. In general, but not always, this is an advantage for plant growth. The effects of a decrease in the free energy of water contained in porous soils or in the tissues of plants include a lowering of the fluid's freezing point and vapour pressure. If the source of the decrease is a matric potential, there is also negative pressure, or suction, that tends to pull all kinds of retaining walls together. The effect of freezing in soils and rocks is worth a brief aside.

As the temperature falls the water in the larger soil pores freezes first and the free-energy gradient is such that water will be withdrawn from the smaller pores. As a result ice lenses form in the coarser pore spaces and the finer pore spaces are exposed to greater shrinkage forces. Because water expands on freezing, the ice lenses have a disruptive effect as they make room for themselves. In rock this is the beginning of one method of soil formation. There tends to be a preferred size for the rock fragments produced by ice disruption. This size is near the optimum for transport by wind and is the dominant size in many of the loses soils that have accumulated in areas near glaciers.

In soil the ice disruption is the source of "frost tilth," which is sought by farmers when they leave land roughly plowed in the fall and hope for a sufficiently frosty winter. There are still some uncertainties with respect to the world's water balance, but agreement was reached on probable values or ranges during an international symposium on the subject held in Britain this summer as one of the activities of the International Hydrological Decade. The figures that follow are taken from the proceedings of the symposium. The world's water exists as liquid (salt and fresh), as solid (fresh) and as vapour (fresh).

There is some uncertainty in the value of the total volume, but it is near 1,500 million cubic kilometres. Estimates of the components are most easily expressed as average depths per unit area of the entire surface of the earth, which has a total area of 510 million square kilometres. Oceans and seas-liquid salt water-make up about 97 percent of all water, with an equivalent depth of between 2,700 and 2,800 metres; the greater part is in the Southern Hemisphere. Of the remaining 3 percent, three-quarters is locked up as solid in the polar ice caps and in glaciers. Here measurement is quite difficult, and a spread in estimates is inevitable. The equivalent depth of ice and snow may be near 120 metres, but at the recent symposium a value of 50 metres was not challenged.

The other large component of fresh liquid water is subject to similar uncertainty: the estimates for underground water may be near 45 metres, but again a value near 15 metres was not challenged. Estimates for surface water, mainly in the great lakes of the world, ranged from .4 metre to one metre. There is general agreement on the average water-vapour content of the atmosphere, at an equivalent in liquid of .03 metre. Although this is a very small fraction of the total, size is no measure of importance. Without water in the atmosphere there would be no weather: Leonardo da Vinci's dictum, "Water is the driver of nature," is justified on meteorological grounds alone. A little detail at this point will be helpful as an introduction to another aspect of the world circulation of water. The amount of water vapour is not the same everywhere, either geographically or seasonally. It is greatest at and near the Equator. If the air there were squeezed dry, it would yield about 44 millimetres of rainfall.

In middle latitudes, say from 40 to 50 degrees, the summer yield would be near 20 millimetres and the winter yield near 10 millimetres, with large variations that depend on geography and weather patterns. In the polar regions the yield ranges from two millimetres in winter to as much as eight in summer. Water vapour enters the atmosphere by evaporation (this term includes transpiration by vegetation), and the main oceanic sources are fairly identifiable. It leaves the atmosphere as rain or snow, and because the precipitation may take place close to the source or thousands of miles away, the residence time may vary from a few hours to a few weeks. A general average is nine or 10 days. The general balance of evaporation and precipitation needs three sets of figures, one set for the entire earth, one for the oceans and one for the land surface.

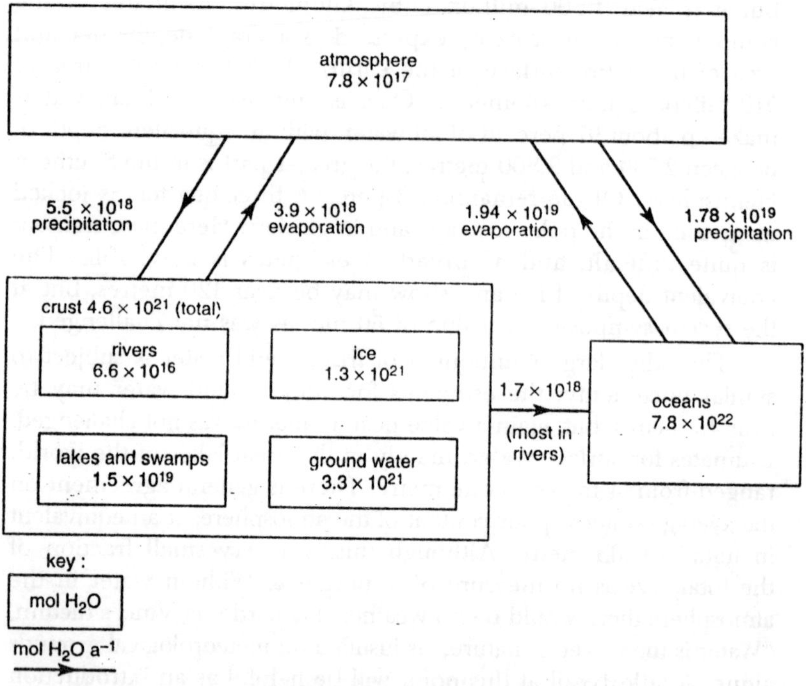

Fig. 10.1. The hydrological cycle.

Here, within a few percent, there is almost complete agreement on values. For the entire earth, average evaporation and precipitation are equal–as they must be–at very nearly 100 centimetres per year. For the oceans, expressed as equivalent depths over the area of the oceans, the average annual precipitation is between 107 and 114 centimetres, the average annual evaporation is between 116 and 124 centimetres and balance is restored by river flow, with an annual very close to 10 centimetres in all estimates. For the land surface the average annual precipitation is near 71 centimetres, the average annual evaporation is near 47 centimetres and the average annual river discharge is near 24 centimetres. (The ocean figure of 10 centimetres corresponds to the 24-centimetre land figure.) Because half of the land surface–ice caps, deserts, mountains, tundra–contributes little or nothing to evaporation, a better evaporation average would take into consideration only the land component of the biosphere where the availability of water is combined with the opportunity for evaporation. Here the average evaporation may total 100 centimetres per year.

The evaporation in high latitudes would of course be far less that the evaporation nearer the Equator. Available measurements support this conclusion. In Finland, at 65 degrees north latitude, the average evaporation is 20 centimetres per year; in southeastern England, at 50 degrees north, it is 50; in North Carolina, at 35 degrees north, it ranges from 80 to 120. On the Equator in the Congo basin the average is 120 centimetres per year; at the same latitude in Kenya it is 150. In the papyrus swamps of the Nile in the southern Sudan, 10 degrees north of the Equator, the average is 240 centimetres per year, but this is a special case. Here the river carries its water into the desert environment of the Sudd; evaporation rates are high not only because of the clear skies and intense sunshine overhead but also because the surrounding desert is a source of hot dry air that augments evaporation. This kind of advective augmentation operates in many places other than the Sudan, particularly in semiarid regions where irrigation is practiced, and not quite enough is known about it. Once in the air, water vapour may circulate locally or become part of the general circulation of the atmosphere.

The general circulation is one of the three important ways of moving water across the earth. Some indication of the worldwide volumes involved is given by the fact that the total annual precipitation over the U.S. comes to some 6,000 cubic kilometres, whereas the liquid equivalent of the water vapour that passes over the U.S. in a year owing to the general circulation of the atmosphere is 10 times that amount. Of the two remaining important ways of moving water across the earth, the major ocean currents comprise one and the discharge of rivers comprises the other. Both have substantial effects on the biosphere.

The ocean currents carry energy surpluses or deficits over great distances; one well-known instance accounts for the extreme contrast between the climates on the west and east sides of the Atlantic in the areas between 50 and 55 degrees north latitude. Without the Gulf Stream northwestern Europe would be a much less pleasant place in which to live and work; indeed, if the cold Lobrador Current had replaced the Gulf Stream, the history of civilization would have been very different. The rivers of the world not only are long-distance movers of water but also serve as conduits for dissolved and suspended material.

Because of its chemical and physical properties, water is a very efficient erosive agent; erosion, transport and deposition have to be recognized as geological processes associated with water in the biosphere. They are the processes that have produced lands and soils, now densely populated and intensively cropped, where annual floods and silt deposition are regarded as the mainstay of life. Elsewhere, notably in the Americas, silt is an embarrassment in the deltas where it settles, and its production is equally unwelcome in river headwaters. Two further points about river waterdeserve mention. First, the salt content of river water differs markedly in composition from that of the oceans. This suggests that the oceanic brine is not merely the accumulation of salts from aeons of land-surface leaching. Second, information about river discharge rates is scanty and not always reliable. As an example, it is only recently that a good estimate of the flow of the Amazon has been obtained. It proved to be twice the best previous estimate and indicates that almost a fifth of the world's river discharge comes from this one stream. It is not possible to do more than guess at the average amount of water the world's plant and animal populations contain.

Considered as the equivalent of rainfall, it may amount to about one millimetre over the entire surface of the earth. This is less by one order of magnitude than the amount of water vapour in the atmosphere, and its distribution is even more varied in space and time. For a fully grown good crop of corn in North America or of sugar beet in northwestern Europe the amount might come to the equivalent of five millimetres of rainfall, and its summer residence time would be two or three days. This is a measure of the rate of water supply needed to maintain optimum conditions for growth. Here, at the point of water untake by the roots of plants, begins the problem with respect to water in the biosphere that makes all other water problems seen trifling. With unimportant exceptions, the basis of all life on the earth is photosynthesis by green plants, a process that involves physics (in the fixation of solar energy) and chemistry (in the union of carbon dioxide and water to form carbohydrates and more complex biochemical compounds).

Water comes into the story in two ways: in transit (as part of the transpiration stream) and in residence (as its hydrogen is chemically bound into the plant structure). The amount that is bound, however, may be less than a fifth of the amount in transit. To give scale the argument that follows, here are some values based on a real crop in a real climate. In producing 20 fresh-weight tons of crop, 2,000 tons of water will pass into the plants at their roots. At harvest perhaps 15 tons of the water supply will be in transit, leaving the crop with a dry weight of five tons. To produce the five tons of dry matter three tons of water will have been fixed and transformed. The energy fixed in the dry matter will be 1 percent or less of the total solar energy received by the crop; nearly 40 percent of the energy will have been used to evaporate the transpired water. The average value of 40 percent for the net solar radiation income retained by a green crop cover varies, of course, with season and climate. The first loss is to reflectivity: of the solar radiation reaching the crop about 30 percent is reflected. There is also an income of long-wave radiation from the sky, but this is outweighed by the outgo of long-wave radiation from the earth to the atmosphere. When the deficit is met by deducting it from the remaining balance of short-wave solar income, the net retained income is decreased to 40 percent of the initial input.

As already noted, when the water is available, very nearly all this energy is used in evaporating water. Here once again water stands at the extreme of a range of physical properties. The volume of water evaporated per unit of energy input is less than it would be for any other liquid. The relevant physical constant, the latent heat of vaporization, is somewhat less than 600 calories per gram at ordinary temperatures, but the rounded figure is adequate for the present purpose. If we let R_1 represent the total radiant income in calories per square centimetre over a period of time, then the net radiation is about $4R_1$ and the evaporation equivalent is near R_1/- 1,500 grams per square centimetre (or centimetres of water depth as the equivalent of rainfall). Consider some real midsummer values to show what this means.

In a humid temperate climate the value of R_1 is close to 450 calories per square centimetre per day. This works out to an evaporation equivalent of three millimetres per day, which is a good estimate for June in southwestern England. For many of the farming areas of the U.S. The R_1 value is close to 650 calories per square centimetre, bringing the evaporation rate up to about 4.5 millimetres per day. The maximum rates known, which are found in irrigated areas, range from 4.5 to 7.5 millimetres per day. It is possible that the higher rates are influenced by advection from surrounding nonirrigated areas, as is the case in the papyrus swamps of the Nile. The most important fact to be considered in connection with this wide range of evaporation rates is that there are only very small variations among the evaporation rates of different kinds of plants. Thus the governing factor in variation is almost exclusively a climatic one. This fact and much other evidence suggest that the supposed water "need" of a crop is dictated not by the plants but by the weather. In this connection the concept of "potential transpiration," which came into use simultaneously and independently in at least two parts of the world, is of great value both in research and in the practical aspects of soil water management.

It is worthwhile seeing how potential transpiration is linked with elementary plant physiology and with some of the physics of soil water already considered. A growing plant takes in water at the roots and, in the absence of immediate replenishment, the process dries the soil so that more and more energy is required for further extraction. The energy requirement is very small, however, compared with the amount of energy needed to evaporate the same quantity of water from the plant's leaves. There can be no serious error in assuming, as Frank J. Veihmeyer of the University of California at Davis does, that all soil water is equally available for transpiration up to the stage marked by the onset of wilting.

The purpose of well-managed irrigation, of course, is to make sure that plants never get to the wilting stage. For maximum growth irrigation may have to consist of frequent small applications of water rather than occasional large ones. Given an adequate supply of water, the chain of consequences is simple. There are maximum values for each of several factors: water content in the plant, hydrostatic pressure in the plant and leaf turgidity. When neither the intensity of the light nor the concentration of carbon dioxide constitutes a limiting factor, maximum leaf turgidity permits maximum opening of the stomatal apertures in the leaf surface, thus affording the best possible opportunity for movement of carbon dioxide into the leaf. The state of the stomatal opening that allows easy inflow of carbon dioxide, however, also allows equally easy outflow of water vapour.

By far the greater part of the water need of plants is actually a "leakage" process that has to be kept goring to ensure continued growth. Given a sufficiently wet soil around the plants' roots, the rate of leakage is dictated not by plant physiology but by the physical factors of temperature, humidity and ventilation. The sole constraint is imposed by the law of energy conservation. In its last stages the transpiration stream undergoes a change of state from liquid to vapour, and the rate of change depends on the rate at which energy can reach the system to supply the necessary latent heat of vaporization. So much for the physics of the process. When the supply of water in the soil approaches exhaustion, plant physiology rather than physics begins to predominate. Plant type, root structure, phase of plant development, soil type, soil depth-these become the important factors. What is available for utilization has more significance than the weather has, particularly in semiarid zones. Because agriculture is most active in the more humid zones of the biosphere, it is useful to estimate how much reserve soil water is available on the average in these zones. Factors already described prevent any exact answer to this question.

Nonetheless, a cautions estimate, advanced with considerable reservation, would be about 10 centimetres of rainfall equivalent. Three examples will suffice to show the need for caution. There are large agricultural areas of North Carolina and neighbouring states where an inert subsoil is covered by no more than 20 centimetres of useful topsoil. Here the entire water reserve available to the agricultural cycle cannot exceed a rainfall equivalent of five centimetres. This is one extreme, the deep volcanic soils of East Africa are at the other. In those soils the roots of many plants go down as much as six metres below the surface. The available water in a profile that deep is equivalent to nearly 50 centimetres of rain, and the plant can extract water throughout a long dry season at something very close to the potential transpiration rate. An example from France falls somewhere in between. There the drying of the soil was observed while a crop of sugar beet transpired at the full potential rate throughout a dry summer.

(a)

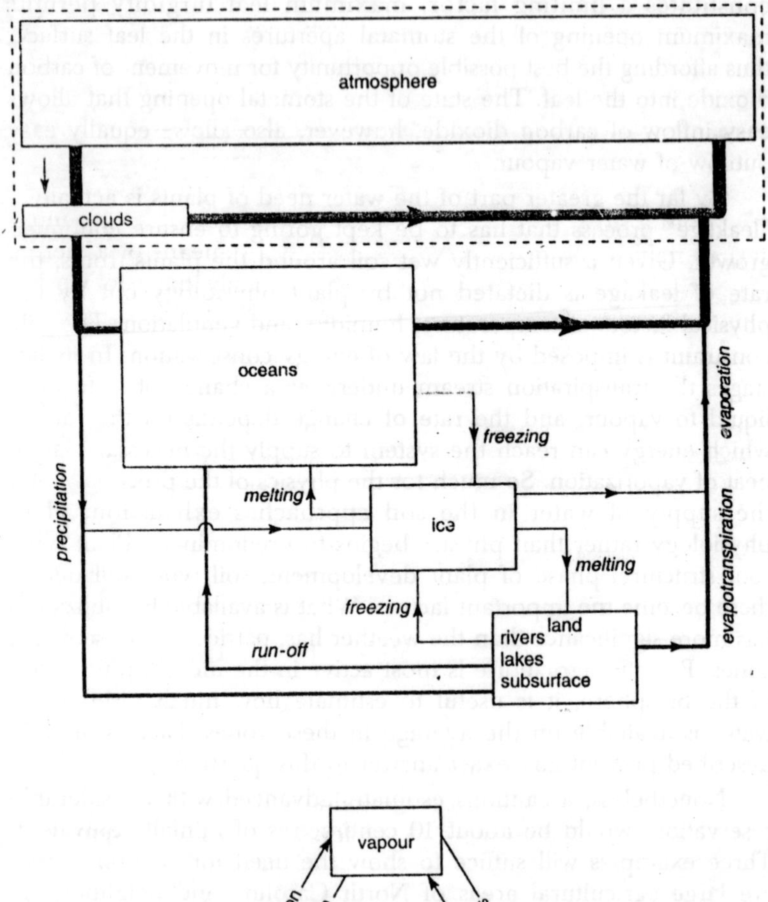

(b)

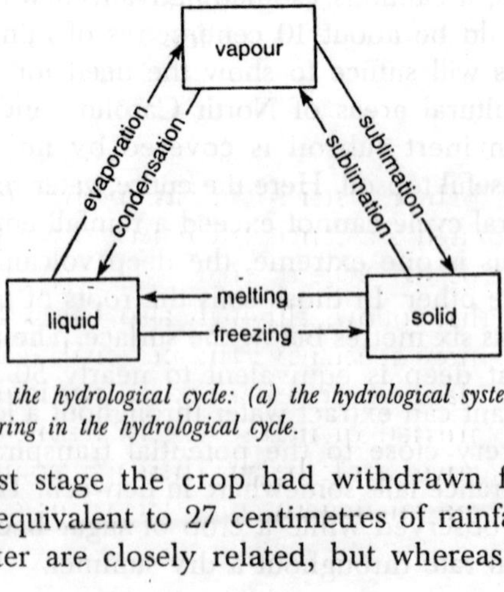

Fig. 10.2. Processes in the hydrological cycle: (a) the hydrological system; (b) changes in phase occurring in the hydrological cycle.

At the driest stage the crop had withdrawn from the soil available water equivalent to 27 centimetres of rainfall. Soil water and ground water are closely related, but whereas soil water is always biologically important, the importance of ground water may range from being trivial to being all that matters. The soil is a kind of buffer between rainfall and ground water. In general any deficit in soil moisture that has built up in a dry period must be completely restored by rain before there is any water surplus available to move down to ground water. This is an important consideration for the water engineer, who may be drawing a water supply from a stream (permanent streams are sustained by ground water) or may be tapping an aquifer directly by means of a well. In the first instance the engineer will presumably have to work within legal constraints on how much river water he can divert.

In the second, if he is to choose a safe aquifer pumping rate, the engineer must (or should!) have some awareness of the current soil-moisture deficit and of the likely rates of rainfall in the months ahead. The river engineer can use the same information for another purpose: the soil-moisture deficit will enable him to estimate the risk of flooding in the event of a heavy storm. In some countries the control of ground water is a major outlet for engineering skill, and its exploitation is the basis of farming technique. One need only think on the one hand of the Netherlands and on the other of such semiarid regions as Iran, where deep tunnels tap the buried aquifers and carry ground water to valley bottoms.

In many semiarid regions the vegetation along transient streams maintains its luxuriance because the groundwater level there is close to the surface and within reach of plant roots. The plants' effective reach depends both on the soil and on the kind of plant, but in general it is seldom more than a few metres. The movements of any water table deeper than that is unaffected by the plant growth or the evaporation processes taking place above it, and its ground water contributes nothing to the biological activity at the surface. What has been said about water so far has involved terms that are generally accepted, and the concepts themselves are supported by good reasoning, good evidence or both.

The remarks that follow, although also based on reasoning and evidence, are more speculative and personal. If the biosphere is taken to be the place where water and energy interact, can the interaction be expressed quantitatively in terms of biological productivity? The answer has to be no. There are too many variable. All the same, by rearranging some of the water quantities and energy quantities that are known, a suggestive relation can be obtained. Start with the fact that, for a good crop, 1 percent or less of the incoming solar radiation is fixed as dry matter (here and in what follows that 1 percent refers to the total botanical yield, irrespective of economic value). We shall give this percentage the symbol e, and express it numerically as 100 per 10,000. This degree of efficiency is achieved only by an experiment station or by an extremely competent commercial farmer.

Based on the statistics of world cereal production, including straw as well as grain, the average achievement in highly mechanized industrial farming shows an efficiency of only about 35 per 10,000. This decreases to roughly 17 per 10,000 in North America, and in tropical Africa and Asia subsistence farming rarely shows an efficiency better than 8, even when the possibility of two crops per year is allowed for. There is obviously room for improvement everywhere and the question in the present context is: Where does water come in, and how? Some evidence is now being accumulated suggesting that, when there are no limitations on water supply, the total crop growth is proportional to the total of potential transpiration over the period of growth. The factor of proportionality depends on many things: plant variety, management, kinds and quantities of fertilizer, pest and disease controls and the like. Hence the negative answer to the earlier question. Still, something can be done with ratios.

There is reason to believe potential transpiration is a fairly constant fraction of the solar radiation income. Combining this fact with the relation between potential transpiration and total crop growth, it is possible to derive a connection between growth rate and utilization of water, assuming that unlimited water is available (by inference, there would be a similar response to timely irrigation). In what may seem too precise a form, the answer is that the increase in yield (t) equals .39 ε. Here ε represents efficiency and t can be read, according to preference, as metric tons per hectare per centimetre of water applied or as tons per acre per inch of water. Considering the fivefold (or perhaps tenfold) world variation in efficiency, some uncertainty in the multiplying factor is unimportant; others may prefer, or find, a different value. Taking the illustration already given, suppose the area involved is one acre and the efficiency is exactly or almost 100 per 10,000.

The predicted increase in yield in response to water, applied when it is needed, is .39 ton of dry matter per acre per inch of water. In terms of fresh weight the gain is about 1.5 tons per acre per inch. This is the kind of response obtained in experiments with irrigated potatoes in Britain. There are some countries where the value of ε is small because of lack of water, but there are many, including several of the rice-growing nations, where the small value of ε is more truly a measure of the inefficiency of the farming system itself. To get the most out of water, whether it comes from irrigation or from rainfall, the standard of performance elsewhere in the system must be improved: better varieties, better soil management, better crop husbandry, better plant hygiene and better pest control. Then water may be the driver of nature in agriculture as well as in the atmosphere.

11

NITROGEN CYCLE

Although men and other land animals live in an ocean of air that is 79 percent nitrogen, their supply of food is limited more by the availability of fixed nitrogen than by that of any other plant nutrient. By "fixed" is meant nitrogen incorporated in a chemical compound that can be utilized by plants and animals. As it exists in the atmosphere nitrogen is an inert gas except to the comparatively few organism that have the ability to convert the element to a combined form. A smaller but still significant amount of atmospheric nitrogen is fixed by ionizing phenomena such as cosmic radiation, meteor trails and lightning, which momentarily provide the high energy needed for nitrogen to react with oxygen or the hydrogen of water. Nitrogen is also fixed by marine organisms, but the largest single natural source of fixed nitrogen is probably terrestrial microorganisms and associations between such microorganisms and plants. Of all man's recent interventions in the cycles of nature the industrial fixation of nitrogen far exceeds all the others in magnitude.

Since 1950 the amount of nitrogen annually fixed for the production of fertilizer has increased approximately fivefold, until it now equals the amount that was fixed by all terrestrial ecosystems before the advent of modern agriculture. In 1968 the world's annual output of industrially fixed nitrogen amounted to about 30 million tons of nitrogen; by the year 2000 the industrial fixation of nitrogen; may well exceed 100 million tons. Before the large-scale manufacture of synthetic fertilizers and the wide cultivation of the nitrogen-fixing legumes one could say with some confidence that the amount of nitrogen removed from the atmosphere by natural fixation processes was closely balanced by the amount returned to the atmosphere by organisms that convert organic nitrates to gaseous nitrogen. Now one cannot be sure that the denitrifying processes are keeping pace with the fixation processes. Nor can one predict all the consequences if nitrogen fixation were to exceed denitrification over an extended period.

We do known that excessive runoff of nitrogen compounds in streams and rivers can result in "blooms" of algae and intensified biological activity that deplete the available oxygen and destroy fish and other oxygen-dependent organisms. To appreciate the intricate web of nitrogen flow in the biosphere let us trace the course of nitrogen atoms from the atmosphere into the cells of microorganisms, and then into the soil as fixed nitrogen, where it is available to higher plants and ultimately to animals. Plants and animals die and return the fixed nitrogen to the soil, at which point the nitrogen may simply be recycled through a new generation of plants and animals or it may be broken down into elemental nitrogen and returned to the atmosphere. Because much of the terminology used to describe steps in the nitrogen cycle evolved in previous centuries it have an archaic quality. Antoine Laurent Lavoisier, who clarified the composition of air, gave nitrogen the name azote, meaning without life.

The term is still found in the family name of an important nitrogen-fixing bacterium: the Azotobacteraceae. One might think that fixation would merely be termed nitrification, to indicate the addition of nitrogen to some other substance, but nitrification is reserved for a specialized series of reactions in which a few species of microorganisms oxidize the ammonium ion (NH_4^+) to nitrite (NO_2^-) or nitrite to nitrate (NO_3^-), When nitrites or nitrates are reduced to gaseous compounds such as molecular nitrogen (N_2) or nitrous oxide (N_2O), the process is termed denitrification. "Ammonification" describes the process by which the nitrogen of organic compounds (chiefly amino acids) is converted to ammonium ion.

The process operates when microorganisms decompose the remains of dead plants and animals. Finally, a word should be said about the terms oxidation and reduction, which have come to mean more than just the addition of oxygen or its removal. Oxidation is any process that removes electrons from a substance. Reduction is

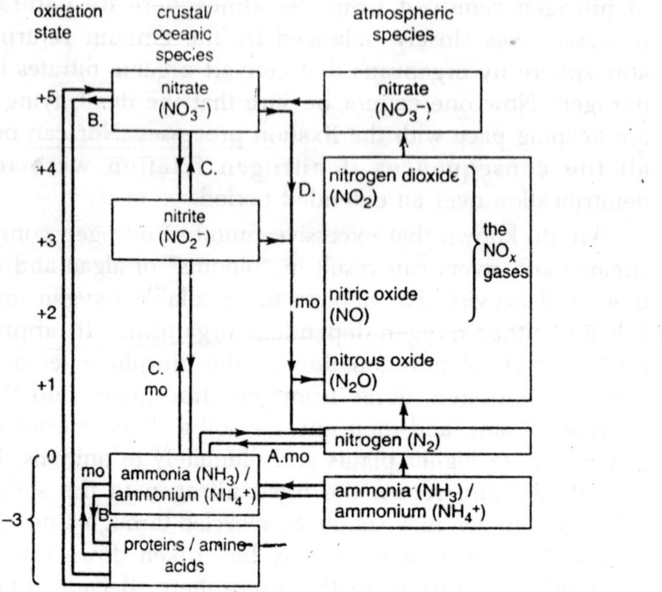

Fig. 11.1. Nitrogen speciation in the environment.

the reverse process: the addition of electrons. Since electrons can neither be created nor destroyed in a chemical reaction, the oxidation of one substance always implies the reduction of another. One may wonder how it is that some organisms find it profitable to oxidize nitrogen compounds whereas other organism-even organisms in the same environment–owe their survival to their ability to reduce nitrogen compounds.

Apart from photosynthetic organisms, which obtain their energy from radiation, all living forms depend for their energy on chemical transformations. These transformations normally involve the oxidation of one compound and the reduction of another, although in some cases the compound being oxidized and the compound being reduced are different molecules of the same substance, and in other cases the reactants are fragments of a single molecular species. Nitrogen can be cycled because the reduced inorganic compounds of nitrogen can be oxidized by atmospheric oxygen with a yield of useful energy. Under anaerobic conditions the oxidized compounds a nitrogen can act as oxidizing agents for the burning of organic compounds (and a few inorganic compound), again with a yield of useful energy. Nitrogen is able to play its complicated role in life processes because it has an unusual number of oxidation levels, or valences. An oxidation level indicates the

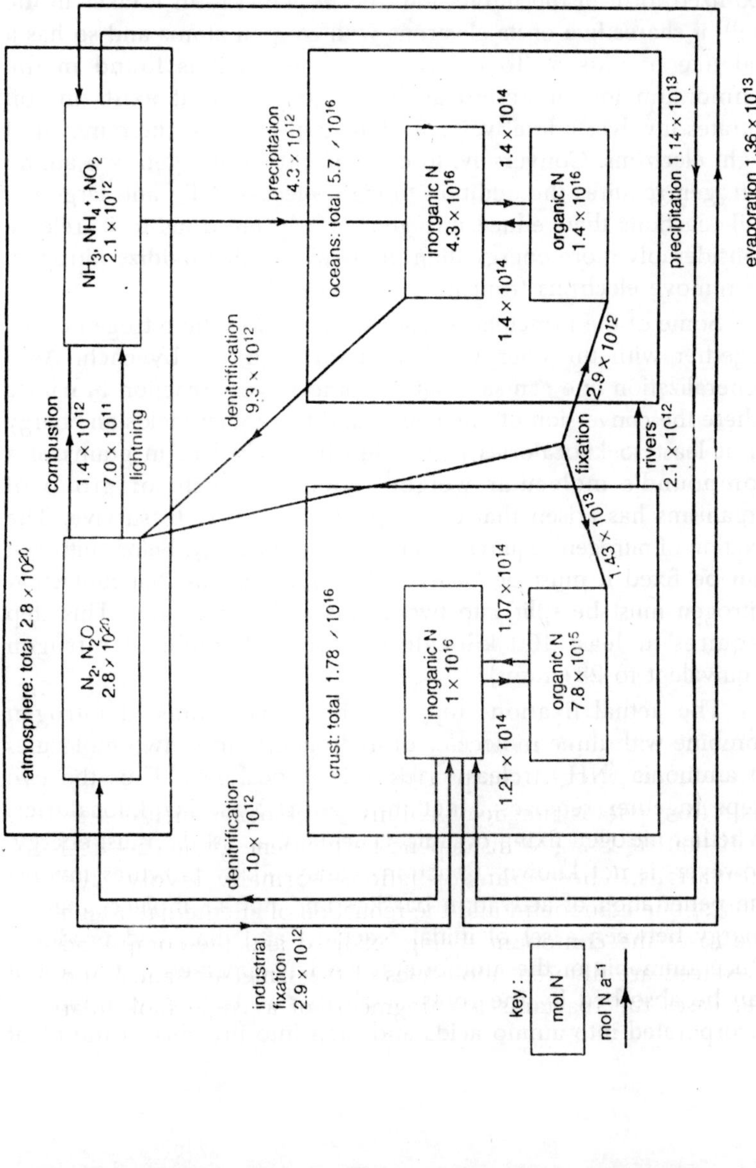

Fig. 11.2. The Nitrogen cycle.

numbers of electrons that an atom in a particular compound has "accepted" or "donated".

In plants and animals most nitrogen exists either in the form of the ammonium ion or of amino ($-NH_2$) compounds. In either case it is highly reduced, it has acquired three electrons by its association with three other atoms and thus is said to have a valence of minus 3. At the other extreme, when nitrogen is in the highly oxidized form of the nitrate ion (the principal form it takes in the soil), it shares five of its electrons with oxygen atoms and so has a valence of plus 5. To convert nitrogen as it is found in the ammonium ion or amino acids to nitrogen as it exists in soil nitrates involves a total valence change of eight, or the removal of eight electrons. Conversely, to convert nitrate nitrogen into amino nitrogen requires the addition to eight electrons. By and large the soil reactions that reduce nitrogen, or add electrons to it, release considerably more energy than the reactions that oxidize nitrogen, or remove electrons from it.

Some of the principal reactions involved in the nitrogen cycle, together with the energy released (or required) by each. As a generalization one can say that for almost every reaction in nature where the conversion of one compound to another yields an energy of at least 15 kilocalories per mole (the equivalent in grams of a compound's molecular weight), some organism or group of organisms has arisen that can exploit this energy to survive. The fixation of nitrogen requires an investment of energy. Before nitrogen can be fixed it must be "activated," which means that molecular nitrogen must be split into two atoms of free nitrogen. This step requires at least 160 kilocalories for each mole of nitrogen (equivalent to 28 grams).

The actual fixation step, in which two atoms of nitrogen combine with three molecules of hydrogen to form two molecules of ammonia (NH_3), releases about 13 kilocalories. Thus the two steps together requires a net input of at least 147 kilocalories. Whether nitrogen fixing organisms actually invest the must energy, however, is not known. Reactions catalyzed by enzymes involve the penetration of activation barriers and not a simple change in energy between a set of initial reactants and their end products. Once ammonia or the ammonium ion has appeared in the soil, it can be absorbed by the roots of plants and the nitrogen can be incorporated into amino acids and then into proteins. If the plant is subsequently eaten by an animal, the nitrogen may be incorporated into a new protein.

In either case the protein ultimately returns to the soil, where it is decomposed (usually with bacterial help) into its component amino acids. Assuming that conditions are aerobic, meaning that on adequate supply of oxygen is present, the soil will contain many microorganism capable of oxidizing amino acids to carbon dioxide, water and ammonia. If the amino acid happens to be glycine, the reaction will yield 176 kilocalories per mole. A few microorganisms represented by the genus Nitrosomonas employ nitrification of the ammonium ion as their sole source of energy. In the presence of oxygen, ammonia is converted to nitrite ion (NO_2^-) plus water, with an energy yield of about 65 kilocalories per mole, which is quite adequate for a comfortable existence.

Nitrosomonas belongs to the group of microorganisms termed autotrophs, which get along without an organic source of energy. Photoautotrophs obtain their energy form light: chemoautotrophs (such as Nitrosomonas) obtain energy from inorganic compounds. There is another specialized group of microorganisms, represented by Nitrobacter, that are capable of extracting additional energy from the nitrite generated by Nitrosomonas. The result is the oxidation of a nitrite ion to a nitrate ion with the release of about 17 kilocalories per mole, which is just enough to support the existence of Nitrobacter. In the soil there are numerous kinds of denitrifying bacteria (for example Pseudomonas denitrificans) that, if obliged to exist in the absence of oxygen, are able to use the nitrate or nitrite ion as electron acceptors for the oxidation of organic compounds. In these reactions the energy yield is nearly as large as it would be if pure oxygen were the oxidizing agent. When glucose reacts with oxygen, the energy yield is 686 kilocalories per mole of glucose.

In microorganisms living under anaerobic conditions the reactions of glucose with nitrate ion yields about 545 kilocalories per mole of glucose if the nitrogen is reduced to nitrous oxide, and 570 kilocalories if the nitrogen is reduced all the way to its elemental gaseous state. The comparative value of ammonium and nitrate ions as a source of nitrogen for plants has been the subject of a number of investigations. One might think that the question would be readily resolved in favour of the ammonium ion: its valence is minus 3, the same as the valence of nitrogen in amino acids, whereas the valence of the nitrate ion in plus 5. On this basis plants must expend energy to reduce nitrogen from a valence of plus 5 to one of minus 3. The fact is, however, that there are complicating factors; the preferred from of nitrogen depends on other variables. Because the ammonium ion has a positive charge it tends to be trapped on clay particles near the point where it is formed (or where it is introduced artificially) until it has been oxidized.

The nitrate ion, being negatively charged, moves freely through the soil and thus is more readily carried downward into the root zone. Although the demand for fertilizer in solid form (such as ammonium nitrate and urea) remains high, anhydrous ammonia and liquid ammoniacal fertilizers are now widely applied. The quantity of nitrogen per unit weight of ammonia is much greater than it is per unit of nitrate; moreover, liquids are easier to handle than solids. Unit the end of the 19th century little was known about the soil organisms that fix nitrogen. In fact, at that time there was some concern among scientists that the denitrifying bacteria, which had just been discovered, would eventually deplete the reserve of fixed nitrogen in the soil and cripple farm productivity. In an address before the Royal Society of London, Sir William Crookes painted a bleak picture for world food production unless artificial means of fixing nitrogen were soon developed. This was a period when Chilean nitrate reserves were the main source of fixed nitrogen for both fertilizer and explosives.

As it turned out, the demand for explosives provided the chief incentive for the invention of the catalytic fixation process by Fritz Haber and Karl Bosch of Germany in 1914. In this process atmospheric nitrogen and hydrogen are passed over a catalyst (usually nickel) at a temperature of about 500 degrees Celsius and pressure of several hundred atmospheres. In a French version of the process, developed by Georges Claude, nitrogen was obtained by the fractional liquefaction of air. In current versions of he Haber process the source of hydrogen is often the methane in natural gas. As the biological fixation of nitrogen and the entire nitrogen cycle became better understood, the role of the denitrifying bacteria fell into place. Without such bacteria to return nitrogen to the atmosphere most of the atmospheric nitrogen would now be in the oceans or locked up in sediments.

Actually, of course, there is not enough oxygen in the atmosphere today to convert all the free nitrogen into nitrates.

One can imagine, however, that if a one-way process were to develop in the absence of denitrifying bacteria, the addition of nitrates to the ocean would make seawater slightly more acidic and start the release of carbon dioxide from carbonate rocks. Eventually the carbon dioxide would be taken up by plants, and if the carbon were then deposition as coal or other hydrocarbons, the remaining oxygen would be available in the atmosphere to be combined with nitrogen. Because of the large number of variables involved it is difficult to predict how the world would look without the denitrification reaction, but it would certainly not be the world be know. The full story of the biological fixation of nitrogen has not yet been written.

One would like to know how the activating enzyme (nitrogenase) used by nitrogen fixing bacteria can accomplish at ordinary temperatures and pressures what takes hundreds of degrees and thousands of pounds of pressure in a synthetic-ammonia reactor. The total amount of nitrogenase in the world is probably no more than a few kilograms. The nitrogen-fixing microorganism are divided into two broad classes: those that are "free-living" and those that live in symbiotic association with higher plants. This distinction, however, is not as sharp as it was one though to be, because the interaction of plants and microorganisms has varying degrees of intimacy.

The symbionts depend directly on the plants for their energy supply and probably for special nutrients as well. The free-living nitrogen fixers are indirectly dependent on plants for their energy or, as in the case of the blue-green algae and photosynthetic bacteria, obtain energy directly from sunlight. Although the nitrogen-fixation reaction is associated with only a few dozen species of higher plants, these species are widely distributed in the plant kingdom. Among the more primitive plants whose symbionts can fix nitrogen are the cycads and the ginkgos, which can be traced back to the Carboniferous period of some 300 million years ago. It is probably that the primitive atmosphere of the earth contained ammonia, in which case the necessity for nitrogen fixation did not arise for hundreds of millions of years. Various kinds of bacterial, particularly the Azotobacteraceae, are evidently the chief suppliers of fixed nitrogen in grasslands and other ecosystems where plants with nitrogen-fixing symbionts are absent. Good quantitative information on the rate of nitrogen fixation in such ecosystems is hard to obtain.

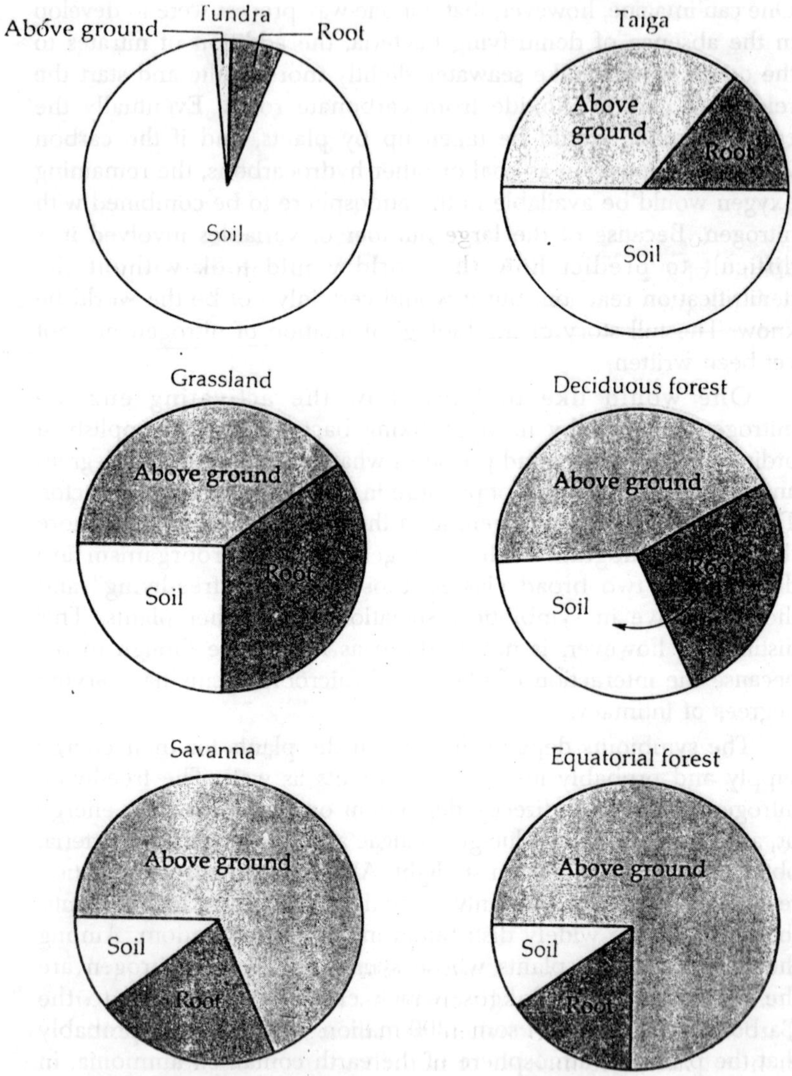

Fig. 11.3. Deposition of nitrogen in the three organic-matter compartments (above ground, root, and soil) for each of six biome types.

Most investigations indicate a nitrogen-fixation rate of only two or three kilograms per hectare per year, with a maximum of perhaps five or six kilogram. Blue-green algae seem to be an important source of fixed nitrogen under conditions that favour their development. They may be a significant source in rice paddies and other environments favouring their growth. In natural ecosystems with mixed vegetation the symbiotic associations involving such plant genera as Alnus (the alders) and Ceanothus (the buckthorns) are important suppliers of fixed nitrogen. For the earth as a whole, however, the greatest natural source of fixed nitrogen is probably the legumes. They are certainly the most important from an agronomic standpoint and have therefore been the most closely studies. The input of nitrogen from the microbial symbionts of alfalfa and other leguminous crops can easily amount to 350 kilograms per hectare, or roughly 100 times the annual rate of fixation attainable by nonsymbiotic in a natural ecosystem.

Recommendations for increasing the world's food supply usually emphasize increasing the cultivation of legumes not only to enrich the soil in nitrogen but also because legumes (for example peas and beans) are themselves a food crop containing a good nutritional balance of amino acids. There are, however, several obstacles of carrying out such recommendations. The first is custom and taste. Many societies with no tradition of growing and eating legumes are reluctant to adopt them as a basic food. For the farmer legumes can create a more immediate problem: the increased yields made possible by the extra nitrogen lead to the increased consumption of other essential elements, notably potassium and phosphorus.

As a consequence farmers often say that legumes are "hard on the soil." What this really means is that the large yield of such crops places a high demand on all minerals, and unless the minerals are supplied the full benefit of the crop is not realized. Symbiotic nitrogen fixers have a greater need for some micronutrients (for example molybdenum) than most plants do. It is now known that molybdenum is directly incorporated in the nitrogen-fixing enzyme nitrogenase. In Australia there were large areas where legumes refused to grow at all until it was discovered that the land could be made fertile by the addition of as little as two ounces of molybdenum per acre. Cobalt turns out to be another essential micronutrient for the fixation of nitrogen.

The addition of only 10 parts per trillion of cobalt in a culture solution can make the difference between plants that are stunted and obviously in need of nitrogen and plants that are healthy and growing vigorously. Although legumes and their symbionts are energetic fixers of nitrogen, there are indications that the yield of a legume crop can be increased still further by direct application of fertilizer instead of depending on the plant to supply all its own needs for fixed nitrogen. Additional experiments are needed to determine just how much the yield can be increased and how this increase compares with the industrial fixation of nitrogen in term of energy investment. Industrial processes: call for some 6,000 kilocalories per kilogram of nitrogen fixed, which is very little more than the theoretical minimum.

The few controlled studies with which I am familiar suggest that the increase in crop yield achieved by the addition of a kilogram of nitrogen amounts to about the same number of calories. This comparison suggests that one can exchange the calories, put into industrial fixation of nitrogen for the calories contained in food. In actually this trade-off applies to the entire agricultural enterprise. The energy required for preparing, tilling and harvesting a field and for processing and distributing the products is only slightly less than the energy continued in the harvested crop. Having examined the principal reactions that propel the nitrogen cycle, we are now in position to view the process as a whole and to interpret some of its broad implication.

One must be cautious in trying to present a worldwide inventory of a particular element in the biosphere in indicating annual flows from one part of a cycle to another. The balance sheet for nitrogen is particularly crude because we do not have enough information to assign accurate estimates to the amounts, of nitrogen that are fixed and subsequently returned to the atmosphere by biological processes. Another source of uncertainty involves the amount of nitrogen fixed by ionizing phenomena in the atmosphere. Although one can measure the amount of fixed nitrogen in rainfall, one is forced to guess how much is produced by ionization and how much represents nitrogen that has entered the atmosphere from the land or the sea, either as ammonia or as oxides of nitrogen. Because the ocean is slightly alkaline it could release ammonia at a low rate, but that rate is almost impossible to estimate. Land areas are a more likely source of nitrogen oxides, and some reasonable estimates of the rate of loss are possible.

One can say that the total amount of fixed nitrogen delivered to the earth by rainfall is of the order of 25 million metric tons per year. My own estimate is that 70 percent of this total is previously fixed nitrogen cycling through the biosphere, and that only 30 percent is freshly fixed by lightning and other atmospheric phenomena. Another factor that is difficult to estimate is the small but steady loss of nitrogen from the biosphere to sedimentary rocks.

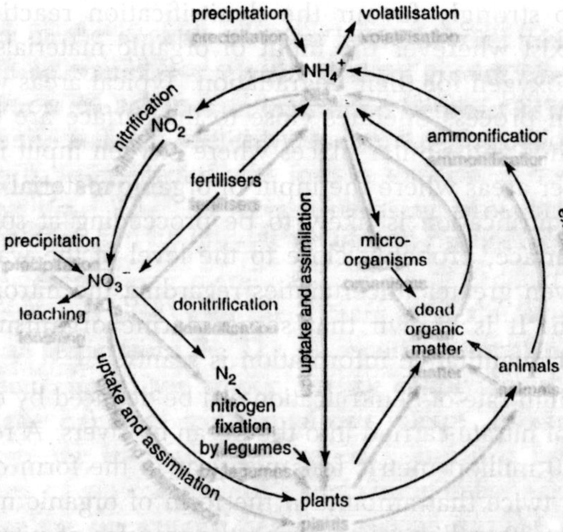

Fig. 11.4. The soil-based nitrogen sub-cycle.

Conversely, there is a continuous delivery of new nitrogen to the system by the weathering of igneous rocks in the crust of the earth.

The average nitrogen content of igneous rocks, however, is considerably lower than that of sedimentary rocks, and since the quantities of the two kinds of rock are roughly equal, one would expect a net loss of nitrogen from the biosphere through geologic time. Conceivably this loss is just about balanced by the delivery of "juvenile" nitrogen to the atmosphere by volcanic action. The amount of fixed nitrogen reintroduced in this way probably does not exceed two or three million tons per year. Whereas late-19th-century scientists worried that denitrifying bacteria were exhausting the nitrogen in the soil, we must be concerned today that denitrification may not be keeping pace with nitrogen fixation considering the large amounts of fixed nitrogen that are being introduced in the biosphere by industrial fixation and the cultivation of legumes. It has become urgent to learn much more about exactly where and under what circumstance denitrification takes place.

We know first of all that denitrification does not normally proceed to any great extent under aerobic conditions. Whenever free oxygen is available, it is energetically advantageous for an organism to use it to oxidize organic compounds rather than to use the oxygen in nitrate salts. One can conclude that there must be large areas in the biosphere where conditions are sufficiently anaerobic to strongly favour the denitrification reaction. Such conditions exist wherever the input of organic materials exceeds the input of oxygen for their degradation. Typical areas where the denitrification process operates close to the surface are the arctic tundra, swamps and similar places where oxygen input is limited. In many other areas where the input of organic material is sizable, however, denitrification is likely to be proceeding at some point below the surface, probably close to the level of the water table. There are even greater uncertainties regarding the nitrogen cycle in the ocean. It is known that some marine organisms do fix nitrogen, but quantitative information is scanty.

A minimum rate of denitrification can be deduced by estimating the amount of nitrate carried into the ocean by rivers: A reasonable estimate is 10 million metric tons per year in the form of nitrates and perhaps twice that amount in the form of organic material, a total of about 30 million tons. Since the transfer of nitrogen into sediments is slight, one can conclude that, at least before man's intervention in the nitrogen cycle, the ocean was probably capable of denitrifying that amount of fixed nitrogen. The many blanks in our knowledge of the nitrogen cycle are disturbing when one considers that the amount of nitrogen fixed industrially has been doubling about every six years. If we add to this extra nitrogen the amounts fixed by the cultivation of legumes, it already exceeds (by perhaps 10%) the amount of nitrogen fixed in nature.

Unless fertilizers and nitrogenous wastes are carefully managed, rivers and lakes can become loaded with the nitrogen carried in runoff waters. In such waterways and in neighbouring groundwater systems the nitrogen concentration could, and in some cases already does, exceed the levels acceptable for human consumption. Under some circumstances bacterial denitrification can be exploited to control the buildup of fixed nitrogen, but much work has to be done to develop successful management techniques. The problem of nitrogen disposal is aggravated by the nitrogen contained in the organic wastes of a steadily increasing human and domestic-animal population. Ideally this waste nitrogen should be recycled back of the soil, but efficient and acceptable means for doing so remain to be developed. At present it is economically sounder for the farmer to keep adding industrial fertilizers to his crops. The ingenuity that has been used to feed a growing world population will have to be matched quickly by an effort to keep the nitrogen cycle in reasonable balance.

12

CARBON CYCLE

The biosphere contains a complex mixture of carbon compounds in a continuous state of creation, transformation and decomposition. This dynamic state is maintained through the ability of phytoplankton in the sea and plants on land to capture the energy of sunlight and utilize it to transform carbon dioxide (and water) into organic molecules of precise architecture and rich diversity. Chemists and molecular biologists have unraveled many of the intricate processes needed to create the microworld of the living cell. Equally fundamental and no less interesting is the effort to grasp the overall balance and flow of material in the worldwide community of plants and animals that has developed in the few billion years since life began. This is ecology in the broadest sense of the word: the complex interplay between, on the one hand, communities of plants and animals and, on the other, both kinds of community and their nonliving environment. We now know that the biosphere has not developed in a static inorganic environment.

Rather the living world has profoundly altered the primitive lifeless earth, gradually changing the composition of the atmosphere, the sea and the top layers of the solid crust, both on land and under the ocean. Thus a study of the carbon cycle in the biosphere is fundamentally a study of the overall global interactions of living organisms and their physical and chemical environment. To bring order into this world of complex interactions biologists must combine their knowledge with the information available to students of geology, oceanography and meteorology. The engine for the organic processes that reconstructed the primitive earth is photosynthesis. Regardless of whether it takes place on land or in the sea, it can be summarized by a single reaction: $CO_2 + 2H_2A$ + light→ $CH_2O + H_2O + 2A$ + energy.

The formaldehyde molecule CH_2O symbolizes the simplest organic compound; the term "energy" indicates that the reaction stores energy in chemical form. H_2A is commonly water (H_2), in which case 2A symbolizes the release of free oxygen (O_2). There are, however, bacteria that can use compounds in which A stands for sulphur, for some organic radial or for nothing at all. Organisms that are able to use carbon dioxide as their sole source of carbon are known as autotrophs. Those that use light energy for reducing carbon dioxide are called phototrophic, and those that use the energy stored in inorganic chemical bonds (for example the bonds of nitrates and sulphates) are called chemolithotrophic.

Most organisms, however, require performed organic molecules for growth; hence they are known as heterotrophs. The nonsulphur bacteria are an unusual group that is both photosynthetic and heterotrophic. Chemoheterotrophic organisms, for example animals, obtain their energy from organic compounds without need for light. An organism may be either aerobic or anaerobic regardless of its source of carbon or energy. Thus some anaerobic chemoheterotrophs can survive in the deep ocean and deep lakes in the total absence of light of free oxygen. There is more to plant life than the creation of organic compounds by photosynthesis.

Plant growth involves a series of chemical processes and transformations that require energy. This energy is obtained by reactions that use the oxygen in the surrounding water and air to unlock the energy that has been stored by photosynthesis. The process, which releases carbon dioxide, is termed respiration. It is a continuous process and is therefore dominant at night, when photosynthesis is shut down. If one measures the carbon dioxide at various levels above the ground in a forest, one can observe pronounced changes in concentration over a 24-hour period. The average concentration of carbon dioxide in the atmosphere is about 320 part per million.

When the sun rises, photosynthesis begins and leads to a rapid decrease in the carbon dioxide concentration as leaves (and the needless of conifers) convert carbon dioxide into organic compounds. Toward noon, as the temperature increases and the humidity decreases, the rate of respiration rises and the net consumption of carbon dioxide slowly declines. Minimum values of carbon dioxide

10 to 15 parts per million: below the daily average are reached around noon at treetop level. At sunset photosynthesis ceases while respiration continues, with the result that the carbon dioxide concentration close to the ground may exceed 400 parts per million. This high value reflects partly the release of carbon dioxide from the decomposition of organic matter in the soil and partly the tendency of air to stagnate near the ground at night, when there is no solar heating to produce convection currents. The net productivity, or net rate of fixation, of carbon dioxide varies greatly from one type of vegetation to another.

Rapidly growing tropical rain forests annually fix between one kilogram and two kilograms of carbon (in the form of carbon dioxide) per square metre of land surface, which is roughly equal to the amount of carbon dioxide in a column of air extending from the same area of the earth's surface to the top of the atmosphere. The arctic tundra and the nearly barren regions of the desert may fix as little as 1 percent of that amount. The forests and cultivated fields of the middle latitudes assimilate between 2 and 4 kilogram per square metre. For the earth as a whole the areas of high productivity are small. A fair estimate is that the land areas of the earth fix into organic compounds 20 to 30 billion net metric tons of carbon per year. There is considerable uncertainty in this figure; published estimates range from 10 to 100 billion tons. The amount of carbon in the form of carbon dioxide consumed annually by phytoplankton in the oceans is perhaps 40 billion tons, or roughly the same as the gross assimilation of carbon dioxide by land vegetation. Both the carbon dioxide consumed and the oxygen released are largely in the form of gas dissolved near the ocean surface.

Therefore most of the carbon cycle in the sea is self-contained: the released oxygen is consumed by sea animals, and their ultimate decomposition releases carbon dioxide back into solution. As we shall see, however, there is a dynamic exchange of carbon dioxide (and oxygen) between the atmosphere and the sea, brought about by the action of the wind and waves. At any given moment the amount of carbon dioxide dissolved in the surface layers of the sea is in close equilibrium with the concentration of carbon dioxide in the atmosphere as a whole. The carbon fixed by photosynthesis on land is sooner or later returned to the atmosphere by the decomposition of dead organic matter. Leaves and litter fall to the ground and are oxidized by a series of complicated processes in the soil. We can get an approximate idea of the rate at which organic matter in the soil is being transformed by measuring its content of the radioactive isotope carbon 14.

At the time carbon is fixed by photosynthesis its radio of carbon 14 to the non-radioactive isotope carbon 12 is the same as the ratio in the atmosphere (except for a constant fractionation factor), but thereafter the carbon 14 decays and becomes less abundant with respect to the carbon 12. Measurements of this ratio yield rates for the oxidation of organic matter in the soil ranging from decades in tropical soils to several hundred years in boreal

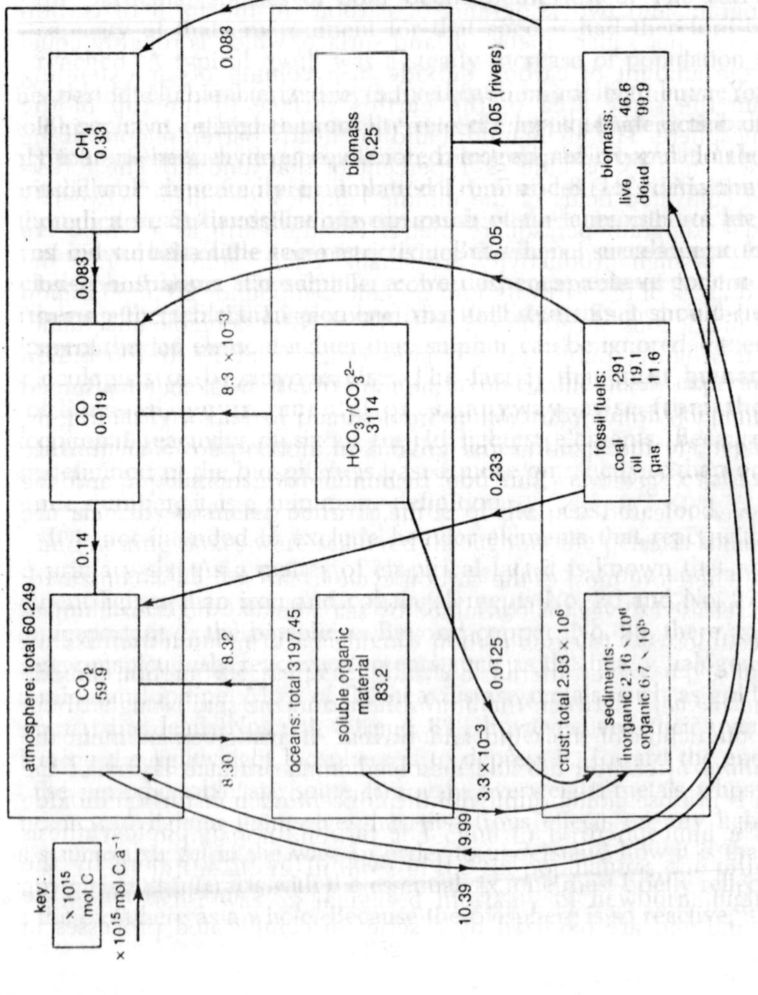

Fig. 12.1. The carbon cycle.

forests. In addition to the daily variations of carbon dioxide in the air there is a marked annual variation, at least in the Northern Hemisphere.

As spring comes to northern regions the consumption of carbon dioxide by plants greatly exceeds the return from the soil. The increased withdrawal of carbon dioxide can be measured all the way up to the lower stratosphere. A marked decrease in the atmospheric content of carbon dioxide occurs during the spring. From April to September the atmosphere north of 30 degrees north latitude loses nearly 3 percent of its carbon dioxide content, which is equivalent to about four billion tons of carbon. Since the decay processes in the soil go on simultaneously, the net withdrawal of four billion tons implies an annual gross fixation of carbon in these latitudes of at least five or six billion tons. This amounts of about a fourth of the annual terrestial productively referred to above (20 to 30 billion tons), which was based on a survey of carbon fixation. In this global, survey the estimated contribution from the Northern Hemisphere, where plant growth shows a marked seasonal variation, constituted about 25 percent of the total tonnage. Thus two independent estimates of worldwide carbon fixation on land show a quite satisfactory agreement. The forest of the world not only are the main carbon dioxide consumers on land; they also represent the main reservoir of biologically fixed carbon (except for fossil fuels, which have been largely removed from the carbon cycle save for the amount reintroduced by man's burning of it).

The forests contain between 400 and 500 billion tons of carbon, or roughly two-thirds of the amount present as carbon dioxide in the atmosphere (700 billion tons). The figure for forests can be estimated only approximately. The average age of a tree can be assumed to be about 30 years, which implies that about 15 billion tons of carbon in the form of carbon dioxide is annually transformed into wood, which seems reasonable in comparison with a total annual assimilation of 20 to 33 billion tons. The pattern of carbon circulation in the sea is quite different from the pattern on land.

The productivity of the soil is mostly limited by the availability of fresh water and phosphorus, and only to a degree by the availability of other nutrients in the soil. In the oceans the overriding limitation is the availability of inorganic substances. The phytoplankton require not only plentiful supplies of phosphorus and nitrogen but also trace amounts of various metals, notably iron. The competition for food in the sea is so keen that organisms

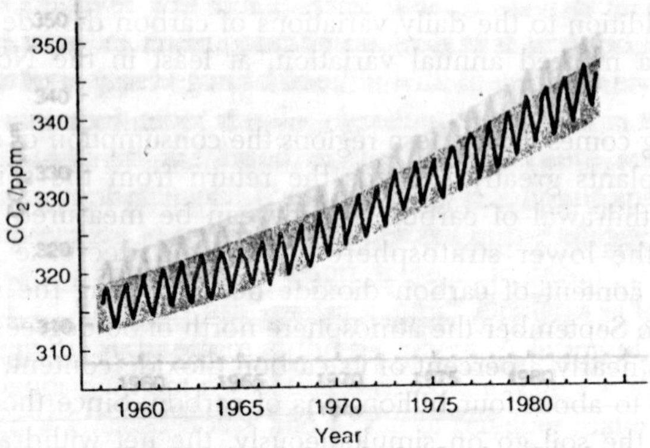

Fig. 12.2. Atmospheric levels of carbon dioxide.

have gradually developed the ability to absorb essential minerals even when these nutrients are available only in very low concentration. As a result high concentrations of nutrients are rarely found in surface waters, where solar radiation makes it possible for photosynthetic organisms to exists.

If an ocean area is uncommonly productive, one can be sure that nutrients are supplied from deeper layers. (In limited areas they are supplied by the wastes of human activities.) The most productive waters in the world are therefore near the Antarctic continent, where the deep waters of the surrounding oceans well up and mix with the surface layers. There are similar upwellings along the coast of Chile, in the vicinity of Japan and in the Gulf Stream In such regions fish are abundant and the maximum annul fixation of carbon approaches. 3 kilogram per square metre. In the "desert" areas of the oceans, such as the open seas of subtropical latitudes, the fixation rate may be less than a tenth of that value. In the Tropics warm surface layers are usually effective in blocking the vertical water exchange needed to carry nutrients up from below.

Phytoplankton, the primary fixers of carbon dioxide in the sea, are eaten by the zooplankton and other tiny animals. These organisms in turn provide food for the larger animals. The major part of the oceanic biomass, however, consists of microorganisms. Since the lifetime of such organisms is measured in weeks, or at most in months, their total mass can never accumulate appreciably. When microorganisms die, they quickly disintegrate as they sink to deeper layers. Soon most of what was once living tissue has become dissolved organic matter. A small fraction of the organic particulate matter escapes oxidation and settles into the ocean depths.

There is profoundly influences the abundance of chemical substances because (except in special regions) the deep layers exchange water with the surface layers very slowly.

The enrichment of the deep layers goes hand in hand with a depletion of oxygen. There also appears to be an increase in carbon dioxide (in the form of carbonate and bicarbonate ions) in the ocean depths. The overall distribution of carbon dioxide, oxygen and various minor constituents in the sea reflects a balance between the marine life and its chemical milieu in the surface layers and the slow transport of substances by the general circulation of the ocean. The net effect is to prevent the ocean from becoming saturated with oxygen and to enrich the deeper strata with carbonate and bicarbonate ions. The particular state in which we find the oceans today could well be quite different if the mechanisms for the exchange of water between the surface layers and the deep ones were either more intense or less so.

The present state is determined primarily by the sinking of cold water in the polar regions, particularly the Antarctic. In these regions the water is also slightly saltier, and therefore still denser, because some of it has been frozen out in floating ice. If the climate of the earth were different, the distribution of carbon dioxide, oxygen and minerals might also be quite different. If the difference were large enough, oxygen might completely vanish from the ocean depths, leaving them to be populated only by chemibarotrophic bacteria. The time required to establish a new equilibrium in the ocean is determined by the slowest link in the chain of processes that has been described. This link is the oceanic circulation; it seems to take at least 1,000 years for the water in the deepest basins to be completely replaced. One can imagine other conditions of circulation in which the oceans would interact differently with sediments and rocks, producing a balance of substances that one can only guess at.

So far we have been concerned only with the basic biological and ecological processes that provide the mechanisms for circulating carbon through living organisms. Plants on land, with lifetimes measured in years, and phytoplankton in the sea, with lifetimes measured in weeks, are merely the innermost wheels in a biogeochemical machine that embraces the entire earth and that retains important characteristics over much longer time periods. In order to understand such interactions we shall need some rough estimates of the size of the various carbon reservoirs involved and the nature of their contents. In the context of the present argument the large uncertainties in such estimates are of little significance. Only a few tenths of a percent of the immense mass of carbon at or near the surface of the earth (on the order of 20×10^{15} tons) is in rapid circulation in the biosphere, which includes the atmosphere, the hydrosphere, the upper portions of the earth's crust and the biomass itself. The overwhelming bulk of near-surface carbon consists of inorganic deposits (chiefly carbonates) and organic fossil deposits (chiefly oil shale, coal and petroleum) that required hundreds of millions of years to reach their present magnitude.

Over time intervals as brief as those of which we have been speaking–up to 1,000 years for the deep-ocean circulation–the accretion of such deposits is negligible. We may therefore consider the life processes on land and in the sea as the inner wheels that spin at comparatively high velocity in the carbon-circulating machine. They are coupled by a very low gear to more majestic processes that account for the overall circulation of carbon in its various geologic and oceanic forms. We now know that the two great systems, the atmosphere and the ocean, are closely coupled to each other through the transfer of carbon dioxide across the surface of the oceans.

The rate of exchange has recently been estimated by measuring the rate at which the radioactive isotope carbon 14 produced by the testing of nuclear weapons has disappeared from the atmosphere. The neutrons released in such tests form carbon 14 by reacting with the nitrogen 14 of the atmosphere. In this reaction a nitrogen atom (${}_7N^{14}$) captures a neutron and subsequently releases a proton, yielding ${}_6C^{14}$. (The subscript before the letter represents the number of protons in the nucleus, the superscript after the letter indicates the sum of protons and neutrons.) The last major atmospheric tests were conducted in 1963. Sampling at various altitudes and latitudes shows that the constituents of the atmosphere became rather well mixed over a period for a few years.

The decline of carbon 14, however, was found to be rapid; it can be explained only by assuming an exchange of atmospheric carbon dioxide, enriched in carbon 14, with the reservoir of much less radioactive carbon dioxide in the sea. The measurements indicate that the characteristic time for the residence of carbon dioxide in the atmosphere before the gas is dissolved in the sea is between five and 10 years. In other words, every year something like 100 billion tons of atmospheric carbon dioxide dissolves in the sea and is replaced by a nearly equivalent amount of oceanic carbon dioxide. Since around 1850 man has inadvertently been conducting a global geochemical experiment by burning large amounts of fossil fuel and thereby returning to the atmosphere carbon that was fixed by photosynthesis millions of years ago. Currently between five and six billion tons of fossil carbon per year are being released into the atmosphere. This would be enough to increase the amount of carbon dioxide in the air by 2.3 parts per million per year if the carbon dioxide were uniformly distributed and not removed. Within the past century the carbon dioxide content of the atmosphere has risen from some 290 parts per million to 320, with more than a fifth of the rise occurring in just the past decade.

The total increase accounts for only slightly more than a third of the carbon dioxide (some 200 billion tons in all) released from fossil fuels. Although most of the remaining two-thirds has presumably gone into the oceans, a significant fraction may well have increased the total number of vegetation on land. Laboratory studies show that plants grow faster when the surrounding air is enriched in carbon dioxide. Thus it is possible that man is fertilizing fields and forests by burning coal, oil and natural gas. The biomass on land may have increased by as much 15 billion tons in the past century. There is, however, little concrete evidence for such as increase. Man has of course been changing his environment in other ways.

Over the past century large areas covered with forest have been cleared and turned to agriculture. In such areas the character of soil respiration has undoubtedly changed, producing effects that might have been detectable in the atmospheric content of carbon dioxide if it had not been for the simultaneous increase in the burning of fossil fuels. In any case the dynamic equilibrium among the major carbon dioxide reservoirs in the biomass, the atmosphere, the hydrosphere and the soil has been disturbed, and it can be said that they are in a period of transition. Since even the most rapid processes of adjustment among the reservoirs take decades, new equilibriums are far from being established. Gradually the deep oceans become involved; their turnover time of about 1,000 years and their rate of exchange with bottom sediments control the ultimate partitioning of carbon. Meanwhile human activities continue to change explosively. The acceleration in the consumption of fossil fuels implies that the amount of carbon dioxide in the atmosphere will kept climbing from its present value of 320 parts per million to between 375 and 400 parts per million by the year 2000, in spite of anticipated large removals of carbon dioxide by land vegetation and the ocean reservoir.

A fundamental question is: What will happen over the next 100 or 1,000 years? Clearly the exponential changes cannot continue. If we extend the time scale with which we are viewing the carbon cycle by several orders of magnitude, to hundreds of thousands or millions of years, we can anticipate large-scale exchanges between organic carbon on land and carbonates of biological origin in the sea. We do know that there have been massive exchanges in the remote past. Any discussion of these past events and their implications for the future, however, must necessarily be qualitative and uncertain. Although the plants on land have probably played an important role in the deposition of organic compounds in the soil, the occans have undoubtedly acted as the main regulator.

The amount of carbon dioxide in the atmosphere is essentially determined by the partial pressure of carbon dioxide dissolved in the sea. Over a period of, say, 100,000 years the leaching of calcium carbonates from land areas tends to increase the amount of carbon dioxide in the sea, but at the same time a converse mechanism–the precipitation and deposition of oceanic carbonates–tends to reduce the amount of carbon dioxide in solution. Thus the two mechanisms tend to cancel each other. Over still longer periods of time–millions or tens of millions of years–the concentrations of carbonate and bicarbonate ions in the sea are probably buffered still further by reactions involving potassium, silicon and aluminium, which are slowly weathered from rocks and carried into the sea.

The net effect is to stabilize the carbon dioxide content of the oceans and hence the carbon dioxide content of the atmosphere. Therefore, it appears that the carbon dioxide environment, on which the biosphere fundamentally depends, may have been fairly constant right, up to the time, barely a moment ago geologically speaking, when man's consumption of fossil fuels began to change the carbon dioxide content of the atmosphere. The greatest disturbances of which we are aware are those now being introduced by man himself. Since his tempering with the biological and geochemical balances may ultimately prove injurious–even fatal–to himself, he must understand them much better than he does today. The story of the circulation of carbon in nature teaches us that we cannot control the global balances. Therefore we had better leave them close to the natural state that existed until the beginning of the Industrial Revolution. Out of a simple realization of this necessity may come a new industrial revolution.

13

ENERGY CYCLE

All life on the earth is of course ultimately powered by the sun, and accordingly it is strongly affected by variations of the incoming solar radiation over the globe. The distribution of sunlight with latitude determines to a great extent the location of the major climatic zones–tropical, temperate and polar–and these zones in turn set broad geographic limits to the different forms of terrestrial life. What is less familiar is the central function of the atmosphere and the oceans in redistributing the incoming solar energy and hence in determining the "macroclimate" of the earth.

The importance of the circulation of the atmosphere and the oceans to the operation of the biosphere becomes apparent when one considers that present forms of life could not endure the harsh climate that would exist if conditions of radiative equilibrium were to prevail at all latitudes (that is, if the incoming solar radiation to a zone were exactly balanced by the outgoing terrestrial radiation from that zone). This article is devoted not only to an examination of the character of the incoming short-wave radiation and the outgoing long-wave radiation but also to an attempt to trace the cycle of the solar energy from the time it enters the atmosphere as sunlight it finally finds its way back into space as heat. At the end of the article I shall take up the question of the possible effects of man's intervention in these vast energy processes. When the sun is over the Equator on March 21 and September 23 (the equinoxes), a maximum amount of solar radiation is received at the Equator. On the same dates the radiation received at the north and south poles is practically zero. This symmetrical decrease of radiation

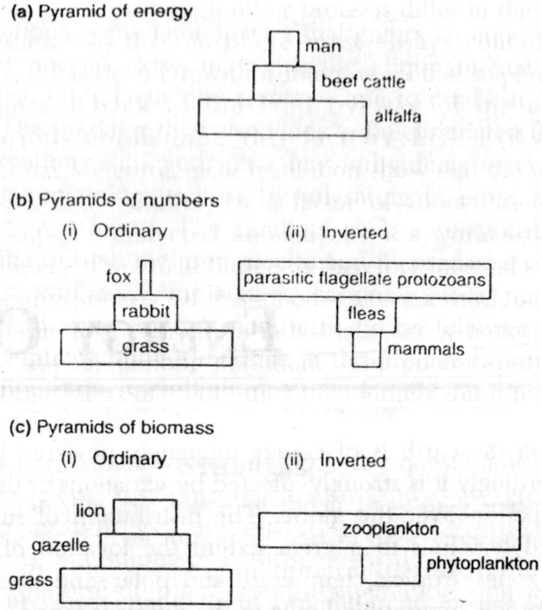

Fig. 13.1. Pyramids of (a) energy, (b) numbers and (c) biomass.

with latitude toward both poles only occurs, however, during the transition seasons, spring and fall. In the summer hemisphere, for example, there is almost no meridional (north-south) heating gradient.

Even more surprising is the fact that the 24-hour average sunshine has a maximum value at the summer pole, not at the subsolar point in the Tropics! This of course is owing to the permanent day-light at the summer pole. From a consideration of these factors alone one would expect significant differences in the large-scale circulation between the summer and the winter hemispheres. In addition one would expect climatic effects related to the variation of incoming radiation with local time in each hemisphere Diurnal changes (that is, changes with a period of 24 hours) in the wind, temperature and humidity are only important, however, close to the ground and at high levels in the atmosphere. Locally the interaction of land and sea breeze effects may also play a role, but the overall atmospheric circulation cannot respond well to such short-period phenomena. This is even truer of the oceanic circulation.

In addition to the imposed diurnal and annual periodicities in the incoming solar radiation, one finds a slight semi-annual variation of insolation in the Tropics, where the sun passes overhead twice during the year. How is the solar energy transmitted and transformed once it enters the top of the atmosphere? Averaged over the globe about 30 percent of the radiation is either scattered back by the constituents of the atmosphere or directly reflected by clouds or the earth's surface. This portion of the solar energy is lost into space and cannot be used to generate atmospheric motions. About 50 percent of the incoming radiation finally reaches the ground or ocean, where it is absorbed as heat. The properties of the surface determine the thickness of the layer over which the available heat is distributed. In the case of an ocean surface wave motions are quite effective in distributing the heat through a thick layers, sometimes extending down to a depth of 100 metres.

The diurnal variation in temperature of the ocean surface itself is thus generally less than 1 degree Celsius. The situation on land depends not only on the diurnal amplitude of the incoming radiation but also on the properties of the soil (for example its wetness) and the presence or absence of vegetation. The energy transfer down into the ground occurs through the slow process of molecular heat conduction. Over bare ground the diurnal temperature range at the surface can amount to several tens of degrees Celsius, but the temperature change is hardly noticeable below a depth of half a metre. What happens to the remaining 20 percent of the incoming solar radiation that is apparently absorbed on its path through the atmosphere? Here it is necessary to consider the spectrum of the incoming and outgoing radiation. The emission spectrum of the sun roughly resembles that of a "black body" radiating at a temperature of 6,000 degrees Kelvin. (A black body is defined as one that absorbs all the radiation falling on it).

In the visible portion of the spectrum (wavelengths between .4 and .7 micron), where the maximum influx of solar energy takes place, the radiation can penetrate almost without loss down to the earth's surface except where clouds are present. High in the atmosphere ordinary oxygen (O_2) and ozone (O_3) molecules absorb an estimated 1 to 3 percent of the incoming radiation. This absorption occurs in the ultraviolet portion of the spectrum and effectively limits the penetrating radiation to wavelengths longer than 3 micron. Although this effect is relatively small, it is important

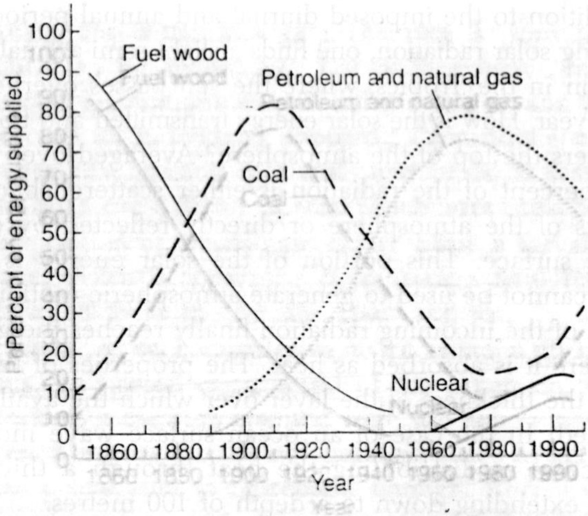

Fig. 13.2. Development of new energy sources in the United States. Coal passed wood as the primary fuel source in 1885, and petroleum passed coal in 1944.

because it is the main source of energy for the circulation above 30 kilometers (see "The Circulation of the Upper Atmosphere," by Reginald E. Newell; Scientific American, March, 1964). Moreover, the absorption at these levels shields the biosphere from the damaging effects of ultraviolet radiation.

At wavelengths longer than one micron most of the atmospheric absorption is due to water vapour, dust and water droplets in clouds. This process operates in the lower troposphere and involves most of the remaining 20 percent of the total incoming radiation. In spite of certain long-term climatic changes climatological records do not show an appreciable net heating or cooling of the earth and its atmosphere. Therefore the earth must emit an amount of radiation equal to the radiation absorbed. A characteristic shift to longer wavelengths does take place, however, since the earth radiates at an effective black-body temperature of about 255 degrees K., a very low value compared with the sun's black-body temperature of 6,000 degrees.

The earth's emission occurs throughout a broad range of wavelengths with a flat maximum at about 12 microns. In this range of the spectrum the atmosphere is not transparent. Water vapour, ozone and carbon dioxide absorb significant amounts of long-wave radiation. If one now calculates the vertical transfer of solar and terrestrial radiation using the observed temperature and humidity structure, one finds that the atmosphere is not in local radiative equilibrium. The net effect due to solar and terrestrial radiation alone would be an intense heating of the earth's surface and a cooling of the atmosphere at the rate of up to two degrees C. per day, depending on the height. In reality the air is prevented from cooling by the vertical transfer of heat directly from the earth's surface and by the release of heat through the condensation of water vapour.

It is at this point that the dynamics of the atmosphere begin to play an important role. The transport of energy upward into the atmosphere forms the major energy supply for the atmospheric heat engine. A large portion of this energy, however, is "latent" (that is, in the form of water vapour), and it is used to raise the atmospheric temperature only when condensation takes place. Close to the surface the energy transport occurs through evaporation of water, through heat conduction and through the transfer of long-wave radiation. At a certain height above the surface turbulent

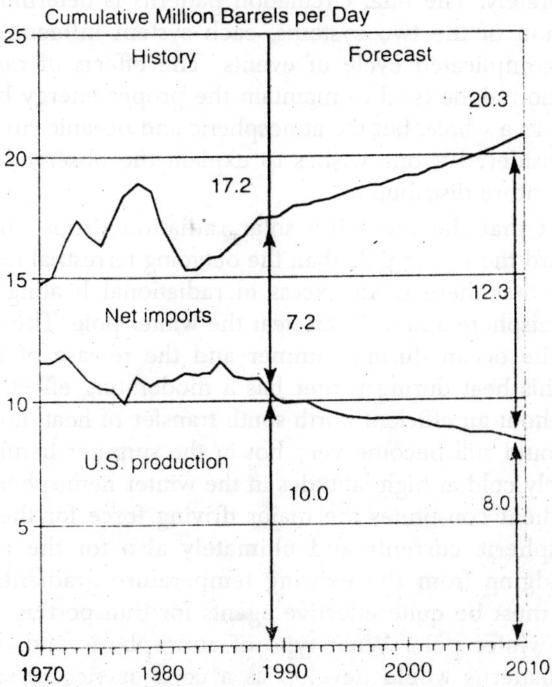

Fig. 13.3. Petroleum production and imports in the United States.

eddies mix the water vapour and heat further upward. The scale of the effective eddies increases with distance from the surface, finally growing to convective clouds of the cumulus type in the free atmosphere. This upward transfer of energy from the surface compensates for the radiative cooling of the atmosphere. A schematic diagram of the average energy cycle in the atmosphere reveals the important role of the earth's surface as the source of latent heat and sensible heat.

There is a significant difference in the character of the heating for the ocean and for the atmosphere. In the ocean the heating is applied at the top, which leads to stable conditions, whereas in the atmosphere the heating is applied at the bottom, giving rise to vigorous convection. The ocean currents, which are driven mainly by the winds, redistribute the absorbed solar heat horizontally and thereby influence in turn the pattern of the heat supply to the atmosphere that finally closes the cycle. The oceans and the atmosphere are strongly coupled systems and cannot very well be treated separately. The final circulation patterns is determined by the interaction of the two systems, each system influencing the other in a complicated cycle of events. The effects of radiation and convection alone tend to maintain the proper energy balance for the earth as a whole, but the atmospheric and oceanic circulation must be considered if one wishes to explain the observed north-south temperature distribution.

The fact that the incoming solar radiation drops off more rapidly toward the winter pole than the outgoing terrestrial radiation does means that there is an excess in radiational heating in the summer hemisphere and a deficit near the winter pole. The storage of heat in the ocean during summer and the release of a large portion of this heat during winter has a moderating effect on the climate. Without an efficient north-south transfer of heat, however, the earth would still become very hot in the summer hemisphere and extremely cold at high latitudes in the winter hemisphere. The heating gradient constitutes the major driving force for the large-scale atmospheric currents and ultimately also for the oceanic currents. Judging from the existing temperature gradient, these circulations must be quite effective agents for transporting energy toward the winter pole. What type of atmospheric and oceanic circulation patterns would develop as a consequence of such an imposed heating gradient? Let us limit the discussion for the time being to the winter situation. It is in this season that the north-south gradient in the solar heating is strongest and that the differences between the radiative equilibrium temperature and the observed temperature are at a maximum. What simple mechanism would suffice to transport heat poleward?

Let us first consider a model atmosphere that has a uniform temperature and rotates at the same rate as the earth. If one starts to heat the air at low levels on the summer side of the Equator, the local temperature will rise and the air column will expand mainly in the vertical direction. This process will create at the upper levels a relatively high-pressure belt located over the "thermal" Equator. Next the north-south pressure gradient will force the equatorial air at all longitudes to move toward the low-pressure zone, mainly into the winter hemisphere, where initially vertical contraction occurred as a result of radiational cooling.

The air will then slowly start to sink over a wide region in the winter hemisphere and will return to the Equator at low levels. The cycle will be closed finally through a rise of the air after it has arrived in the vicinity of the thermal Equator. A simple cellular circulation of this kind, called a mean meridional circulation, would be completely symmetrical with respect to the earth's axis of rotation. The existence of one such cell in each hemisphere with rising warm air near the Equator, and sinking cold air near the poles was originally postulated by the English meteorologist George Hadley in 1735. Such a cell is called a direct cell since it release potential energy and converts it into kinetic energy. Later investigators, notably the 19th-century American meteorologist William Ferrel, showed that one actually needs three cells in each hemisphere to explain the important climatological features at the earth's surface. This picture has been confirmed by many observational and theoretical studies and seems to represent rather well the annual mean conditions in the atmosphere. Recent and more detailed observations, however, have shown that only during the transition months in fall and spring in such an idealized circulation symmetrical with respect to the Equator realized.

The asymmetry in heating during most of the year appears to favour the development of only one strong cell in the Tropics: the one in the winter hemisphere. This cell circulates on the average about 2×10^8 metric tons of air per second. At the same time the "summer" Hadley cell has shrunk to an insignificant size. Let us consider in somewhat greater detail what energy transformations take place in the tropical Hadley cell. In its lower branch subtropical air flows close to the earth's surface toward the summer hemisphere with an average velocity of one or two metres per second. During the long journey equatorward heat and moisture are absorbed from the warm underlying ocean. This is the region of the trade winds.

Near the Equator the air starts to rise in a fairly narrow region called the intertropical convergence zone, where intense precipitation occurs. In that zone a powerful conversion from sensible and latent heat into potential energy occurs as the air expands and the water vapour condenses, the net effect being a cooling of the equatorial atmosphere. The upper branch of the Hadley cell now transports the air, which has become relatively cold but which has a high potential energy, into the winter hemisphere. In the rather wide downward branch in the subtropics the subsiding cold air is strongly heated by compression, and the potential energy supplied to the air in the equatorial convergence zone is converted back into heat. One would expect that the sinking air in the subtropics would be dry, since almost all the moisture was rained out in the upward branch of the Hadley cell. That expectation is confirmed by the location of the continental deserts and the small amount of rainfall over the oceans in this latitude. The large overturning in each hemisphere of the kind Hadley envisioned is not adequate to transport enough energy poleward to counteract the externally imposed heating gradient.

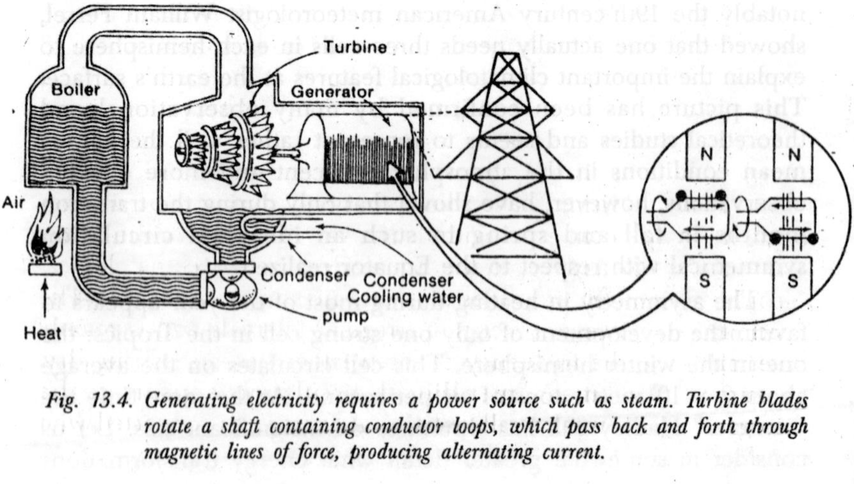

Fig. 13.4. Generating electricity requires a power source such as steam. Turbine blades rotate a shaft containing conductor loops, which pass back and forth through magnetic lines of force, producing alternating current.

In such a situation temperatures near the Equator would start to rise above the observed values, and near the Pole they would start to drop. This would continue until a critical value of the north-south temperature gradient was reached, at which point zonal (east-west) asymmetries would start to develop (this process is called baroclinic instability). Theoretical models indicate that the maximum instability would tend to develop with atmospheric waves a few thousand kilometres long. In the middle latitudes, where the strong meridional temperature gradients are found, these waves can grow very fast and can take over the task of transporting energy poleward from the Hadley cell. The familiar travelling cyclones and anticyclones, which can be found on every weather map in the middle latitudes, are a manifestation of this instability process. They form an extension of the large waves in the middle and upper troposphere. These systems mix heat in an efficient way through horizontal processes.

At the same level one finds warm, humid air flowing poleward and cold, dry air flowing equatorward. On the average these flows are equivalent to a net poleward transport of sensible and latent heat. Waves with lengths of a few thousand kilometers and with time scales of a few days to a week appear to be mainly responsible for the transfers. The typical "variable" climate of middle latitudes is determined to a great extent by such large-scale waves. These waves are more intense and frequent in winter than they are in summer, since they generally develop in regions of strong horizontal temperature contrasts. In middle and high latitudes the mean meridional circulation is weak. One can probably interpret the reverse, mid-latitude Ferrel cell as a circulation that is being driven, or forced, by the atmospheric waves. The net effect of this "indirect" cell is the sinking of relatively warm air and the rising of cold air!

At high latitudes near the Pole there is some suggestion of a direct polar cell; in the Northern Hemisphere this call is very weak. The slow rising motion between roughly 50 and 60 degrees north latitude is connected with the upward branches of the Ferrel and polar cells, and its effects are evident in the climatological records. In this belt one finds a second maximum in rainfall. The precipitation, however, occurs at irregular intervals and is mainly determined by the frequency of the weather systems passing dry. In summer the mean meridional circulation appears to be disorganized and weak at all latitudes. Asymmetries connected with the distribution of continents and oceans dominate the circulation.

The land generally acts as a heat source and the colder water at middle and high latitudes as a heat sink. One apparent asymmetry is the Asian monsoon, which carries warm and humid air far north into the Asian continent. At most other longitudes in the northern part of the Tropics the air still moves equatorward. In the season studying the asymmetries in the circulation is probably more relevant than studying the mean meridional circulation, which has rotational symmetry around the earth's axis. Up to this point I have discussed only some mean features of the climate as they have been derived from the observations of the past 20 to 30 years. Paleontological and even meteorological records show that the climate has changed slowly but significantly in the past and probably is changing now.

On the other hand, it appears unlikely that either the total influx of solar energy or the rotation rate of the earth have changed drastically during the period in which life developed on the earth. Therefore it is probably safe to assume that the basic circulation regime has not changed. However, relatively minor changes in the strength of the north-south energy exchange, through either the mean meridional circulations at low latitudes or the large-scale horizontal waves at middle latitudes, can cause deviations from the present climate. Needles to say, any such deviations could be very significant as far as the living organisms are concerned. The most likely way the climate could be influenced by either natural or artificial means seems to be through a trigger mechanism that ultimately changes the radiation balance. For example, if the cloud cover or dust content of the air were changed at high latitudes, the amount of reflected radiation would increase and consequently less solar radiation would be available to heat the atmosphere and the earth's surface at these latitudes.

The resulting increase in the north-south heating gradient would presumably lead to more violent and more frequent disturbances the middle latitudes. These circulations would in turn affect the original cloud cover. From that point on it seems hopeless to predict with any degree of certainty what the additional effects on the climate would be. Changes in the reflectivity or absorptivity or the earth's surface can also alter the climate. It has been suggested that if the snow and ice fields near the North Pole were to be covered with black carbon, the Arctic Ocean might become ice-free through the increased absorption of solar radiation in summer. Again the present radiation balance and the climate over the earth would be affected.

Still another possibility would be a change in the relative proportion of the atmospheric gases. For example, the measured slow increase in the carbon dioxide content of the air due to the burning of fossil fuels would presumably lead to more absorption of long-wave terrestrial radiation in the atmosphere and consequently to extra heating. One can think of many other ways in which changes in the earth's macroclimate might occur. One must also consider the possibility of external influences such as natural variations in the amount and the spectral distribution of the incident solar radiation itself associated with variation in the level of solar activity. For instance, during periods of high solar activity the solar output mainly increases in the ultraviolet portion of the spectrum. Since most of that radiation is absorbed above 30 kilometers, one would expect to find the largest dynamic response at those levels. It is still an open question whether or not such variations in solar emission–either in the form of increased ultraviolet radiation or possibly also in the form of showers of energetic particles particles–affect the weather deep down in the atmosphere.

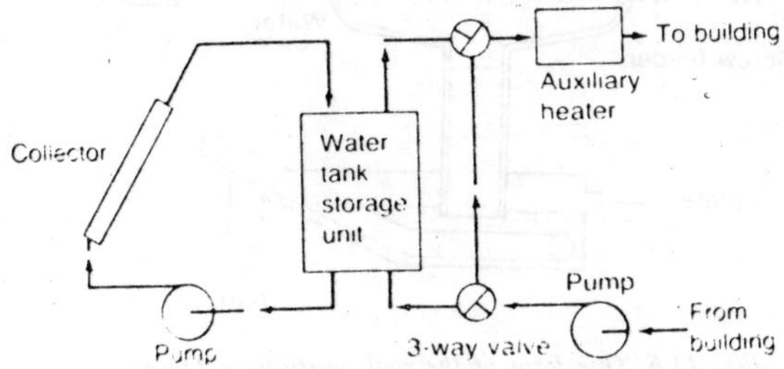

Fig. 13.5. An active solar system that uses water to transfer heat to the building.

The meteorological evidence for direct climate variations caused by such changes in solar output is inconclusive, but variations of this kind certainly cannot be ruled out. As an example one can imagine that through a change in the radiation balance the general regime of atmospheric and oceanic circulation could be brought to settle in a new quasi-equilibrium state that would be slightly different from the state it is in now. According to some recent studies conducted by M. I. Budyko in the U.S. S. R and William D. Sellers in the U.S., even the present state of the atmosphere might not be as stable as one would like to think: these authors suggest that comparatively small changes in the present state can lead to a new ice age. Another consideration is that an initial disturbance of the circulation might, if maintained for a long enough time, excite a long-term climatic fluctuation. The period of this, fluctuation would probably be related to the natural turnover time of the oceans (of the order of a century of more) because of their capacity to store large amounts of heat. Some reassurance that our present climate is not too unstable may be gained from the fact that during the past few centuries the climate has not fluctuated widely.

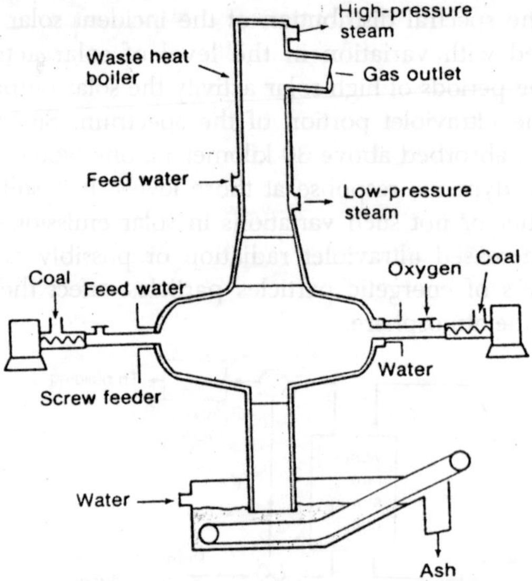

Fig. 13.6. One form of the coal gasification process.

An important complicating factor is the highly interactive nature of the different processes that operate in the ocean and the atmosphere. This makes it practically impossible to deduce through simple reasoning or even by using a simple model what would happen if one could bring about a slight, but more or less permanent, changes in the radiation budget. A high-priority task in the examination of these problems is to establish from observation how the present general circulation in both the ocean and the atmosphere is maintained. Here important contribution have been made in the past 20 years by three groups, one led by Victor P. Starr of the Massachusetts Institute of Technology, the second by Jacob Bjerknes of the University of California at Los Angles and the third by Erik H. Palmen of the university of Helsinki.

The next tasks are to determine whether or not slow changes in the general circulation are taking place, and to try to establish possible cause-and-effect relation. The limited records and the tremendous job of reducing the observation to meaningful parameters, however, severely restrict this direct approach. With the arrival of large electronic computer a powerful new method for studying the climate has been developed. The thermal structure and the dynamics of the atmosphere are simulated through numerical integration in time of the equations that govern the behaviors of the atmosphere. The basic equations are the equations of radiative energy transfer, the equations of motion and the thermodynamic equation. Starting from certain initial condition (for example a uniform-temperature atmosphere at rest), the integration is carried out in time steps of the order of 10 minutes, and new values for the meteorological parameters (wind components, temperature, pressure and humidity) are calculated at each point in a three-dimensional grid covering the global atmosphere.

The most realistic numerical experiments to date have been conducted at the Geophysical Fluid Dynamics Laboratory of the Environmental Sciences services Administration by Joseph Smagorinsky, Syukuro Manabe and Kirk Bryan, Jr. In their experiment the number of grid points in space is of the order of 50,000 (about 10 vertical levels and a horizontal grid distance of about 300 kilometers). After the atmosphere has settled down and recovered from the unrealistic initial conditions, the relevant general circulation statistics are calculated in the same way they would be for the real atmosphere. With the present grid the large-scale weather system seem to be rather well resolved. On the other hand, the smaller-scale phenomena such as cumulus convection cannot be simulated explicitly. They have to be incorporated in a different way.

In spit of these uncertainties and others (such as the lack of knowledge of how to properly incorporate the exchange processes near the earth's surface), the result so far are encouraging. Although the ability to forecast the exact location and intensity of the important weather system degrades rather quickly in the course of a week, the predicated statistics of the average behaviour taken over a much longer period appear to reproducer the observed statistics rather well. The numerical experiment seems to be the most promising road to a better understanding of the present climate. In addition they provide a powerful tool for evaluating the effects on the climate of natural or man-made disturbances.

14

ENERGY CYCLE IN ECOSYSTEMS

THE ENERGY IN ECOSYSTEMS

Productivity

Productivity is the amount of organic substance acquired by an individual, a population, or a system per unit time. The ecological concept, then, differs from that of agriculture, in which "production" is considered to be only that material utilizable by humans in some form. In the agricultural sense, any roots or above ground parts that cannot be used as food or fodder are ignored. Many misunderstandings in discussions between ecologists and agronomists are based on this discrepancy. A fundamental distinction must be made between the numbers of organisms in a habitat and the amount of organic matter they produce per unit time. A large population by no means implies high production.

Indeed, the suggestion has been widely defended that a large population, a large biomass per unit area, is a typical strategy by which organisms adapt to low nutrient availability; a large population is better able to bridge periods when food is scarce than a small one. The turnover–that is, the production–of such a population is comparatively small. It follows from this hypothesis that small biomass is characteristic of habitats with fairly large amounts of available nutrients, and this small biomass should show a high turnover. Habitats with large biomass and limited available nutrients are thought to contain very many species, whereas those with little biomass and high turnover can house but a few species. We can be quite certain that this hypothesis, in this simple form, is not generally applicable. It is presented here in order to underline

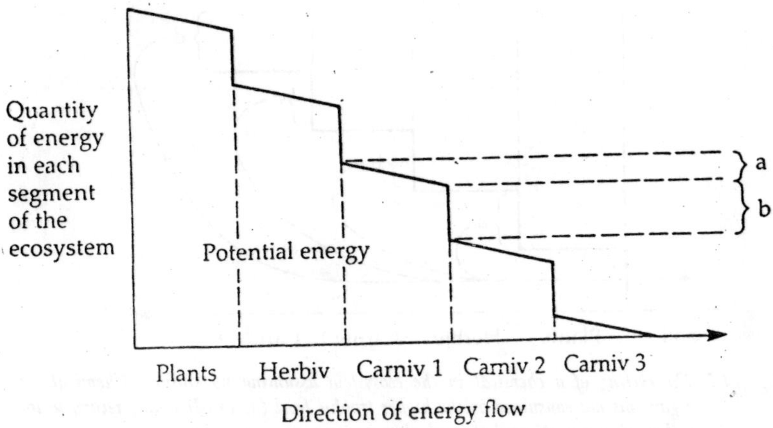

Fig. 14.1. Energy flow through the ecosystem. At each trophic level maintenance energy is lost as heat (a); energy is also lost as heat in each transformation from one trophic level to the next (b).

the difference between biomass and production in a habitat, and because it is definitely worth considering, at least, for a number of habitats. In testing its validity, one will have to ask which kind of organism (plant, animal, microorganisms) the large population comprises, and here, again, divide the system into compartments. The amount produced, as a rule, is smaller than the mass of the producing organism. Consider a forest, for example.

The amount of organic matter contained in the stand of plants has, under some circumstances, been developed over centuries; the production in a single year is distinctly lower. But where small organism are concerned the relationships can be different. This is the case with bacteria, for example, and is an especially familiar feature of planktonic algae. The production of new algae per unit time is considerably higher than the stand of algae present in the body of water at any time. Production falls into two major categories.

One is primary production, by green plants and photosynthetically active bacteria, which produce organic matter from solar energy and inorganic matter. The other is secondary production, by organisms not capable of photosynthesis–bacteria fungi and animals, which convert the organic matter of the plants into the substance of their own bodies. Each of these categories can be subdivided into gross and net production. Using light, a plant produces organic from inorganic matter. This is gross

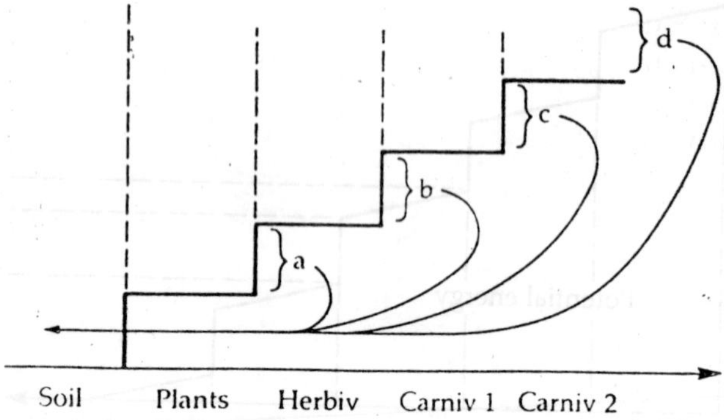

Fig. 14.2. The cycling of a chemical in the ecosystem assuming no erosion. Chemicals in organisms not consumed by the higher trophic level (a) to (d) slowly return to the soil as they are released through decay.

production. But part of this matter is respired, used up by the plant's metabolism. Gross production minus respiratory losses is called net production. It is only the net primary production that we can realistically measure; gross primary production is usually simply a logical construct. Gross production and net production by animals are easier to distinguish. The respiratory losses of animals are relatively accessible to measurement, and the factors determining them are relatively well known.

Net production is of course the most important to ecosystem research, for it is this that can be passed on to the next trophic level. To humans, the animals of especial interest are those with net production not much lower than gross production. The ecological efficiency of animals has therefore been widely studied. For the reason the net primary production called simply primary production in the following discussion) is largely independent of the species of plant. Primary production of a system is affected chiefly by the duration of the growing season and by the water supply. In natural ecosystems it is usually a valid starting assumption that nutrients are present in adequate quantities. The nutrient poverty of many present-day cultivated areas has come about through decades or centuries of overuse.

Fertilization in many cases does no more than restore these areas to their original fertility. Water supply and length of the growing season, then, are a sufficient basis for a first-approximation estimate of primary production in land ecosystems. It is impossible to make direct comparison of this primary production of a system with a the secondary production. Secondary production by a primary consumer–a herbivore–is necessarily higher than that by a secondary consumer that feeds on this herbivore. And production by a tertiary consumer is necessarily still lower. The secondary production of an ecosystem therefore in every case approaches zero–a self-evident fact that need not be discussed further.

The significant point with regard to the human nutritional situation is that as each successive trophic level is passed secondary production decreases drastically. From an energetic point of view it would be more sensible to use fish and flesh refuse directly as human food than to return them to the cycle as food for animals. A number of different methods have been used to estimate the level of net primary production. The simplest is the harvest method, in which the whole stand of plants, together with the roots, is collected. When the plants are annuals, the year's production can thus be measured directly. With plants that live more that one year, one faces the problem of distinguishing the organic matter originally present from that added during the year under study. This distinction is extraordinarily difficult in the case of the root system. For this reason the net primary production given is often that of the parts above-ground alone.

Furthermore, the above-ground net primary production is the most important in general for zoological and ecological purposes. One way of estimating the annual growth is to harvest the stand both at the beginning and at the end of the growing season. More precise data are provided by the gas-exchange measurements. Part of a plant, or the whole plant, is enclosed in a cuvette within which the air is kept at the same humidity and temperature as that of the surroundings. Air is passed through the chamber and the concentration of carbon dioxide in the entering air is compared with that at the exit. The difference in the carbon dioxide that has been incorporated into the plant tissues.

The method is very exact, but because of the accuracy with which the microclimate within the cuvette must be controlled it is technically complicated and laborious. Production in water is measured chiefly by the dark-bottle method. Two bottles are filled with the water to be studied, and they are hung at the desired depth in the water. One of the bottles is made light-tight by covering it with aluminium foil. In the lighted bottle the planktonic algae continue to take up carbon dioxide, whereas in the dark bottle they only respire. The oxygen content of the two bottles is then determined by the Winkler method, and the difference between the two indicates the rate of photosynthesis and thus the amount of carbon dioxide incorporated. In bodies of water with abnormal chemistry the method is not very accurate; the milieu within the bottle changes. Moreover, the endogenous diurnal rhythm of the algae can affect the results.

In many cases, therefore, a more elaborate method using radioactive labelling has been adopted. Here the water is injected with $NaH^{14}CO_3$. Under the assumption that the planktonic algae assimilate $^{14}CO_2$ to the same extent as $^{12}CO_2$ and do not deposit ^{14}C in a special way, measurement of the amount of ^{14}C taken up by the end of the experiment can be used to determine the total amount of carbon assimilated. An objection to this method is, for example, the face that C_4 plants, because of the higher affinity of PEP for carbon dioxide, incorporate ^{14}C to a greater extent than do C_3 plants, so that errors can be introduced. Furthermore, N, P, or Si deficiency can rapidly develop in closed vessels. When this happens protein synthesis is largely blocked, while photosynthesis continues. These cells excrete a considerable percentage of their primary production in the form of soluble compounds; this does not appear in the incorporated ^{14}C activity.

The diversity in the methods of production measurement is necessarily accompanied by a variety of units for production. Sometimes it is expressed in g dry weight, and sometimes in the amount of carbon dioxide taken up. Because inorganic material is often deposited in plants, it is customary to indicate the amount of energy bound in the organic matter (Chart 14.1). Very extensive summaries of the energy content of organic matter are now available. From these data, knowing the general possibilities of error in the methods used, one can derive rules that allow one to omit certain measurements. On the average we can take it that 1g (ash-free) animal substance corresponds to about 5.6 kcal, and 1g of plant substance has about 4.6 kcal. The values for seeds are considerably higher. There is thus no single best method for determining primary production; under different conditions, different methods will be required. All of these are subject to error, and they are not directly comparable. However, the errors have now been reduced to an order of magnitude low enough that estimates of worldwide production is possible.

Chart 14.1. Data for the Conversion of Various Units in Production Ecology

Production is commonly indicated by a variety of units. So that approximate orders of magnitude can be estimated for purposes of comparison, a number of conversions are listed here. The comparisons themselves, of course, can also be only approximate.

1 g C corresponds to	2.2 g	organic plant dry matter
	3.3 g	dry matter of phytoplankton (i.e., including inorganic components)
	42 g	fresh weight of phytoplankton
	1.7 g	organic animal dry matter
	8.3 g	fresh weight of zooplankton

1 g plant dry matter, ash-free, corresponds to	4,000–5,000 cal
1 g animal dry matter, ash-free	5,000–6,000 cal
1 g carbohydrate	3,700–4,200 cal
1 g fat	9,500 cal
1 g protein	3,900–4,150 cal

1 g oxygen uptake corresponds to 3,280 cal
= 0.345 g fat
= 0.728 g plant matter
= 0.596 g animal matter

1 carbon-dioxide release corresponds to 1,800 cal
1 ml O_2 uptake corresponds to 4,687 cal
1 ml CO_2 release corresponds to 3,558 cal
1 cal=4.186 joule 1 joule=0.23889 cal

Although the determination of primary production, risky though it remains, is in principle so well founded methodologically that "recipes" can be given, the same cannot be said of secondary production. Here the difficulties begin as soon as one wishes to transfer the growth rate of a single animal, measured in the laboratory, to field conditions. In the field mortality in the population complicates the issue. Two opposed processes are operating, the growth–the production–of the individuals and the continual decrease in number of individuals by mortality. As has been discussed, the two process do not occur uniformly in time. In insects, for example, production is highest during larval development; the highest mortality occurs in the youngest stages. With species that have sharply separated generations or clearly distinguishable stages in the life cycle, so that they can be subdivided into cohorts, regular monitoring of the population (concerning the difficulties of measuring a population) allows production to be estimated. But with species having many intermingled generations with ill-defined stages, and within which by no means all individuals proceed to reproduce–as appears to be the case among most mammals and birds, at least–things are considerably more difficult. Special procedures must be adapted to meet each case. An especially problematic task is that of establishing the number of microorganisms. Counts made with different straining methods in many cases give discrepant results.

Furthermore, such counts give no information about the species (and thus the special capabilities) of the organism, nor can one discern whether the cells counted were still alive when the sample was taken. Methods employing vital stains such as acridine orange have also proved not to be reliable enough. Recently researchers have turned to a method in which the content of ATP (a compound that breaks down almost immediately when an organism dies) is used to infer at least the biomass of the living microorganisms. This procedure assumes a constant relation between biomass and

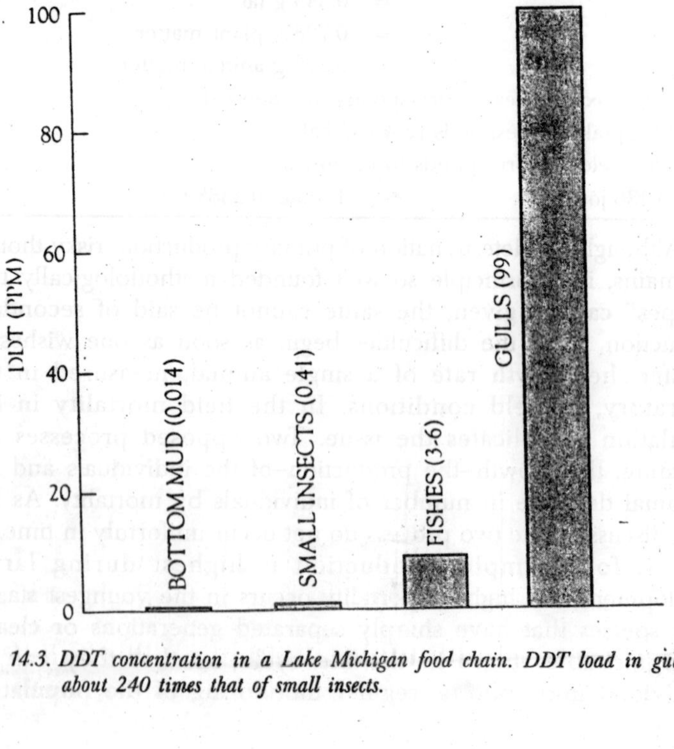

Fig. 14.3. DDT concentration in a Lake Michigan food chain. DDT load in gulls was about 240 times that of small insects.

amount of ATP, which seems to hold only very approximately. In many cases no attempt has been made to estimate biomass; rather, the activity of certain key enzymes is used as the sole indicator of the activity of the microorganisms. Wieser and Zech (1976), for example, showed that there was particularly high dehydrogenase activity at the surface of sand grains on the ocean shore; in the interstitial water the activity of this enzyme, an important part of the electron-transport system, was much lower (only about 10%-20% of the total activity). And yet another problem is generally encountered. This is evident from the fact that, with even a minimal rate of division of the microorganism under the prevailing conditions, the consumption of matter ought to be much greater than it actually is (assuming that the number of bacteria derived by various methods are minimal values.) One can only conclude that the majority of microorganisms in the soil are inactive, whereas others seem to be excessively active.

Animals of the ocean floor have been shown to engage in "gardening." They accumulate bacteria in front of themselves by special secretion and stimulate them to rapid growth. Perhaps this observation is comparable to the fact that in most bird population a large fraction of the individuals do not reproduce–they are not involved in production. But serious research into these problems has only just begun, and the field is still in a state of flux. Normally one would expect a close relationship between the level of primary production and the density of animal populations. Such a relationship, on a general scale, can be demonstrated from marine habitats.

The regions of the ocean with highest primary production to a large extent coincide with those where the fish catch is largest and where the bird population are most dense. Similarly, Hinz showed that on Spitsbergen the regions of highest plant productivity gave the largest numbers of animal captures. But here we encounter a number of problems. No animal-capture numbers are absolute indicators, for they refer selectively to particular groups. If the number of mammals is compared with the primary production of forested and treeless regions, no sensible correlation emerges; the number of productivity of mammals in forests are considerably lower than in any region lacking trees, from the tundra to the steppes at tropical latitudes. In part this simply results from the facts that large mammals cannot move freely in the dense jungle and that most of the production of a primeval forest occurs in the crown stratum, out of reach of the large mammals. (Table 14.1) This interpretation accounts for the fact that a poor arctic tundra sometimes supports a hundred times as many mammals as a primeval forest in central Europe.

At the moment it is impossible to tell whether taking all animal groups into account would give appreciably different results. There is another reason why the level of primary production cannot be closely correlated with animal productivity; from one year to the next there can be extreme differences in the amount of accumulated plant substance, depending on the differences in the weather conditions from year to year. The number of animals present is not necessarily controlled by the same factors and in the same way as the productivity of plants. Thus there are differences in the rate of utilization of the primary production in different years. In central Europe a year very favourable to grasshoppers is not noticeable as such until the following year, when the new generation emerges from the abundantly produced eggs and begins to grow. Plants, by contrast, tend to respond immediately to the prevailing conditions. The situation becomes quite difficult when comparisons over large geographical regions are attempted. Because there are no herbivorous ectothermic animals in the far north, these regions cannot be directly compared with stepes with respect to the ectothermic fauna. Nor can marine and terrestrial regions be compared.

On land the primary production is used by animals of direct interest to man (plant-eating birds and mammals). In the ocean, on the other hand, the fish interesting to humans are all secondary, tertiary, or quaternary consumers. They feed on large clams or fish; only very few species are able to live on planktonic crustaceans–that is, primary consumers. For this reason the production of animals important to man in the ocean must be well below that on land, even for a given level of primary production. Finally, special features may be overlooked when large-scale comparisons are made. Though primary production is about the same in the two bodies of water, the western Baltic Sea provides far higher yields of food fish than does the Skagerrak. The reason is that the two differ in salinity. The Skagerrak has normal marine salinity, with the normal marine fauna.

In the western part of the Baltic the salt concentration is reduced, and a great many animal forms have disappeared. These–

Table 14.1. Degree to Which Primary Production is Utilized by Warm-blooded Herbivores

Type	*Harvestable primary roduction per ha per yr. kg dry matter*	*No. sheep per ha per year this energy can support*	*No. of animals per ha*	*Degree of utilization in % (rough estimate)*
Tundra Spitsbergen	(270-500 kg ha)	1	Reindeer 0.006 Wild geese (90 days) Ptarmigans Musk oxen	1-2
Steppe, Africa (Serengeti)	7,500 kg/ha ($30x10^6$ kcal)	20	0.5 animals of many species ($2.2x10^6$ kcal/ha/yr)	8
European mountains	9,000 kg/ha	24	Red deer 0.006 Roe deer 0.02 Hares	0.02
European mountains, herb and shrub strata only	500 kg/ha	1.4	Red deer 0.006 Roe deer 0.02 Hares	1.0
Hirta St. Kilda, maritime mountain meadow	2,000 kg/ha	5	Soay sheep 1.72	31
Carpathian mountain meadow	$6x10^6$ kcal/ha/yr	4	15 Muridae	1

irregular sea urchins, sea stars, and large gastropods–are competitors of other marine animals, especially clams. Clams (Cyprina islandica above all) are excellent food for the cod, whereas the animals that have vanished from the Baltic would not have been eaten by the fish in any case. Furthermore, there is the following consideration. Two bodies of water may exhibit identical levels of primary production, although in one the producers are very small planktonic algae and in the other very large species.

The very small algae in the first habitat can be utilized only by very small planktonic animals, which in their turn are the food of the larger planktonic animals on which fish feed. By contrast, the larger planktonic algae in the second habitat can be eaten directly by large planktonic animals; in this case the food chain is shorter by one link. At a rough estimate, this shortening increases fish production by a factor of 10. Where vascular plants on land are concerned, it is self-evident that only part of the production is normally used by animals, or that different parts of the plant tissue produced are used by different animals–some eat roots, others leaves, other seeds, and still others the nectar of flowers. Here the central issue is the part of the plant by which production is represented; the overall level of production is not the first consideration. For all these reasons, it is inappropriate to try to find a direct relationship between the level of primary production and the number of animals–especially if one wants anything but a very rough indication, and if one is interested primarily in the animals important as food for humans.

The differences in production within a system can be well illustrated by a food pyramid, in which the individual trophic levels are set one above another. But this form of presentation conceals awkward problems. Should the numbers of individual organisms be indicated, or their biomass? A great many Orchestes fagi can live on a single beech tree, but if the biomasses of the green leaves and of the beetles are compared, the picture looks quite different. And if production is judged entirely on the basis of physiological experiments, still another picture emerges. For example, a certain amount of hay can feed a certain number of rabbits for a certain period of time.

In theory, a certain number of martens can feed on this number of rabbits for the same length of time, and a certain number of wolves can live on these martens for the some period. Finally, it

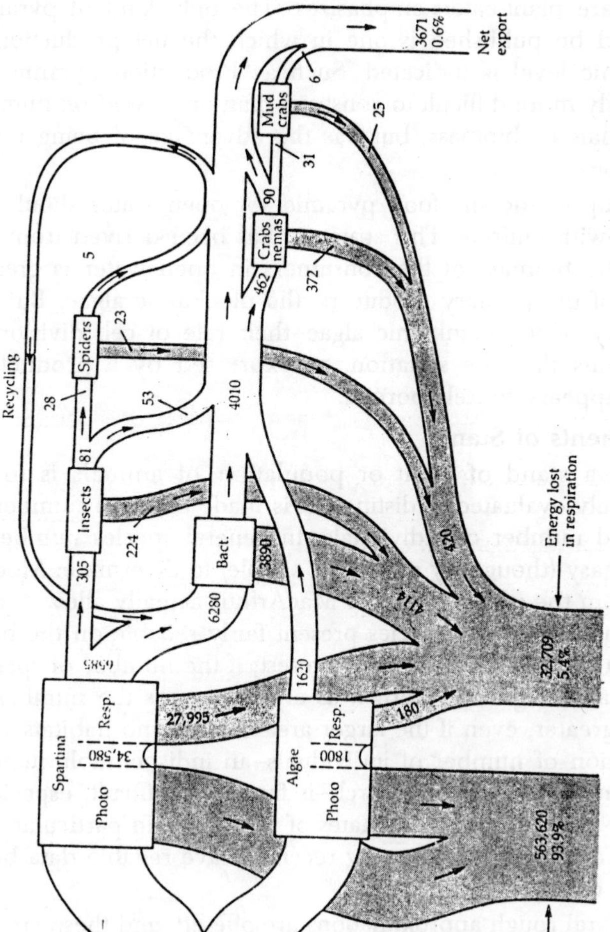

Fig. 14.4. Energy-flow diagram for a Georgia salt marsh. Nemas=nematodes. Units are kilocalories per square meter per year.

has unfortunately become accepted practice to limit such a pyramid to systematic groups; a certain number of plant-eating birds is associated with a certain number of carnivorous birds, and these with a certain number of birds that prey on the carnivores. This approach is unreasonable, as the example of the birds–hardly any of which are plant-eaters-emphasizes. The only kind of pyramid that should be published is one in which the net production at each trophic level is indicated. Such a "Production pyramid" is considerably more difficult to construct than one based on number of individuals or biomass, but has the advantage of being really unambiguous.

The "upside-down" food pyramid for open water should be dispensed with entirely. This structure has been derived from the fact that the biomass of the consumers in open water is greater than that of the primary producers, the planktonic algae. But the productivity of the planktonic algae–their rate of cell division–is so enormous that the situation, as expressed by a production pyramid, appears entirely normal.

Measurements of Stand

When a stand of plant or population of animals is to be quantitatively evaluated, a distinction is made between number of species and number of individuals. In general species number is relatively easy (though by no means simple) to determine. Models derived from the island theory of MacArthur actually allow one to estimate the number of species present fairly reliable on the basis of the area covered by region of interest, if the number of species in comparable regions is known as area increases the number of species is greater, even if the larger area contains no habitats. The determination of number of individuals–an indispensable item of information in ecosystem research–is far more difficult, especially in the case of land plants. Estimates of root mass, in particular, are frequently challenged. Only very recently have reliable data been published, chiefly by Kummerow.

In general rough approximations are offered, and these are not dependable; under different conditions a given plant can have quite different ratios of root-system size to phytomass above ground. There is therefore also a high probability of error in constructing a simple energy pyramid based on numbers of individuals per unit area (or their biomass). So far, there is no method for the quantitative determination of animals on land; enormous errors can occur. But we must know precisely the number of animals in an area if we want to say anything quantitative about the ecosystem. We also need to know the number of offspring that the animals in the system produce per unit time, and the mortality of the progeny (when bird's nests are deserted or robbed, for example, or caterpillars are infested by parasites), but this information is almost never available.

Indeed, it is practically impossible to obtain a simple count of the adult population of large, relatively conspicuous animals. A discouraging example is the recent attempt to quantify the roe deer population in the cultivated areas of central Europe. Danish studies, in which the area investigated was fenced off, showed that the real population was at least three times as large as all the other counts by experts had indicated. This fundamental discovery–that the true number is three to five times greater than the best estimates–may well apply to all the roe deer populations in central Europe. Even enclosed populations in which every individual is marked can never be completely counted by the professionals in charge of them. There is an annual cycle in "observability." If it is so difficult to obtain a reliable count of the roe deer in central Europe, a region relatively easy to inspect, one can imagine how far wrong the number given for other animals and birds may be–apart from animals so extremely rate that they deserve protection, such as the peregrine falcon, golden eagle, and eagle owl. The counts of the various mouse species per unit area so far available are exceedingly unsatisfactory. Almost all data on mass reproduction of mice are based on relative studies; traps are set up at the same spots year after year, according to a maintained pattern. The differences in number of animals caught are used as indicators of fluctuations in population density.

A method that ought in theory to work is to catch as many members of a species as possible, mark them, and then let them go. Then there is a second trapping session, and one can compute the population size from the ratio of marked and unmarked animals. But not animals are caught so easily a second time, not all can be permanently marked, and marking can put the animals at additional risk. And special difficulties arise where the animals are clustered, territorial, or randomly distributed–that is, everywhere. None of the techniques for counting insects gives anything like a satisfactory result. The most reliable counts seem to be obtained for terrestrial insects, those of the soil as well as in the herbaceous, shrub, or canopy strata, by collection by hand from a test area. It is necessary to distinguish precisely among the different species, and in many cases to distinguish the young and adult stages of a single species. Remember that animals of different size can vary greatly in metabolic rate and thus in the amount of food eaten per unit time and in productivity.

It is therefore pointless from the outset to lump animals at the same trophic level in a single biomass. Bearing all this in mind, we must regard the data in Table 14.1 as very rough indicators. There have been an especially large number of studies on the density of bird populations. But here, too, the variations are huge and there is a very high probability of error. Nevertheless, at present these bird counts are probably the most reliable data of all on the density of terrestrial animals. Because birds have been counted over a very long time in very many parts of the world, and because there are large numbers of extremely knowledgeable amateurs in the field of ornithology, estimates of bird population size have attained a special significance. Birds seem to be useful as "bioindicators;" routine monitoring of the bird population can provide evidence about the state of the environment. For this reason programmes have been instituted in several countries (the "Common Bird Census" in the British Isles, for example) in which quantitative determinations of the bird population are regularly carried out.

Despite the likelihood of error this method still gives us the best information about changes occurring on our planet. The situation in water seems to be simpler. Here the level of primary production evidently depends chiefly on the availability of nutrients and light. When larger amounts of nutrients are provided, productivity can be decisively increased. The number of algae in a body of water rises sharply, but at the same time the transmission of light through the water is reduced. At the deeper levels, where photosynthesis had been possible when minerals were scarcer, autotrophic algae can no longer exist. Now oxygen is used up at these depths, but no more is liberated. As a result, in the deeper parts of nutrient-rich waters anoxia kills off both animals and plants. The eutrophication of lakes and regions of oceans is dangerous because of the associated reduction in available oxygen. On land, where oxygen is always available, overfertilization does not have such consequences. The process in water can be followed closely by filtering a certain amount of water through a simple plankton net.

The number of individual planktonic algae and animals in a whole body of water can be readily derived from these results.

Even planktonic microorganisms are now relatively easy to count. Bottom-dwelling organisms are also accessible; grabs can be used to bring to the surface a piece of bottom material of known area, and with enough samples of this sort the number of animals colonizing the bottom can be determined. As always in biology, the identification of species can be difficult, but here again the problems in the aquatic realm are slight compared to those encountered on land. For these reasons there have been many more studies of aquatic populations and ecosystems than of terrestrial ones. In many cases, the results have been applied directly to land communities; but it has turned out that such application is almost always incorrect. The situation on land is not only much more difficult methodologically, but the general functional principles of populations and ecosystems on land are more complicated than in the water.

Food Chains and Food Webs

The distributions of plants and animals derived from such measurement procedures can be arranged in terms of function; the ordering most commonly used is based on the flow of food. We can distinguish food chains, food webs, and–associated with these-trophic levels. Food chains are found primarily in habitats with very few species. In the eastern African Lake Nakuru, a soda lake, one species of blue-green algae [Oscillatoria (Spirulina) platensis] is the producer for two consumers, a copepod and flamingo. When the systems are more complicated, the simple food chain tends more and more to become a food web. This tendency can be followed very well in ponds of different size. In very complex habitats the networks become so complex that they cannot be traced in detail. Whereas the food chains consist of readily distinguishable links–primary producer, consumer I (which feeds on the primary producer), consumer II (which feeds on consumer I), and sometime still higher "trophic levels" –such distinctions are no longer possible in food webs. One can make at most an approximate subdivision into different trophic levels. Such food webs have been presented as descriptions of the "community nexus" in a wide variety of aquatic and terrestrial habitats. They have profound implications for the stability of habitats.

It is a priori clear that a system comprising a network has greater stability than one composed of independent chains, However, no such web has so far indisputably been shown to exist. Proof of

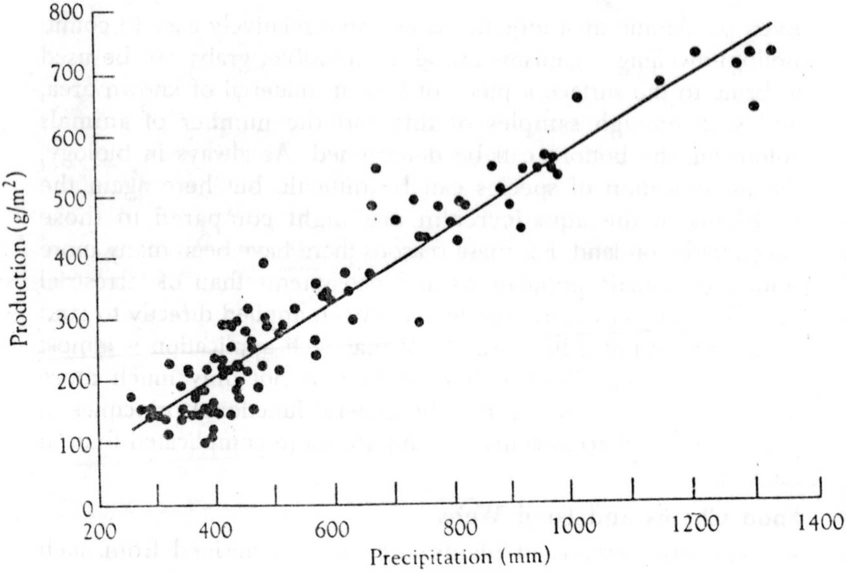

Fig. 14.5. Relationship between mean annual precipitation and mean above-ground net primary production for 100 major land resource areas across the central grassland region.

its existence would require that the quantitative relationships be known. A hypothetical food web as it is classically presented, and with it–hypothetically!–the precisely quantified food relationships. The quantification reveals that in actually we are dealing not with a web, but with chains. In reality, the netlike arrangement plays no role at all. The frequently drawn conclusion that web structure implies stability, then, is not reliable in the absence of quantification. Because no ecosystem on earth has been described in really quantitative terms, there is as yet no confirmation of the web structure. Before the question of stability can be argued further, we must apply ourselves to quantification. Such studies as are available do show that stands of cultivated plants are less vulnerable to mass outbreaks of "pests" when they are not monocultures, but comprise two or more species (for example, the maize and sweet-potato plantations in Costa Rica).

The leaf beetle Phyllotreta cruciferae, which is a severe pest, on cabbages in the USA, did considerably less damage and occurred in considerably less dense population when the cabbage fields also contained other plants, such as tomatoes, tobacco or Ambrosia

ambrosia artemisiifolia). The reason was not a decrease in the number of predators or parasites, but the fact that the insects find their food plants by following the chemical stimuli the plants release; these are not as easily recognizable in the polycultures. At the same time, this example shows that associations based on food relationships and thus on energy flow are only one facet of the complex ecosystem; in this case food relationships are drastically altered by parameters of sensory physiology.

Energy Flux

The ecological significance of food supply, we considered the flow of energy through a single organism. Now we must turn to the flow of energy through an ecosystem. Solar energy is converted to chemical energy by the green plants; this chemical energy is passed on through some number of stages, and in the process is converted in a stepwise manner to heat energy. Eventually, the chemical energy ought to fall to zero. But in asking ecosystem research to verify this expected result, we are making a demand comparable to that for quantification of the community nexus. The only data available so far are more or less applicable estimates for overall systems (chiefly aquatic) and quite detailed studies of the flow of energy through individual populations. Even though this is not a very satisfactory base of departure, a relatively consistent pattern results. The herbivores, consumers 1, use up only a fairly small part of the food theoretically available to them.

All the plant-eating insects in a beech forest together consume at most 10 percent – 15 percent of the substance produced by the beeches in a year; other animal groups can be neglected as herbivores in this habitat. Grasshoppers in the meadows of central and northern Europe consume no more than 10 percent of the plant tissue formed. The warm-blooded herbivores of the tundra–reindeer and musk ox above all–utilize only 5 percent–10 percent, and the large grazing animals on the African steppes barely exceed 10 percent utilization. Still less of the plant matter in the forest is consumed by mammals, for the main productive stratum, the canopy, is out of their reach. Small mammals in general consume barely 1 percent of the net primary production; only in a very few cases can this figure rise to as much as 6 percent. However, particular parts of the plants, such as the seeds, can be consumed in higher proportions -30 percent, 75 percent or even 90 percent.

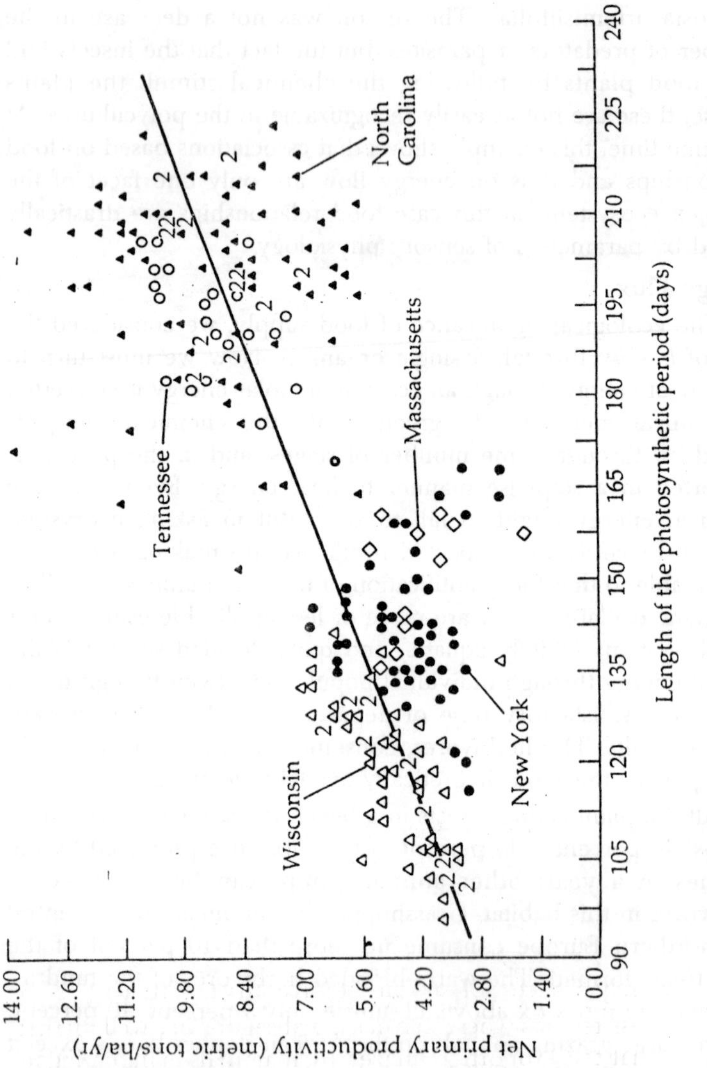

Fig. 14.6. Relationship between net primary production and the length of the growing season for stations in the eastern deciduous forest region of North America.

It is generally true that on land a maximum of 10 percent–20 percent of the substance formed by plants is directly consumed by herbivores. All the rest eventually (very much later, in the case of trees) enters the food chain responsible for the decomposition of dead matter. By far the greatest energy turnover on land, therefore, as chemical energy is used up, occurs on the soil layer. One must bear in mind the excrement, exuvia, and other things derived from the bodies of the first consumers (hair, feathers, shed antlers) are also passed on directly to the food chain in the soil. Now the question arises, how large a percentage of the net production of the first consumers enters the secondary consumers–predators and parasites? Here, again, the value seems in general to vary between 5 percent and 15 percent. The greater part of the substance produced by the first consumers, then, also enters the soil stratum, from the dead bodies.

Three consequences follows from these considerations.

1. We must consider the soil in greater detail, for it is the chief site of turnover inter-restrial ecosystems.
2. We must ask whether the picture we have drawn is really sufficient, or whether, a finer subdivision is required.
3. We must inquire into the causes and the ecological effects of the surprisingly low rates of utilization observed-in the utilization of primary production by herbivores as well as that of the net secondary production of any animal population by animals at the next trophic level.

We shall begin with the third question. In regions where there is a marked change of seasons, the annual production is of course not always at hand; there are times of very high productivity, and other times in which little or practically no food is available. The minimum, during the dry season or in winter, determines the density of the large-mammal population during the main production season, and thus the degree of utilization. The longer or more severe the winter or dry season, the smaller the percentage of production likely to be utilized. This tendency is to seem extent compensated by migrations, like those which used to bring red and roe deer down from the mountains in winter. In the tropical rainforests there are no seasons. But here the productive zone, the crowns of the trees, is largely inaccessible to the herbiverous mammals.

In treeless regions where persons are absent or not well marked, it may be that the proportion utilized by mammals is much higher. This could happen in temperate-zone heaths near the coast, where because of the prevailing winds trees cannot grow. The situation is similar on high-altitude steppes near the equator; here the very small territories cf the vicuna suggest that their rate of utilization is very high. Furthermore the small mammals and insects with high rates of reproduction appear unable to "keep up" with the rapid new growth of vegetation in the spring. They are directly dependent on plant food, and cannot survive warm periods without food. Their lives would be endangered if they were to emerge before the leaves sprout in the spring. If they appear after leafing-out has begun–even immediately after–as a rule the plants have gained an advantage in development that cannot be overcome. But if leafing-out and the appearance of the insects are simultaneous, the trees can easily be stripped bare; this happens, for example, during mass outbreaks of June beetles. A yet we cannot specify a percentage utilization in such cases. All the trees so damaged in temperate zone can put out new leaves and exhibit full production after the June beetles have dies out. However, another example. shows that mass outbreaks in natural regions of vegetation can have extremely dangerous results.

In the Finnish and Swedish birch forests, mass outbreaks of the moth Oporinia autumnata cause widespread destruction; the forest boundaries in many cases are pushed back by 30 km. Ultimately we must admit that the causes of the low rate of utilization of the permanently available food reserves in the tropical rainforest by herbivorous insects are not known. Though low utilization is the rule, we must always be prepared for surprises. In the two examples that follow, plant-eating animals use up nearly 100 percent of the primary production. The first of these is the only native herbivore of central Iceland, the pink-footed goose. The contemporary populations of this bird here are so dense that they consume nearly all of the vegetation of the few oases during the breeding season. This is possible only because the geese can migrate so rapidly; the arrive only after the vegetation has sprouted and leave after they have eaten everything up.

Another likely contributing factor is the very favourable conditions under which this population spends the winter. The losses that probably were once routinely suffered in the overwintering region no longer occur. The second example is provided by the studies of Reichholf on the lakes produced by damming the lower reaches of the Inn River. He showed that the large flocks of ducks that overwinter there consume almost all the aquatic plants. This happens only when the animals are not disturbed during the winter, so that they can stay in the region all day and night. Here, again, the high rate of utilization is possible only because the birds migrate, arriving after the vegetation has grown tall and leaving the lakes when the water plants have been grazed off. The consequences are

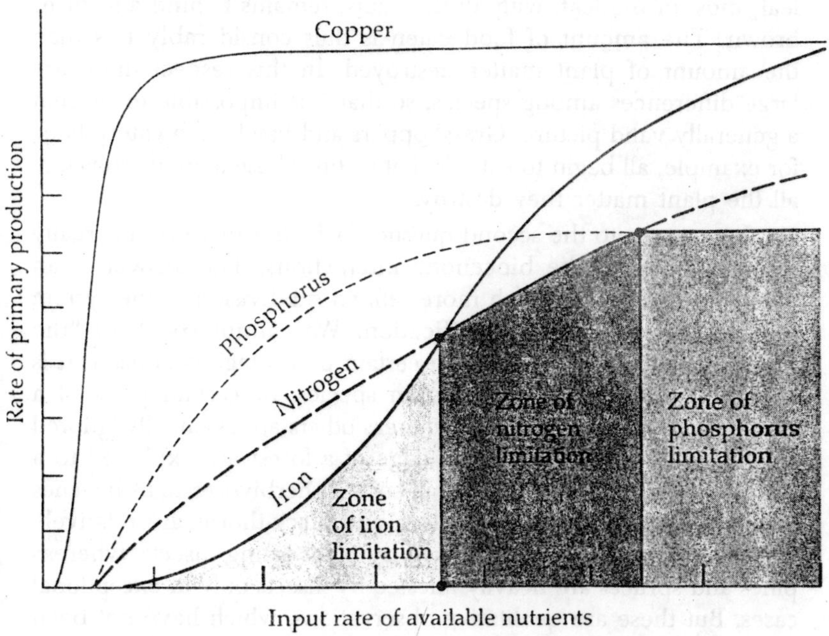

Fig. 14.7. Hypothetical sequences of nutrient factors that limit primary productivity. The rate of primary production follows the heavy line and is limited first by iron, then (as more iron becomes available) by nitrogen, and finally by phosphorus. Some nutrients, such as copper, may always be present in superabundant amounts.

important ecologically. The energy contained in the plants is respired above the water by the ducks. If it were not for the ducks, the plants would sink to the bottom in the winter and rot there, withdrawing oxygen from the water in the process. Without the ducks anoxic conditions would develop; with them, they do not. In this sense the ducks are responsible for the large stocks of fish in the lakes.

In evaluating energy flow, one encounters a special problem with regard to the food not used by the animals. Here, we do not mean the part that leaves the gut undigested–although this element is frequently underestimated. For example, the dipteran larvae feeding in acid beech forests chew up about 30 percent of the litter from the stand, whereas they actually utilize only a few percent. What is meant by "unused food" is the amount destroyed, but not consumed, by animals during feeding. The larvae of the alder-leaf beetle eat only the photosynthetically active layer of the leaf; most of the leaf, with all the veins, remains behind and turns brown. The amount of food eaten is thus considerably less than the amount of plant matter destroyed. In this respect there are large differences among species, so that it is impossible to present a generally valid picture. Grasshoppers and hawkmoth caterpillars, for example, all begin to eat a leaf at its tip. These animals consume all the plant matter they destroy.

Let us turn to the second question, whether these results really suffice to qualify the biocenotic interactions. The answer is an unequivocal "no". A much more refined subdivision of the system is necessary for such quantification. We cannot speak of "the primary production" of which a certain percentage is utilized. It is possible, for example, that certain species, or certain parts of a plant, are 100 percent utilized, whereas others are essentially ignored by the herbivores. Oaks at the edge of a forest are exploited to a considerably greater extent by all sorts of herbivores than beeches similarly located. Willows, poplars, and black-thorns are relatively frequently and obviously attacked by plant-eating insects, whereas pines and spruces are heavily infested by insects only in exceptional cases. But these are qualitative observations, which have not been quantified. Among the large ungulates of Africa, as we have seen, the different species exhibit strict specialization to the different strata of vegetation. But even if we distinguish among species we have gained relatively little.

The individual plants are also differentially vulnerable to herbivores. These differences are in part genetically based, and in part associated with the habitat. Alders or spruces in dry habitats (Or in dry years) tend to be attacked by herbivores more extensively than they are under moist conditions. And finally, the separate parts of a plant are used as food by herbivores to widely differing degrees. So far we have been discussing the parts of plants that grow above ground, disregarding the fact that there is about the same phytomass underground. We know nothing about the degree to which this part is utilized. Moreover, different above-ground organs are formed in different years. The irregular production of seeds by forest trees is an example of this effect. In some years the accumulated production of many years is evidently channelled into seed formation. As yet we cannot say whether more plant matter is consumed by herbivores in a seed year than in those when no seeds are produced.

Presumably consumption is higher in a seed year, but it would make sense to give a percentage figure only if a mean value for many years could be cited; voluminous acorn production is possible only because reserves have accumulated in the oak tree during many previous year. Most of the organic matter formed by the plants is not consumed by herbivores, but eventually reaches the ground. Normally, these substances are continuously decomposed and worked into the soil. In terrestrial habitats, therefore, by far the highest energy turnover occurs in the soil stratum. In spruce and pine forests the amount of litter falling per year has been estimated as 1,500–3,000 kg per hectare, and in good mixed deciduous forests of central Europe as 4,000 kg/ha or more. In a balanced system, then, this amount must be decomposed each year; otherwise organic matter would accumulate on the forest floor. This is not to say that a particular beech leaf is decomposed in the course of a year.

In general, a fallen leaf in an acid beech forest requires about 5 years before it is no longer visually identifiable. This decomposition is not a rapid and complete breakdown into CO_2, H_2O, and minerals. The process is slow and involves complicated resynthesis, in the course of which humus is produced. Humus substances, because of their great stability, are of very special significance in the preservation of soil structure. As far as the participation of organisms in the breakdown of litter is concerned, there are both qualitative and quantitative considerations. In view of the enormous number of soil-dwelling animals it is a priori likely that they play a central role in decomposing plant matter.

Generally, for practical purposes, macrofauna are distinguished from microfauna; the former is understood to include chiefly the earthworms, enchytraeids, isopods, diplopods, and large insect larvae that in Europe are especially important in the breakdown of litter. Evidently the litter is first eaten by the larger animals, which can attack even the less digestible elements with the aid of microbes. Isopods, snails, and dipteran larvae are equally important. The significance of these primary decomposers is evident; in their absence, the rate of decomposition is almost halved. During this first passage through a digestive system, little energy is withdrawn from the material. All animals take up very large amounts of fallen leaves and produce very large amounts of excrement; they utilize the litter relatively poorly. Nevertheless, this first passage is of fundamental importance to the system, as follows:

1. The water-binding capacity is increased. The excreta of the primary decomposers holds moisture more uniformly than the original litter.
2. Having been comminuted, the material becomes accessible to a large number of so-called secondary decomposer–primarily springtails, mites, and nematodes. The excreta of some primary decomposers are more readily consumed by the secondary decomposers than those of others, so that if the process of decomposition is to proceed evenly there must be an undisturbed, well-balanced community of animals in the soil.
3. The excreta of both primary and secondary decomposers provide particularly good conditions for the development of microorganisms, above all actinomycetes and bacteria. Their presence can be inferred by comparison of the metabolic activity (CO_2 release) in fallen leaves with that in excrement.

This first passage through the gut is followed by repeated passage, through the same or other animals. Wieser (1965) showed that isopods flourish particularly well when they can eat their own excreta over and over again. The microorganisms break down those plant parts that are hard to digest, and the animals are nourished primarily by these microorganisms. Would decomposition proceed more rapidly if only the microorganisms were present? After all, the activity of the animals appears to cause a considerable reduction in the numbers of microorganisms! This is the sort of false conclusion that is so commonly encountered in analysis of predator-prey systems.

If the animals were not present, the microorganisms would very rapidly reach an extremely high population density and then almost stop reproducing; their cultures would be in the stationary phase. By virtue of the fact that their population density is kept low, they maintain maximal production-and thus optimal performance–over the long term. The importance of the comminution of litter is also shown by certain experiments done in Russia, and others by Herlitzius and Herlitzius (1977). Completely intact beech foliage on acid forest soil is very resistant to attack and hard to break down. The excrement of leaf-eating beetles that falls down from the canopy and the dead bodies of these beetles provide nuclei in which microorganisms can begin to grow. During the repeated passage through animal bodies, easily decomposable substances are withdrawn from the substrate.

The carbohydrates are predominantly broken down into carbon dioxide and water. The percentage of the most important plant nutrients (nitrogen, phosphorus, potassium) rises, and lignin and tannins–products of the polymetization of phenols–accumulate in the excreta. In this neutral or weakly alkaline milieu these substances come onto close contact with nitrogenous metabolites produced by the animal decomposers. The conditions thus favour the microbial breakdown of the lignins to phenols, at a rate that depends on the amount of nitrogen available to the microorganisms. The analysis of humus fractions from different soils, together with model studies, shows that in the milieu described, when nitrogen and sufficient oxygen are present, phenolic substances are oxidized to oxyquinones, which polymerize to form larger molecules.

The process is accelerated by phenoloxidases of the microorganisms. The end result is the production of humic acids–polymers of complicated structure which are extraordinarily stable. Breakdown of humic acids in the soul occurs very slowly; it is likely that the chief agents of their decomposition are mycorrhiza fungi. Humic acids, then, are an extremely stable element of the soil, especially because they can form clay-humus complexes by combining with aluminium and iron hydroxides. These complexes ultimately determine the fine structure of the soil, the friability required for good ventilation and water retention. The proportion of clay-humus complexes in the soil is thus responsible for soil fertility–at least in the majority of soils. The quantitative division of the labour of remineralization, between herbivores and soil organisms, has probably resulted from a many-faceted coevolution that has led to optimization of the system. When the decomposition of litter proceeds as described, the result is the ideal form of humus, mull. But it does not take the same course in all habitats; there are wide differences in rate of decomposition in different types of forest.

The quality of the humus, the rate of breakdown of leaf litter, and the number of litter consuming animals in a habitat are closely related. Comparison of three forest habitats with decreasing humus quality (mull on lime soil, mull on acid soil, and mor on acid soil–the last is a very unfavourable form of humus) showed that the rate of decomposition decreased in this sequence, and the number of animals was related about as 3:1-5:1. Soil pH and the organisms supported are frequently correlated, and even though the dominant plant species are in many cases the same (beech on chalk and variegated sandstone; pines on sand and chalk, birch forest on quite different geological substrata in Greenland, Iceland, and Scandinavia) the differences are striking–particularly with respect to the density of the animal species present. A number of other factors are associated with pH, which together make acid soils generally poorer in nutrients. Humus, as an organic substance, does not continue to exist indefinitely. It in turn is eventually broken down by microorganisms. Naturally, this process occurs especially rapidly under hot, humid conditions (in a tropical rainforest). Although humus is not a plant food, in most soils it is essential to plant nutrition. The humic acids can absorb molecules that otherwise would soon be leached out by the rain. Moreover, humus substances can retain large amounts of water.

Humus therefore guarantees the plants a steady supply of minerals and water. The destruction and disappearance of humus in many soils is responsible for a decline in primary production. In principle, the same relationships are found in water; only a relatively small fraction of he producing planktonic algae is consumed as living cells by herbivores. Most of the algae die, sink toward the bottom, and are eaten in deeper water by consumers in the detritus food chain. On land as in water, then, we have what has been called a Y-shaped model of energy flow. It is biocenotic interrelationships which, while unquantified, appeared to be a very complex food web, and found that quantification produced two nearly independent food chains; although the latter was a purely hypothetical model, the above considerations show it to be quite realistic. Most of the production in water is by phytoplankton rather than by sessile plants.

It is therefore restricted to water levels near the surface. The more transparent the water, the deeper the producing layer can be. The higher the nutrient content, the more dense the phytoplankton , so that light penetrates less deeply and the producing layer is shallower. In this sense a large body of water can be compared with a forest, where production is also restricted to the upper surface; in the shaded region near the ground the chemical energy formed is used up. In deep bodies of water where oxygen is abundant the rain of detritus is essentially all consumed before reaching the bottom. This is generally true in the oceans; in the Baltic Sea, for example, it has been documented by analyses of protein content.

In such waters the bottom contains very little organic matter, and as a result the fauna is relatively diverse and in many cases consists of fairly large animals which presumably develop very slowly. In waters with little oxygen at depth consumption occurs anaerobically or not at all; the bottom becomes covered with decaying mud in which H_2S is formed.

Oxygen deficiency occurs in the depths when the water becomes opaque because the nutrient content is too great, and/ or when the deeper water of an ocean or take is never exchanged with the more superficial water because autumn and spring storms fail to churn the water sufficiently. Under such conditions the water becomes increasingly uninhabitable down to a certain depth. At great depths more and more organic material accumulates. These large-scale relationships probably can serve as models of the origin of petroleum. Conditions are like this in the Black Sea, and it is feared that they will become so in the Baltic. But the situation in the Baltic is complicated by irregular inputs of oxygen-rich, high salinity water from the North Sea; these masses of water, being more saline and thus heavier that the Baltic water, move into the Baltic along the bottom, displacing the water that is low in oxygen and salts. So far this process has repeatedly brought about complete decomposition of the deposited organic matter. In shallow bodies of water the organic matter produced reaches the bottom before it is decomposed.

The abundance of this food source is evident in the large numbers of filter-feeding animals that inhabit fresh water and the shallow parts of the oceans–barnacles (Balanus), mussels, bryozoans, a variety of polychaetes, and many others. Let us consider these relationships in detail. Primary production is entirely performed by unicellular algae of several sorts–peridinaeans, diatoms, and the very small flagellates of the nannoplankton. In cooler regions or seasons the diatoms predominate, whereas under warmer conditions they are less in evidence. In warm oceans primary production depends heavily on the nannoplanktonic flagellates. The significance of this fact is enormous, even though the level of primary production may be identical in the two cases.

The organism of the nannoplankton are so small that they escape not only ordinary plankton nets but also most of the animals that eat plankton; only highly specialized animals, usually very small, are capable of utilizing the nannoplankton. The best known example is the mantle of the planktonic tunicate Oikopleura, an extremely complicated structure within which the animal encloses itself. The mantle presents a very fine-meshed lattice to the flow of water generated by the tunicate, and the nannoplankton that accumulate on it are used by the animal as food. In fact, it was this observation that first led to the discovery of the nannoplankton, which until that time had been overlooked. The nannoplankton, then, serve as food for small to very small planktonic animals, and these in turn are eaten by larger zooplankton. But if the primary production is by relatively large phytoplankton, the large planktonic animals can utilize it directly. By the rule of thumb that only about 1/10 of the energy is retained at each transition from one trophic level to another, it is clear that this extension of the food chain when the nannoplankton are the primary producer must have a drastic effect on secondary production by larger inhabitants of the open water; the necessary consequence is a paucity of fish.

Recent studies indicates that the poverty of tropical oceans is not necessarily all a matter of mineral deficiency; to some extent it results entirely form the fact that the nannoplankton are responsible for primary production. (Evidently this nannoplankton population is extraordinarily resistant to a great variety of influences, so that if it were to replace the primary producers in other parts of the ocean fish production, and thus the world's fisheries, would be endangered). For this reason the precise analyses of the requirements and reproductive rates of phytoplankton currently underway should have high priority. The ocean's planktonic animals comprise two fundamentally different groups. The animals in one group spent their entire lives in the open water, whereas those in the other are the larval forms of bottom-dwellers. The latter are naturally seldom encountered in the large oceans.

Planktonic larvae are characteristic of the smaller seas–North Sea, Baltic Sea, and Mediterranean, for example-but are almost nonexistent in fresh water. The planktonic animals in lakes spend their whole lives in the plankton, with at most certain temporary stages (as happens among the water fleas and their relatives) resting on the bottom. There are probably great differences in the food requirements of the zooplankton, as has been suggested by the studies of Lampert. But we know almost nothing about them. The next trophic level is represented by small or some what larger plankton-eating fishes, such as herring (Clupea) and whitefish

(Coregonus), and these in natural systems are eaten by large predatory fishes, mammals and birds. Here, again, we may take it that in general only about 10 percent of the algae are directly ingested and digested by animals (ingested algae, if undamaged, are often passed out again in the excreta and then simply continue to grow).

Yet another factor is involved. Planktonic algae "lose" a large part of their photosynthates (though there is no reliable, quantitative estimate of how large it is) into the water, where they drift about as relatively short-chain carbohydrates, as fatty acids, and as amino acids. These losses are greater if, while photosynthesis is proceeding rapidly under strong light, certain substrate materials fall below a critical concentration; then subdivision can no longer occur. If the nitrogen concentration is below the critical level no more amino acids and protein molecules can be synthesized, but carbohydrates continue to be produced, and these must be eliminated.

As a result, the water accumulates large amounts of free organic molecules. The same thing happens when planktonic organisms die or are injured. For this reason the amount of dissolved organic compounds in the ocean is considerably higher than that of all organic compounds bound in plants and animals. The compounds dissolved is fresh water are all taken up, incorporated and broken down by bacteria; in the ocean there are many organisms that are either obligatory or facultative feeders on such dissolved organic compounds. In water, then, the formation of a soil such as occurs on land is neither necessary nor possible.

The principle of energy flow in water is based on complete breakdown of organic matter. There is no question of humus formation. If organic matter is deposited on the bottom and not decomposed, the first result is a decaying sludge; then firm deposits of organic matter are formed which eventually can become fossil fuels (peat, coal, petroleum). These regions are quite unlike humus soils-they are extremely hostile to life, and practically devoid of animals. Bacteria and protozoans are the sole inhabitants of such reducing sediments.

The Significance of Animals in an Ecosystem

Animals are much more conspicuous in water than on land. On land they are surrounded by higher plants and are barely apparent, whereas on the bottoms of bodies of water they play a leading role. This is especially true of the ocean; the subdivision of marine bottom-dwelling animals has been done according to principles very like those used to subdivide terrestrial regions on the basis of plant communities. The large, well-known associations in the ocean are the pure animal communities of the coral reef; the benthic animal and plant communities hardly enter the macroscopic picture. On land, by contrast, animals in general contribute little to the flux of energy and the cycling of materials. Rarely do herbivores consume more that 10 percent of the energy flux by the plants. For this reason, many ecosystem analyses have given only cursory treatment to the animals. Not a few researchers–particularly botanists interested in productivity and system analysts interested in the system as such–are inclined to deny the animals any appreciable role in the ecosystem. But nothing could be further from reality.

An animal like the rhoe deer, which preferentially consumes the buds of trees, can dramatically reduce the production of a forest; the rabbit, an equally selective feeder, in population of normal density can keep the stands of many special plants down the extent that they are no longer ordinarily encountered. After myxomatosis had drastically reduced the rabbit population in the British Isles certain wildflowers appeared in all sorts of places where people had been looking in vain for years. Beavers in a ecosystem can back up large lakes behind their dams; if the dams break extensive meadows appear in otherwise continuous woodland. There is not much quantitative data available, and that has been taken from only a few animals.

Indeed, effects such as those described can hardly be expressed quantitatively or included in treatments of energy flow or of the cycling of matter. Cycling and energy flow are but two facets in the intricate pattern of an ecosystem. We shall now present a few examples that illustrate the complex involvement of animals on ecosystems. These must be anecdotal and in some cases have not been completely worked out, but they reveal the difficulties inherent in this line of research.

Leaf-eating beetles evidently cause no decline in production in a deciduous forest. Because the beetles eat holes in the upper leaves more light reaches those below, so that these can proceed with photosynthesis at a greater rate. Here loss and profit are in balance. In fact, there is an added benefit. The excrement of the beetles, which contains finely ground up leaves and in many cases is full of minerals, as well as the dead beetles, laden with nitrogen and phosphorus, fall to the ground and act like nuclei of crystallization–

centres from which the decomposition of the leaf litter proceeds must more rapidly and efficiently than it would without these highly nutritive particles. The leaves shed from the trees contain very few nutrients even, for bacteria and fungi, and are excessively hard to digest. Only by way of the excreta and bodies of animals are they vulnerable to attack. With these aids, decomposition can occur rapidly; without them it would take so long that natural rejuvenation of the woods would be nearly impossible, for a thick layer of undecomposed leaves would pile up. The beetle is thus a fundamental element in the system. But its influence can be detached only by laborious and extremely meticulous research; without such research its effects might be preserved only when it was 80 or 100 years too late. Moreover, the effect varies on different types of soil.

On chalky soils, where the decomposition of organic matter proceeds relatively rapidly, the contribution of the beetle corpses is less important than on acid sand and silicate soils. Because plant-eating insects preferentially infest trees in which the sap flow is not completely intact, aging trees with lowered productivity, but which are still highly competitive, tend to be invaded and killed. Competition is thus reduced for the younger trees, and in the long term productivity of the system is increased.

Aphids withdraw from their host plants large amounts of fluid containing sugar. These sucking insects also require nitrogen, present in much smaller concentrations in plant sap; by the time they have acquired enough nitrogen they have taken in far too much sugar, and this is excreted in the form of "honeydew". Are the plants seriously damaged by this loss? Experiments have shown that nitrogen-fixing bacteria can be brought to peak performance by aqueous solutions of sugar; after application of aqueous sugar solution the fixation of atmospheric nitrogen increases sharply. It is possible that the following system has developed by coevolution. The plants pass on the surplus carbohydrates they have formed to the sapsuckers, which pass most of these carbohydrates on to the soil and thus activate the bacteria to produce more of an element that is ordinarily a limiting factor for plant development. That is, the plant releases a surplus product in order to obtain adequate amounts of a factor in minimum. In such a case we would be dealing not with a plant and a pest, but with a system which-as a system-has been optimized.

Spittlebugs and their relatives can occupy tropical forests in such crowds that a continual fine rain drips from them onto the ground. They withdraw huge amounts of water from the plants. With what results? The soil of tropical rainforests contain almost no nutrients. Only the uppermost layer, the leaf litter, contains minerals that can be used by the plants, but in general this layer is too dry for the plant roots to penetrate.

Because of the continual sprinkling by spittlebugs, some roots can invade even this top parts of the soil and thus provide the plant with nutrients, whereas the greater part of the root mass, deeper in the soil, is responsible exclusively for the provision of water. In this case, too, plant and sucking insect together would represent an optimized system. Studies of soil animals and litter decomposers have shown that they feed chiefly on bacteria, which carry out the acute decomposition. Without these animals the bacteria would be much more numerous. It would appear, then, that the soil animals inhibit decomposition of the litter. But the opposite is actually true. By their constant eating of bacteria, as has clearly been shown for Orchestia, these animals keep the population of bacteria in the exponential phase of growth. If the soil animals were not present the bacteria would rapidly increase to capacity and enter the stationary phase, in which they become practically inactive.

By keeping the population density low the animals ensure the highest productivity of the bacteria–that is, the greatest effectiveness in breaking down the litter. In the last analysis this example is simply another expression of the predator-prey relationship we found for the red grouse. In that case, too, the activity of the predator kept the rate of grouse reproduction as high as possible; without predation, the grouse reproduction rate falls off dramatically.

The lemmings in the arctic tundra show oscillation in population size, with mass reproduction bringing a peak every 3-5 years. The plants over large areas of tundra can be killed when the animals cut through their roots. But without the turnover of the soil that occurs at regular intervals by the large numbers of burrowing lemmings, breakdown of the ground litter in these arctic regions would take an exceedingly long time. The outbreaks of lemmings are necessary if organic material is not to accumulate to excess. The dung produced by the cattle introduced into the Australian steppes could not be naturally re-mineralized there. It accumulated on the ground, often drying into a hard crust, and destroyed the vegetation.

Moreover, it provided a haven for stinging flies that had also been introduced by accident and proved severely annoying, and it gave the parasitic worms passed out by the cattle in their dung ample opportunity to re-infect them. In 1963 the Australian government began a massive effort to get rid of this nuisance. Many dung beetles were subjected to the most through tests, and finally–from April of 1967 on–about 275,000 beetles of four species were released. One species (Onthophagus gazella) reproduced and spread at a spectacular rate. Wherever this beetle settled in well, the cattle dung was worked into the soil in about two days (there the beetle larvae continued to feed on it). Two other African species have reproduced to a similar extent in other parts of Australia, with similar effects.

The danger of parasites is very much less, and the troublesome flies have almost disappeared. But at the moment the additional release of mites is under consideration. All dung beetles normally carry mites with them; these move into the dung, reproduce there, and then attach to the next beetle to be carried to the next pile of dung. The European burying beetle Nicrophorus is also associated with mites. Its larvae as a rule are subordinate in competition with fly maggots, but the mites of the beetle attack the fly's eggs–the competitors of the beetle larvae are eliminated. The relationship between the dung beetles and their mites is thought to be similar, and therefore experiments are currently underway to provide the dung beetles with normal mites. No one pays much heed to dung beetles encountered in the countryside; but they can be vital to the functioning of the system, as the Australian example shows. For reasons obvious in these examples, Mattson and Addy (1975) do not regard the animals of an ecosystem in isolation., but rather consider the whole as a system for long-term optimization, with uniform operation over centuries. The example presented, in which the ducks consumed the primary production almost completely and thus prevented a lake from reverting to land, is an argument in this direction.

But the principle of optimization becomes especially apparent when we consider the implications of animals as flower pollinators and seed-dispersing agents. Here they determine the structure of the plant community, altogether their function could not be adequately comprehended in the context of energy flow or the cycling of matter. Remarkably, these functions of animals–despite their outstanding significance to plant communities–are often neglected in botanically oriented analyses of ecosystems. Plants growing in stands comprising a few species, or in natural monocultures, usually are wind-pollinated.

The greater the number of species in a stand, and the more nearly random the distribution of the individuals of each species, the more the species are forced to adopt a better-aimed method of pollination, by animals. A logical consequence of pollination by animals is seed dispersal by animals. This connection has rarely been noted; Regal (1977), in particular, has called attention to it. When many species are distributed in a mosaic throughout a system, individuals have a good chance of survival only if the seeds can be transported over considerable distances. Thus, in quite general terms, we find that wind pollination prevails in the northern tundras and in the forests of northern and temperate latitudes, where the number of species are small, as well as in windswept steppes. Even in the temperate-zone forests pollination by insects and seed transport by birds gains in importance, and in subtropical and tropical forest regions animals are indispensable to most of the plants.

It is true that many plants, whether pollinated by wind or animals, can exist for quite long times (sometimes for several generations, by parthenogenesis or self-pollination) without normal pollination. But in the long run an exchange of genes is necessary, as it is in animals. Just how necessary is evident in the complicated coevolution that has occurred between plant and insect. Examples have been presented in many research reports and handbooks. The enormous energy expenditure by the plants is understandable only in this context. Similar instances of coevolution to ensure seed transport are not known to such a great extent, and studies of this aspect have been fewer. Evidently real specialization, so common where pollination is concerned, is rare in the case of seed disposal. But the fact that many plant seeds (mistleote, tomato) germinate with difficulty or not at all unless they have passed through the body of an animal points in the direction of coevolution. Under heavy grazing pressure most dicotyledonous plants disappear and are replaced by grasses and their relatives. When cattle are allowed into a wood to feed, it becomes overgrown with grass.

It is quite certain that grasses, with their ability to grow continually, become so widespread on as a result of the activity of grazing mammals. Grazers can convert a plant community containing many species into a quite different, much less diverse community. A good example is given by the meadows in coastal regions which are sometimes flooded by the sea–the North Sea, for instance. Here many species grow naturally, but when sheep are kept on the meadows most of them disappear. Only meadow-grass (Puccinnellia), with its persistent growth, can withstand the pressure of grazing. The very dense grass vegetation that is produced under the influence of the sheep holds the soil together much more firmly than did the original salt-meadow vegetation. For this reason sheep are a valuable aid in converting these coastal fields into land that can be used for agriculture. The natural vegetation allows channels to open in the ground from time to time, so that conversion to dry land is much less likely; the system is maintained.

Another outstanding way in which plant-eating animals can affect ecosystems is by transmitting plant diseases (fungus and virus diseases). Bark beetles, long-horned beetles, and sap-sucking insects are disease carriers. Previously this fact has been viewed entirely as an injurious aspects of insects, so that at present nothing can be said about its significance in a balanced ecosystem. An ecologist would infer that such diseases do the most damage to already weakened plants, thus making room for other species that find conditions more favourable at that location. The spread of such pathogens would then be regarded as a mechanism by which turnover in the ecosystem is accelerated. But this is pure speculation, unsupported by any data. The examples discussed so far have illustrated the significance of animals to the ecosystem as a whole. The importance of many animal species to the existence of other animals has been known for much longer, and indeed is self-evident.

Woodpeckers that dig out holes provide breeding places for stock doves, owls, hoopoes, rollers, and tits, as well as for the dormouse and many insects. By their burrowing activities foxes in the forest, marmots and prairie dogs in the steppe, and rabbits in the savanna and at the edges of forests provide a continual supply of fresh soil; this is colonized by pioneer plants that otherwise could not exist in such mature stands. The rich fauna in the Florida Everglades results from the presence of alligators. These animals dig deep pits for themselves, where all sorts of other water animals–fish, turtles, insects-can shelter during the dry season, while at their margins even sensitive water plants can survive hot dry periods. If It were not for the alligators the water birds would find no food during the dry seasons, and the aquatic animals for which the Everglades are famous would die out.

In the mud flats of the North Sea the factor restricting colonization by sessile marine animals is lack of an adequate substrate. Only a few species are capable of living directly on the mud. One of these is the oyster. Where oysters appear, they are followed by a great number of sessile species which can now attach to the oyster shells. It was the observation of this relationship that led to the discovery by Mobius of the phenomenon of the biocenosis, and thus to the actual beginning of the science of ecology. These examples should illustrate the point sufficiently, but let us mention just a few more of the many others that could be cited.

The big Chilaces typhae lives in cattails, but can enter the reeds only if the caterpillars of a moth have bored passages into them. Some parasitic wasps cannot find their host by themselves; they follow the trails of other species that prefer the some host and, having found their victims in this way, proceed to lay their eggs. Their larvae develop more rapidly than those of the first wasps, and thus win the competition against the actual discoverer of the host. Given all these possibilities, it is quite inappropriate to limit an ecosystem analysis to energy flow and the cycling of matter. To do so would be like making a physiological study of an animal in which nothing but feeding and digestion were considered.

The functions of animals described here determine the composition of the ecosystem, the level of primary production (of which they than consume 10%), and the form in which this primary production is delivered (i.e., by which species and which parts of the plant). The action of animals in an ecosystem can be compared with that of switches and amplifiers in an electronic system, or with the sense organs and nervous system of an individual organism. But these effects are very difficult to represent numerically. It is just as futile to try to express in numbers the way a system is affected by the removal of a pollinator or seed transporter as it would be to find a numerical expression for the loss of an individual's vision, hearing, or sense of smell. On land, too, animals can determine the entire structure of their system.

Particularly striking examples are the islands colonized by sea birds, which build up large deposits of guano in arid regions and can thus be an economic factor of major importance. Seabird colonies situated in trees can kill the trees with their droppings; cormorants have been known to do this. In the water, animals dominate the structure of the habitat as the higher plants ordinarily do on land.